The European Union in Africa

MANCHESTER
1824
Manchester University Press

The European Union in Africa

Incoherent policies, asymmetrical partnership, declining relevance?

Edited by Maurizio Carbone

Manchester University Press

Published by Manchester University Press
Altrincham Street, Manchester M1 7JA, UK
www.manchesteruniversitypress.co.uk

British Library Cataloguing-in-Publication Data is available

Library of Congress Cataloging-in-Publication Data is available

ISBN 9781 7894 9387 0 *paperback*

First published by Manchester University Press in hardback 2013

This edition first published 2016

Printed by Lightning Source

Contents

Part III Policies and partnerships

Part IV Conclusion

List of tables

Notes on the contributors

Maurizio Carbone is Professor of International Relations and Development and Jean Monnet Chair of EU External Policies at the University of Glasgow.

Gordon Crawford is Professor of Development Politics and Director of the Centre for Global Development at the University of Leeds.

Mary Farrell is Reader in European and International Politics and Jean Monnet Chair at the University of Greenwich.

Toni Haastrup is Teaching Fellow in International Security in the Politics and International Studies, University of Warwick.

Amelia Hadfield is Professor in European Affairs and Senior Research Fellow at the Institute of European Studies, Vrije Universiteit Brussel.

Simon Lightfoot is Senior Lecturer in European Politics at the University of Leeds.

Alan Matthews is Professor Emeritus of European Agricultural Policy, Department of Economics, Trinity College, Dublin.

Gorm Rye Olsen is Professor in Global Politics and Head of the Institute of Society and Globalisation at Roskilde University.

Jan Orbie is Professor of European Union Politics at the Centre for EU Studies at Ghent University.

Michael Smith is Professor of European Politics and Jean Monnet Chair at Loughborough University.

Fredrik Söderbaum is Associate Professor at the School of Global Studies, University of Gothenburg, and Associate Senior Research Fellow at the United Nations University-Comparative Regional Integration Studies (UNU-CRIS), Bruges.

Christopher Stevens is a Senior Research Associate at the Overseas Development Institute (ODI), London.

Ian Taylor is Professor in International Relations and African Politics at the University of St Andrews.

Tine Van Criekinge is Fellow in International Relations at the London School of Economics.

Richard Whitman is Professor of Politics and International Relations at the University of Kent.

Preface

This book originates from a workshop which was held at the University of Glasgow in December 2009. In that context, I was encouraged to draft a proposal for a book, whose original title was 'One Europe, One Africa'. The idea back then was to concentrate on the 2007 Joint Africa–EU Strategy and its attendant Africa–EU partnerships. Increasingly, it became evident that this type of approach would be too narrow to fully capture the tensions and contradictions in the relationship between the two 'partners'. For this, I asked a few more contributors to join the project. The completion of the book, thus, took longer than anticipated. The aim was not to focus only on the more established areas of the EU–Africa cooperation – aid, trade, security – but to discuss new areas of cooperation, such as migration, climate change, energy and social policies. Moreover, the ambition was not to treat Africa as a passive actor, an empty space where the EU could just test its international ambitions. Thus, this volume looks at the evolving relationship between the EU and Africa from both the EU's and Africa's perspectives. It should be noted however that, except very brief analyses in some chapters, this volume does not take into account the EU's (lack of a) role in North Africa in the 2011 Arab Spring.

My gratitude goes to a number of people who have contributed to the successful completion of this volume. First, I am indebted to the contributors, who have respected deadlines and reacted to comments in a very constructive way. Second, I would like to thank the team at Manchester University Press. They have all supported and encouraged me at various stages, from the initial proposal, throughout the peer review processes, to the final editing stage. Third, I am grateful to my colleagues and friends in Glasgow, who make my life at the university easier and enjoyable. Lastly, my thoughts are always with my family in Italy. I have now been away for more than 15 years, but my roots are still in my small village, Ceramida.

This book is dedicated to the people of Africa. Over the past decade

I have travelled to Botswana, Djibouti, Egypt, Ethiopia, Gambia, Kenya, Malawi, Senegal, South Africa and Uganda, spending between one week and two months in each country. In these trips I met politicians, civil servants and members of civil society for research purposes and ordinary citizens for my personal growth. Every time I came back to Europe I felt enriched and couldn't wait to start a new trip. Only those who have been in Africa understand what the *mal d'Afrique* entails. I have been lucky enough to suffer from this many, many times.

Ceramida and Glasgow, 30 July 2012

Author's note

The term 'European Union' is generally used to refer to the *sui generis* entity that was created with the Treaty of Maastricht in 1993, whereas for the period between 1957 and 1992 the term 'European Community' is often preferred. In this book, we have used the term 'European Union' for both periods.

Please note that references to the various EU treaties (including Rome, Maastricht, and Lisbon) can be found at this webpage: http://europa. eu/about-eu/basic-information/decision-making/treaties/index_en.htm (accessed 10 July 2012). References to the various Lomé Conventions can be found at this webpage: http://aei.pitt.edu/ (accessed 10 July 2012).

Part I
Introduction

EU–Africa relations in the twenty-first century: evolution and explanations

Maurizio Carbone

With the adoption of the Joint Africa–EU Strategy (JAES) (Council of the European Union, 2007) at the second summit between the European Union (EU) and Africa held in Lisbon in December 2007, it was announced that a new era in relations between the two parties was about to start. Some clashes, nevertheless, had occurred before the summit, when a heated debate took place over the participation of Zimbabwe's President, Robert Mugabe: several EU member states were opposed to it, while all African countries defended their right to decide who should (and who should not) attend the meeting. More generally, the two partners seemed to have diverging agendas: for Europeans, the priorities were security and migration; for Africans, they were aid and trade. Tensions further escalated when the discussion touched upon the new free trade agreements introduced by the Cotonou Agreement (European Union, 2000), the so-called Economic Partnership Agreements (EPAs): several African leaders stated that they were no longer willing to accept any imposition by the EU. This resentment was further alimented by the rise of interest in Africa of 'emerging' powers, most notably China but also the other BRIC countries (Brazil, Russia, India), which offered new partnership alternatives to several African countries.

Against this background, this volume seeks to explain how the relationship between the EU and Africa has evolved in the first decade of the twenty-first century. For this, it treats the EU as both a 'bilateral donor', focusing in particular on the new partnership agreement between the EU and the African, Caribbean and Pacific (ACP) group of countries, and a 'collective actor', paying special attention to the JAES and a number of EU policies that affect African development beyond aid. More specifically, this introductory chapter sets the context for the remainder of the book. The first section sketches the evolution of EU–Africa relations, between the adoption of the Cotonou Agreement in June 2000 and the third Africa–EU Summit held in Tripoli in November 2010. The second section presents some contending explanations, drawing on studies of

EU external relations as well as offering a perspective of Africa. The third section introduces the structure of the volume, giving an overview of the various chapters including the conclusion which summarises the key findings and more generally discusses tensions and contradictions in the EU's policies towards Africa.

An evolving relationship: from Cairo to Tripoli via Cotonou

The evolution of EU–Africa relations should be set against two tracks. The first track concerns the programme managed by the European Commission. In this case, the most important change is certainly the adoption of the Cotonou Agreement, which marked a fundamental departure from the principles of the long-standing Lomé Convention.[1] The second track concerns the attempt to create a continent-wide policy towards Africa, under the slogan 'one Europe, one Africa', which started with the first Africa–EU Summit held in Cairo in April 2000.

The EU–ACP Partnership Agreements

When it was adopted in the early 1970s, the Lomé Convention was greeted as one of the most progressive agreements for North–South cooperation in that it established a contractual right to aid and gave ACP countries non-reciprocal trade preferences. But the disappointing performance of almost all ACP countries, the gradual inclusion of economic and political conditionalities by the EU, and the changing international context led to a profound re-thinking of the EU–ACP cooperation framework. The adoption of the Cotonou Agreement became part of a process of 'normalisation' between the EU and its former colonies. In fact, it was established that: aid allocation would be made conditional not only on recipient needs but also on their performance; new free trade agreements would replace the previous preferential trade regime after an interim period (2000–07); political dialogue, meant to include issues that previously fell outside the fields of cooperation (that is, peace and security, arms trade, migration, drugs and corruption) would be reinforced (Babarinde and Faber, 2005; Flint, 2009).

Less than two years after it came into force in April 2003 (because of the protracted process of ratification), the Cotonou Agreement was revised in February 2005. Its overall structure was not altered, but the few changes reflected the EU's preferences. Security became a more important priority, and the new provisions in this area – such as combating terrorism, countering the proliferation of weapons of mass destruction (WMD), preventing mercenary activities and committing to the International Criminal Court (ICC) – were strongly criticised by African countries (Hadfield, 2007). The second revision, concluded in March 2010, introduced some changes with the view to strengthening cooperation in regional integration and climate change, and acknowledging the role of ACP national parliaments as actors of cooperation (Carbone, 2013). Interestingly, as soon as the second revision was concluded, the relevance of the EU–ACP framework started to be challenged by various parties, and not only in Europe. The issue, however, was not only whether the EU wanted to renew the Cotonou Agreement when it expires in 2020, but also whether the ACP continued to see the EU as its privileged partner. Some observers still pointed out how a number of common interests or shared values called for the revamping of the relationship between the two parties in a less asymmetrical fashion.

At the same time, besides these formal negotiation episodes, EU–ACP relations have been affected, if not compromised, by three issues. Firstly, the European Commission's excessive emphasis on trade liberalisation over development in the negotiations of the EPAs was strongly criticised by African countries (as well as by some EU member states and the NGO (non-governmental organisation) community). Since no EPA had been signed with any of the African regions by December 2007, the negotiations continued after the agreed deadline. Eventually, several countries initialled (that is, they committed to signing) an interim EPA, but at the end of 2010 none of them agreed to fully ratify it (Ngangjoh-Hodu and Matambalya, 2010). Secondly, the risk of a securitisation of EU development policy became more evident in April 2004 when the funds for the newly created African Peace Facility (APF) were taken from the 9th European Development Fund (EDF), which until then had mainly been used for social and economic development. These concerns were partially mitigated by the fact that the APF was used for missions carried out by the African Union (AU) – in Sudan, the Comoros, Somalia and the Central African Republic – and to support capacity-building programmes of the AU Commission and other sub-regional organisations (Sicurelli, 2010). Thirdly, the implementation of the incentive-based approach to democratic governance, which was introduced by the 10th EDF and was meant to reward those countries which took concrete governance commitments, was disappointing. The EU assumed that all ACP countries would engage in reforms in order to get additional funds, but not only did most African countries perceive it as simply making resources conditional on good governance, but the actual disbursement patterns showed that most countries received the same amount of incentives (Molenaers and Nijs, 2009; Carbone, 2010).

The Joint Africa–EU Strategy

The real novelty in EU–Africa relations since the turn of the century has been the EU's attempt to pursue a common (for member states and European institutions) and continent-wide (beyond North and sub-Saharan) approach to Africa. This process started with the first Africa–EU Summit held in Cairo in April 2000, but despite the adoption of a declaration and a plan of action it was clear that European representatives placed more emphasis on political aspects, most notably human rights, democracy and conflict prevention, while African representatives concentrated on economic issues, most notably aid, debt relief and trade opportunities. The second meeting planned for Lisbon in April 2003 was postponed owing to a disagreement over the participation of President Robert Mugabe and other Zimbabwean leaders. Because of the earlier imposition of a travel ban, several EU member states sought to prevent Mugabe from entering the EU area. African leaders saw this as an inappropriate interference, arguing that it was not possible to hold a meeting without all African states being represented (Olsen, 2004).

Meanwhile, several events called for a consolidation of EU–Africa relations. At the international level, global inequality and poverty eradication became a priority for the international community, as witnessed by the adoption of the Millennium Development Goals (MDGs) and the launch of the Doha Development Agenda (DDA). At the African level, the adoption of the New Partnership for Africa's Development (NEPAD) provided a platform for foreign donors and investors to work together in support of African development, while the setting up of the African Union further reassured the international community that African leaders wanted to take ownership of their future. At the EU level, the new commitments to boost the volume of aid (including doubling that for Africa), enhance the quality of aid and promote better policy coherence for development, were complemented by the adoption of the European Security Strategy (European Council, 2003) and the European Consensus on Development (European Union, 2006). With all these initiatives, the EU clearly sought to enhance development effectiveness, but at the same time it was attempting to strengthen its profile in the international arena (Carbone, 2007).

Against this background, the EU adopted a strategy for Africa in December 2005 (Council of the European Union, 2005). This document represented the convergence of two views. On the one hand, the European Commission (2005), in a paper produced by Directorate General for Development (DG Development), focused on foreign aid and policy coherence for development as key tools to achieve the MDGs.

On the other hand, the Council, in a paper produced by the High Representative for the CFSP, Javier Solana (in 2005), saw the promotion of peace and security in Africa as central to the CFSP.[2] The EU's Strategy for Africa received mixed evaluations. True, it represented a remarkable novelty in that it set a framework for a more consistent EU policy towards Africa. Yet, it did not make a qualitative leap because it simply reiterated existing commitments on aid and trade and was adopted with little consultation of relevant stakeholders. Because of this, at the EU–AU Ministerial meeting in Bamako in December 2005 it was agreed to transform the EU's Strategy for Africa into a partnership between Africa and the EU (European Commission, 2007).

This time the drafting process was more participatory, but the second EU–Africa Summit in December 2007 was once again preceded by a division over the presence of President Mugabe: some EU leaders, including the British Prime Minister Gordon Brown, were against his participation, and when it was decided to allow Mugabe to be involved, these leaders decided to pull out. The new Joint Africa–EU Strategy is a much longer and comprehensive document than its predecessor (Council of the European Union, 2007). The starting point was the idea of a 'new strategic partnership', in which the two parties would cooperate not only because of a 'Euro-African consensus on values, common interests and common strategic objectives', but also because 'each one expects to benefit from the other' (Olivier, 2011: 59). The targets were both Africa's development and the global governance architecture; in particular, the two parties strived to promote peace and security, democratic governance, human rights and people-centered development in Africa, and to address issues of common interest in the international arena through effective multilateralism and a more representative system of global governance. To meet these objectives, a detailed action plan for 2008–10 was adopted, which included eight 'Africa–EU Partnerships': peace and security; democratic governance and human rights; trade and regional integration; MDGs; energy; climate change; migration, mobility and employment; science, information society and space. In sum, it seemed that there was hardly any field in which EU and Africa were not meant to cooperate (Schmidt, 2008).

The third Africa–EU Summit held in Tripoli in November 2010, which attracted limited interest by political leaders and the media, recorded a number of disappointments. While a number of observers pointed to the fact that the EU was losing ground *vis-à-vis* the BRIC group, particularly China (Wissenbach, 2009; Bach, 2010), the emergence of these new actors did not seem to affect the EU's overall strategy toward the African continent. In fact, the EU did not make new commitments to increase foreign aid. Moreover, the voices of African leaders complaining that trade barriers and agriculture subsidies 'shamefully' protected European markets from African products, hindering their ability to trade as equals, remained unheeded (Carbone, 2011). The initial implementation of the First JAES Action Plan produced only cumbersome institutional frameworks meant to consolidate dialogue between the two parties, with a large number of meetings and technical activities often seen as the only indication of success. The issue of peace and security was singled out as one of the few areas in which there was some progress: the EU has been among the key sponsors of Africa-led peace-keeping missions and other mechanisms promoting the AU's role as a mediator in political and security crises.

Contending explanations

Until the end of the 1990s, relations between the EU and Africa were conceived mainly in development terms and within the prism of EU–ACP relations. Some scholars argued that the Lomé Convention represented a prime example of successful cooperation between developed and developing states, the latest step in a historical process from colonialism to mutual cooperation and equality. By contrast, more critical observers maintained that the Lomé system perpetuated inequalities between the two parties; in fact, it guaranteed Europe raw materials and a vast market for its manufacturing goods and investment, while African countries faced several obstacles when they tried to exports their products.[3] While the Lomé period has been the object of contentious debate, the post-Lomé period and the EU's attempts to pursue a 'continent-to-continent' approach have attracted limited academic attention.

EU internal factors

This volume seeks to shed light on three different debates within EU studies. The first debate relates to the EU policy-making process (and more generally EU integration theories), particularly the balance between the role of the member states and that of EU supranational institutions. On the one hand, some scholars argue that the EU's Africa policy is the result of the convergence of the preferences of key member states, most notably France and the United Kingdom. This is mostly evident in the case of security policy (Chafer and Cumming, 2011), partially in the case of development policy (Olsen, 2008b) and less so in trade policy (Elgström and Pilegaard, 2008). On the other hand, a group of scholars maintains that the EU's Africa policy reflects the bureaucratic interests of the European Commission, and this has ensured continuity more than change (Babarinde and Faber, 2005). In the field of trade, the Commission has used informational and procedural advantages to increase its autonomy *vis-à-vis* the EU member states in the EPA negotiations (Elgström and Frennhoff Larsén, 2010). In the field of security, it has 'used' Africa to steal competences away from the Council: by framing security in Africa as a development issue not only did it push the Council to take some decisions in security policy (Krause, 2003), but it even replaced it in the management of resources (Sicurelli, 2008).[4]

The second debate is interested in the relations between the EU and Africa as a subset of the EU's external action. In the late 1990s, not only was it clear that the traditional tools used for development had become inadequate, but the EU saw its involvement in conflict prevention and management in Africa as vital in becoming a more significant player in international politics (Olsen, 2008a; Bagayoko and Gibert, 2009; Sicurelli, 2010). This, however, did not necessarily mean that the profile of Africa in the EU's external relations was augmented; for Bayart, in fact, 'EU Member States use Africa policy as a way of creating a united Europe on the cheap, precisely because the stakes in their relations with Africa are so very low' (Bayart, 2004: 454). Similarly, for Hadfield (2007), the injection of securitised principles has transformed EU development policy from a mainstay of EU external relations into a low-profile aspect of EU external relations. By contrast, others argue that the rise of international terrorism has not significantly altered the EU's approach to Africa and most of the initiatives taken in the early 2000s on security policy have actually contributed to raising the profile of the developing world in the EU's overall external affairs (Olsen, 2008b; Carbone, 2011).

The third debate concerns the kind of power the EU is, or wants to be, in the international arena. On one side, there are those who claim that the EU is a realist power. They argue that the EU has pursued its interests, attempting to guarantee itself access to the continent's emerging markets and preserve its interests in international security (Farrell, 2005; Faust and Messner, 2005; Storey, 2006). On the other side, there are those who have investigated the role of the EU as a normative power. They suggest that the EU has sought to promote a wide range of norms, notably liberalisation and democratisation (Elgström, 2000), sustainable development (Flint, 2009), regional integration (Farrell, 2009; Söderbaum and Stålgren, 2010) and the empowerment of African countries, particularly in the fields of environmental politics and human rights (Scheipers and Sicurelli, 2008). In the middle, there are those who maintain that it is actually not possible to clearly separate interests from norms (Langan, 2011), or that the EU has used its relations with Africa as an indirect instrument of structural power by imposing development and general trade liberalisation rather than its specific interests (Holden, 2009).

EU external factors and Africa's perspective

While the existing literature on EU–Africa relations has focused on the EU, increasingly more attention has been paid to Africa's perspectives – which also reflects a general trend in the field of international relations (Smith, 2009). For many years, the prevailing view seemed to be that 'to a very large extent, Africa's position in the global system depend[ed] on the outside world's, mainly the OECD states' assessment of its importance to the affluent and strong countries' (Engel and Olsen, 2004: 15–16). If few Africanists in the 1990s claimed that 'the discourse on Africa's marginality is a nonsense' (Bayart, 2004: 267), following the establishment of NEPAD and the AU, a consensus started to emerge that Africa was taking a more prominent role in world politics (Taylor and Williams, 2004; Taylor, 2011).

While acknowledging, and even supporting, these processes, the EU has not always acted accordingly. In spite of the rhetorical emphasis on partnership, the relations between the EU and Africa in the new century seem to have reinforced the patterns of asymmetrical relations of the previous century. A first group of scholars, who draw on international political economy, have shown that, while the Lomé Convention (at least initially) attempted to incorporate some of the South's claims for special treatment in the international system, the Cotonou Agreement has completely re-designed the EU–ACP relationship by instilling neo-liberalism principles and even re-designing the development strategies of various African states. The EU, by blending ideas of consent (for example, dialogue, partnership, ownership) and coercion (for example, trade liberalisation and aid conditionality), has managed to impose its material (and even normative) interests on weaker partners. The African 'partners', because of their dependence on European markets, had no other choice but to accept (Brown, 2002; Hurt, 2003; Nunn and Price, 2004). A second group of scholars has reached similar conclusions by focusing on the processes that led to the Cotonou Agreement and the JAES. Of course, there are some structural problems but at times the EU has willingly made decisions that have negatively affected Africa's performance in the negotiations. In the EU–ACP framework, the ACP has a small secretariat in Brussels, supported (largely) by the EU itself. In the case of the JAES, the AU Commission, the privileged interlocutor of the European Commission, was at an embryonic stage when the JAES was negotiated and initially implemented (Sicurelli, 2011). But more significantly, the EU has contributed to undermining the idea of group solidarity that bridged the differences between the ACP constituent parts, by proposing sub-regional trade agreements (that is, EPAs) and by adopting three separate regional strategies for Africa, the Caribbean and the Pacific (Slocum-Bradley, 2007).

Another strand of the existing literature has examined the role of the EU in comparative perspectives. In the past, the EU's approach was compared to that of the US, in the fields of foreign aid, trade and security policies (Olsen, 2008b), as well as in relation to more normative issues such as human rights and climate change (Scheipers and Sicurelli, 2008). The conclusion was that the EU generally tended to favour a multilateral approach and (often rhetorically) emphasised the idea of partnership, while the US adopted a more unilateralist approach, with limited consideration for local ownership. Increasingly, a number of scholars have concentrated on the rise of China, at times comparing its approach to that of the EU. In this case, the view that has emerged is that China has promoted South–South dynamics, on the basis of the principles of non-interference in political affairs and mutual benefits. By contrast, the EU has propagated North–South dynamics and supported a comprehensive model for economic, social and political development, with particular emphasis on democratic governance and human rights (Wissenbach, 2009; Carbone, 2011; Men and Barton, 2011). Similar conclusions come from the literature on the external perceptions of the EU. In particular, the EU is clearly perceived by African policy-makers as being driven by its commercial interests, and increasingly so to face competition from China and other BRIC countries (Elgstrom, 2007). More generally, it seems that the colonial legacy still affects the relationship between the two parties, in that it contributes to building expectations of compensation (through aid) and recording disappointments when those expectations are not met, particularly as a result of the EU's tough stance on aid conditionality and economic protectionism (Sicurelli, 2011).

Outline of the book

Following this introductory chapter, this volume is divided into two parts: one on 'actors and contexts', and the other on 'policies and partnerships'. The overall aim is to challenge three widely held assumptions about the role of the EU in Africa, encapsulated in the subtitle of the book (hence the choice of a question mark): incoherent policies, asymmetrical partnership, declining relevance. More specifically, are the

EU's policies towards Africa coherent? Is the EU–Africa relationship asymmetrical in spite of the emphasis on partnership and ownership? Is the EU's relevance in, and for, Africa in the decline (*vis-à-vis* new and more established powers)? Each of the chapters tackles at least one of these three questions, though the various contributors do not necessarily reach similar conclusions.

With the risk of oversimplifying, the three questions can be answered as follows. Firstly, the EU has made more progress than could be expected in making its policies towards Africa more coherent. Tensions between EU member states and supranational institutions still persist in a number of areas, which ultimately challenge the EU's international legitimacy and its self-proclaimed identity as a champion of the interests of the developing world. Secondly, the EU–Africa relationship is less asymmetrical than it is generally believed. However, this is not necessarily because of the EU's rhetorical commitments to partnership and ownership, but mainly because Africa has become a more assertive actor and has used any opportunity to widen its policy space in negotiations with external actors. Thirdly, despite the fact that the EU has increasingly faced competition from a number of 'emerging powers', it is still a major player in Africa – and it is still perceived by its counterparts as such. This position of privilege is due to its comprehensive approach, which goes beyond the pursuit of only material interests and engagement with a selected number of strategic countries. To be sure, comprehensive approach does not equate with coherent approach, as argued in the answer to the first question, nor does it mean that the EU has managed to take full control of its paternalistic reflexes. It simply means that the EU has cooperated with Africa in a large number of fields, or at least more than any other established and emerging power. Of course, not all of the contributors share these conclusions, but this interplay of coherence, partnership and relevance should be kept in mind when readers look at the two parts that comprise this book.

Actors and contexts

The first part of the volume seeks to unravel some the forces that have shaped the evolution of EU–Africa relations since the beginning of the new century. The first three chapters focus on the EU. Fredrik Söderbaum (chapter 2) concentrates on some of the 'internal' challenges the EU faces to implement a coherent approach towards Africa. In particular, he shows a significant variation across policies, which is due to the resistance of the member states to pull sovereignty in some areas, but also to the disunity of the European Commission which has prevented it from exercising leadership. These problems have an impact, of course, on the EU's internal and international legitimacy: genuine actorness, in fact, cannot be built around symbolic presence and declarations emphasising partnership which are detached from sound policy and real performance. Gorm Rye Olsen (chapter 3) questions the uniqueness of the EU as an actor in Africa while discussing the two key 'external' threats to its long-standing privileged position on the continent. The EU may have been more altruistic and responsive to Africa's needs than any other actor, however China has slowly emerged as a potential alternative because of its emphasis on 'mutual interest' and its attempts to project itself as a 'responsible global power'. The US approach by contrast has been largely driven by the pursuit of material interests and Africa has not been not involved in most of its initiatives. A significant difference between the EU and China, however, lies in the fact that the EU tends to emphasise a continent-to-continent approach together with the promotion of regional integration, while China prefers to engage with a few strategic countries. Richard Whitman and Toni Haastrup (chapter 4) argue that sub-Saharan Africa has proved a crucial component in the evolution of an embryonic strategic culture for the EU. By analysing strategic declarations and the various civilian and military operations deployed by the EU in Africa, they find that three different frames have informed the EU's behaviour in Africa: the development–security nexus, the human security imperative and a preference for local enforcement.

The other two chapters in this first part of the volume concentrate on the African side of the EU–Africa relationship. In chapter 5, Ian Taylor discusses the interplay of internal challenges and external opportunities in Africa's posture in world politics. Contrary to the prevailing view of the continent's marginality and irrelevance, he claims that Africa has been inextricably entwined in world politics; in this sense, the establishment of the AU with all its limitations, and before that NEPAD, has only strengthened African agency. Because of this assertiveness, Africa has been able to successfully engage with the new external actors – China and India of course, but also Russia, Iran, Mexico, Turkey – and international corporations searching for economic opportunities, particularly in the area of oil. The novelty in the contemporary rush for resources *vis-à-vis* the colonial scramble, thus, is African agency. In chapter 6, Mary Farrell shows how, amidst a flurry of activity at the sub-regional and continental levels, African countries have taken a number of steps in the direction of closer cooperation with their neighbours, and have launched several initiatives to support regional cooperation and integration. The EU, on its part, has attempted to promote regional integration, but its approach has met with the resistance of African countries and regions, as witnessed by the refusal to sign the highly controversial EPAs and the increased competition in the infrastructure sector coming from China. Similarly, at a more continental level, despite the rhetoric, the two 'partners' have increasingly offered different perspectives on several issues, particularly in the areas of security, migration and environmental policies.

Policies and partnerships

The second part of this volume examines a number of policy areas, ranging from more established areas of cooperation – most notably aid, trade, agriculture and fisheries – to new areas of concern – such as migration, energy, climate change and social policies. The first two chapters look at some interesting dynamics in the area of development cooperation. Maurizio Carbone (chapter 7) finds that in its ambitious agenda on aid effectiveness, the EU has (not necessarily willingly) emphasised donor coordination and division of labour over recipient ownership. More specifically, in the supranational aid programme, the EU has significantly improved its development record by increasingly targeting the poorest countries and delivering aid more efficiently. Yet, it has often imposed its agenda, at times in the guise of governance and increased security, and has failed to engage with recipient governments, parliaments and non-state actors. The EU's 'federator' role has been resisted not only by the EU member states, but also by African countries, who have feared the imposition of stricter conditionalities. Along similar lines, Gordon Crawford (chapter 8) finds a chasm between the rhetoric of policy statements and the reality of implementation in the EU's attempt to promote democracy and human rights in Africa. The EU's self-proclamation of normative power, therefore, is not substantiated in practise; not only have limited resources been allocated, but the EU has been silent in countries where it has strategic and commercial interests. The ironic conclusion is that, if there is any new partnership at all in EU–Africa relations, it is in the rhetorical commitment to values and the lack of pressure to undertake political reforms in the direction of democratisation.

The subsequent two chapters discuss economic relations. In chapter 9, Christopher Stevens argues that in the negotiations of the EPAs, the European Commission has embarked on a losing strategy, which not only has attracted strong criticism from EU member states, civil society organisations and African countries, but has also resulted in a failure to shape policy outcomes – in fact, none of the African regions has signed a full EPA by the agreed deadline, which highlights the limitations of the

EU's 'hard power' even in negotiations with a group of weaker developing countries. Moreover, the EU has jeopardised its efforts to promote regional integration and has backpedalled on other objectives such as enforceable environmental, social and labour provisions. In chapter 10, Alan Matthews maintains that there have been some improvements in the coherence of EU agriculture and fisheries policies with African development. In his view, the number of reforms conducted in the 1990s and 2000s in both policies have brought some benefits to Africa, particularly to African farmers in terms of market access, and to coastal communities with the adoption of the fisheries partnership agreements. At the same time, he shows that a number of incoherencies still persist, such as the EU's resort to export subsidies, the increased number of non-tariff barriers, the strict rules of origin, the disappearance of the commodity protocols and the potential negative impact that preference erosion in both agriculture and fisheries policies may have on food security.

The two chapters that follow focus on two issues that have found increasing space in the global policy agenda: energy and climate change. Amelia Hadfield (chapter 11) examines the emergence and evolution of an energy-development policy nexus in EU–Africa relations. She shows how two broad policy dynamics in Africa, that at first sight appear to have little in common with each other, have established an early set of synergies. New producers in sub-Saharan Africa may contribute to easing the EU's traditional fossil fuel reliance on Russian and Gulf exporters, but these same countries are also far older recipients of EU development aid, having for decades been blighted with underpowered state capacity, poor quality governance and chronic levels of poverty. In chapter 12, Simon Lightfoot points to the importance of climate change in EU–Africa relations. Firstly, climate change has a developmental dimension, hitting Africa earlier and harder than many other countries. Secondly, the EU has attempted to gain support from developing countries, particularly in Africa, in its attempt to shape the post-Kyoto Climate Agreement. However, the EU's efforts to project a vision of normative power and offer global leadership has been significantly weakened by the lack of policy coherence, as well as the limited amount of funding made available to meet any of the priorities identified in the EU–Africa partnership on climate change.

The last two substantial chapters deal with the more social aspects of the EU–Africa partnership: migration and labour rights. In chapter 13, Tine Van Criekinge argues that the issue of migration, which has increasingly become more prominent in both Africa's political agenda and EU's external agenda, has created tensions between the two parties. On the one hand, African governments have attempted to make sure that migration contributes to development, most notably through remittances, 'brain gain' and by attracting external assistance to address its roots causes. On the other hand, the EU has attempted to promote a comprehensive and balanced approach, but has ultimately used a combination of repressive measures and incentives with the view to inducing African countries to comply with re-admission and migration control measures. Besides this preference for a security-oriented approach, the EU external migration policy has suffered from the lack of adequate funding, policy expertise and coordination between EU member states and supranational institutions. In chapter 14, Jan Orbie argues that since the early 2000s the European Union has followed international trends and has explicitly committed itself to promoting the social dimension of globalisation in its external relations, including in Africa. Nevertheless, after meticulous analysis of political agreements, budgetary commitments and trade arrangements, he found that the EU's implementation record has been poor, and social issues have been largely overshadowed by other considerations. This failure, however, has not strained the EU–Africa partnership: in fact the EU's overall emphasis on labour standard has generally been seen by African countries as a neo-colonial agenda, interfering with their sovereignty. By contrast, African policy-makers have generally favoured a paradigm which prioritises growth and development over social issues.

Conclusion

In chapter 15, Michael Smith reviews the main findings of the volume and draws some more general conclusions on EU–Africa relations. Broadly in line with this introduction, he identifies four general themes coming from the various chapters: intra-EU institutional dynamics, tensions in policy implementation, impact, links with the broader world order. Firstly, internal institutional forces have major shaping effects on the formulation and the direction of the EU's Africa policies. The European Commission and the EU member states are at the core of the multilevel policy-making dilemmas faced by the EU. But the chain extends upwards to the institutions of global governance and downwards to the role played by the (post-Lisbon Treaty) EU delegations. Unsurprisingly, the results are a fragmented policy-making process and the related problems of a lack of coherence and consistency. Secondly, there is a highly differentiated set of relationships between the EU and Africa, both in terms of countries and issues. This generates a number of tensions and contradictions between normative and material interests. The EU's commitment to poverty eradication and African development, therefore, is only one part of the story. The other part is represented by the hard-nosed trade negotiations, the resistance to African migration and the increased securitisation of development policy. Thirdly, beyond the rhetoric of equality and partnership, some interesting dynamics are at play in the EU–Africa relationship. On the one hand, the EU's approach often evokes distinctly coercive forms of behaviour. On the other hand, there is evidence that African countries find ways to avoid, resist or re-configure EU policies, because of the agency available to African countries or because of the benefits they manage to reap from the uncoordinated character of EU polices. Fourth, the EU–Africa relationship can be seen as part of the EU's search for order. Granted, the EU may no longer be in a position of dominance as a partner for African countries, because of the growing interest in the continent on the part of China, the USA and other emerging powers. Nevertheless, one of the truly distinctive aspects of the EU's relationship with Africa is the attempt to construct and institutionalise a comprehensive inter-regional partnership and to build into it both the normative and material dimensions of the EU's engagement with the international order.

Notes

1 Besides the relations with the members of the ACP group, the EU has developed formal relationships with North Africa through the Euro-Mediterranean Partnership (EMP) and European Neighbourhood Policy (ENP), and with South Africa through the Trade and Development Cooperation Agreement (TCDA). This introduction, like most of the contributions, focuses only on sub-Saharan Africa.

2 The European Parliament (2005) endorsed the idea of having a 'common' strategy towards Africa in November 2005, but strongly emphasised the need to use conditionality to promote human rights and democratic practices.

3 For a review of EU–Africa relations between the 1950s and the 1990s, which is beyond the scope of this volume, see for instance, Ravenhill (1985), Lister (1988), Grilli (1993), Brown (2002) and Arts and Dickson (2004).

4 In addition to member states and EU institutions, other actors have at times played a role, so multilevel governance approaches have been proposed to explain the evolution of EU–Africa relations (Holland, 2002).

References

Arts, K. and Dickson, A.K. (eds) (2004) *EU Development Cooperation: From Model to Symbol*, Manchester: Manchester University Press.

Babarinde, O. and Faber, G. (eds) (2005) *The European Union and Developing Countries: The Cotonou Agreement*, Leiden: Brill.

Bach, D. (2010) 'The EU's "Strategic Partnership" with Africa: Model or

Placebo?', in O. Eze and A. Sesay (eds), *The Africa–EU Strategic Partnership: Implications for Nigeria and Africa*, Lagos: Nigerian Institute of International Affairs.

Bagoyoko, N. and Gibert, M. (2009) 'The Linkage between Security, Governance and Development: The European Union in Africa', *Journal of Development Studies*, 45 (5): 789–814.

Bayart, J.-F. (2004) 'Commentary: Towards a New Start for Africa and Europe', *African Affairs*, 103 (412): 453–8.

Brown, W. (2002) *The European Union and Africa: The Restructuring of North–South Relations*, London; New York: I.B. Tauris.

Carbone, M. (2007) *The European Union and International Development: The Politics of Foreign Aid*, London: Routledge.

Carbone, M. (2010) 'The European Union, Good Governance and Aid Coordination', *Third World Quarterly*, 31 (1): 13–29.

Carbone, M. (2011) 'The European Union and China's Rise in Africa: Competing Visions, External Coherence, and Trilateral Cooperation', *Journal of Contemporary African Studies*, 29 (2): 203–21.

Carbone, M. (2013) 'Friends with Benefits: The EU, the ACP Group, and the Renegotiation of a Special Relationship', unpublished paper.

Chafer, T. and Cumming, G. (eds) (2011) *From Rivalry to Partnership? New Approaches to the Challenges of Africa*, Farnham: Ashgate.

Council of the European Union (2005) *The EU and Africa: Towards a Strategic Partnership*, 15961/05 (Presse 367), 19 December.

Council of the European Union (2007) *The Africa–EU Strategic Partnership: A Joint Africa–EU Strategy*, 16344/07 (Presse 291), 9 December.

Elgström, O. (2000) 'Lomé and Post-Lomé: Asymmetric Negotiations and the Impact of Norms', *European Foreign Affairs Review*, 5 (2): 175–95.

Elgström, O. (2007) '"Outsiders" Perceptions of the European Union in International Trade Negotiations', *Journal of Common Market Studies*, 45 (4): 949–67.

Elgström, O. and Pilegaard, J. (2008) 'Imposed Coherence: Negotiating Economic Partnership', *Journal of European Integration*, 30 (3): 363–80.

Elgström, O. and Frennhoff Larsén, M. (2010) 'Free to Trade? Commission Autonomy in the Economic Partnership Agreement Negotiations', *Journal of European Public Policy*, 17 (2): 205–23.

Engel, U. and Olsen, G.R. (eds) (2004) *Africa and the North: Between Globalization and Marginalization*, London: Routledge.

European Commission (2005) *EU Strategy for Africa: Towards a Euro-African Pact to Accelerate Africa's Development*, COM (2005) 489, 12 October.

European Commission (2007) *From Cairo to Lisbon: The EU–Africa Strategic Partnership*, COM (2007) 357, 27 June.

European Council (2003) *A Secure Europe in a Better World: European Security Strategy*, Brussels: European Council, 12 December.

European Parliament (2005) *A Development Strategy for Africa*, Resolution, 17 November.

European Union (2000) 'Partnership Agreement between the Members of the African, Caribbean and Pacific Group of States of the One Part, and the European Community and its Member States, of the Other Part', signed in Cotonou on 23 June 2000, *Official Journal of the European Communities*, L 317/3, 15 December.

European Union (2006) 'Joint Statement by the Council and the Representatives of the Governments of the Member States Meeting within the Council, the European Parliament and the Commission on European Union Development Policy: "The European Consensus"', *Official Journal of the European Union*, C 46/1, 24 February.

Farrell, M. (2005) 'A Triumph of Realism over Idealism? Cooperation between the European Union and Africa', *Journal of European Integration*, 27 (3): 263–83.

Farrell, M. (2009) 'EU Policy towards Other Regions: Policy Learning in the External Promotion of Regional Integration', *Journal of European Public Policy*, 16 (8): 1165–84.

Faust, J. and Messner, D. (2005) 'Europe's New Security Strategy: Challenges for Development Policy', *The European Journal of Development Research*, 17 (3): 423–36.

Flint, A. (2009) *Trade, Poverty and the Environment: The EU, Cotonou and the African–Caribbean–Pacific Bloc*, Basingstoke: Palgrave.

Forwood, G. (2001) 'The Road to Cotonou: Negotiating a Successor to Lomé', *Journal of Common Market Studies*, 39 (3): 423–42.

Grilli, E. (1993) *The European Community and the Developing Countries*, Cambridge: Cambridge University Press.

Hadfield, A. (2007) 'Janus Advances? An Analysis of EC Development Policy and the 2005 Amended Cotonou Partnership Agreement', *European Foreign Affairs Review*, 12 (1): 39–66.

Holden, P. (2009) *In Search of Structural Power: EU Aid Policy as a Global Political Instrument*, Aldershot: Ashgate.

Holland, M. (2002) *The European Union and the Third World*, New York: Palgrave.

Hurt, S. (2003) 'Co-operation and Coercion? The Cotonou Agreement between the European Union and ACP States and the End of the Lomé Convention', *Third World Quarterly*, 24 (1): 161–76.

Kingah, S. (2006) 'The New EU–Africa Strategy: Grounds for Cautious Optimism', *European Foreign Affairs Review*, 11: 527–53.

Krause, A. (2003) 'The European Union's Africa Policy: The Commission as Policy Entrepreneur in the CFSP', *European Foreign Affairs Review*, 8 (2): 221–37.

Langan, M. (2012) 'Normative Power Europe and the Moral Economy of Africa–EU Ties: A Conceptual Reorientation of "Normative Power"', *New Political Economy*, 17 (3): 243–70.

Lister, M. (1988) *The European Community and the Developing World: The Role of the Lomé Convention*, Aldershot: Avebury.

Mangala, J. (2011) *Africa and the New World Era: From Humanitarianism to a Strategic View*, New York: Palgrave Macmillan.

Men, J. and Barton, B. (2011) *China and the European Union in Africa: Partners or Competitors?*, Farnham: Ashgate.

Molenaers, N. and Nijs, L. (2009) 'From the Theory of Aid Effectiveness to the Practice: The European Commission's Governance Incentive Tranche', *Development Policy Review*, 27 (5): 561–80.

Ngangjoh-Hodu, Y. and Matambalya, F.A.S.T. (2010) *Trade Relations between the EU and Africa: Development, Challenges and Options beyond the Cotonou Agreement*, London and New York: Routledge.

Nunn, A. and Price, S. (2004) 'Managing Development: EU and Africa Relations through the Evolution of the Lomé and Cotonou Agreements', *Historical Materialism*, 12 (4): 203–30.

Olivier, G. (2011) 'From Colonialism to Partnership in Africa-Europe Relations?', *The International Spectator*, 46 (1): 53–67.

Olsen, G.R. (2004) 'Challenges to Traditional Policy Options, Opportunities for New Choices: The Africa Policy of the EU', *The Round Table*, 93 (375): 425–36.

Olsen, G.R. (2008a) 'Coherence, Consistency and Political Will in Foreign Policy: The European Union's Policy towards Africa', *Perspectives on European Politics and Society*, 9 (2): 157–71.

Olsen, G.R. (2008b) 'The Post September 11 Global Security Agenda: A Comparative Analysis of United States and European Union Policies Towards Africa', *International Politics*, 45: 457–74.

Ravenhill, J. (1985) *Collective Clientelism: The Lomé Conventions and North–South Relations*, New York: Columbia University Press.

Scheipers, S. and Sicurelli, D. (2008) 'Empowering Africa: Normative Power in EU–Africa Relations', *Journal of European Public Policy*, 15 (4): 607–23.

Schmidt, S. (2008) 'Towards a New EU–African Relationship: A Grand Strategy for Africa?', *Foreign Policy in Dialogue*, 8 (24): 8–18.

Sicurelli, D. (2008) 'Framing Security and Development in the EU Pillar Structure: How the Views of the European Commission Affect EU Africa Policy', *Journal of European Integration*, 30 (2): 217–34.

Sicurelli, D. (2010) *The European Union's Africa Policies: Norms, Interests and Impact*, Farnham: Ashgate.

Sicurelli, D. (2011) 'Regional Partners? Perceptions and Criticisms at the African Union', in S. Lucarelli and L. Fioramonti (eds), *External Perceptions of the European Union as a Global Actor*, London and New York: Routledge, pp. 180–94.

Smith, K. (2009) 'Has Africa Got Anything to Say? African Contributions to the Theoretical Development of International Relations', *The Round Table*, 98 (402): 269–84.

Söderbaum, F. and Stålgren, P. (eds) (2010) *The European Union and the Global South*, Boulder, CO: Lynner Rienner.

Storey, A. (2006) 'Normative Power Europe? Economic Partnership Agreements and Africa', *Journal of Contemporary African Studies*, 24 (3): 331–46.

Slocum-Bradley, N. (2007) 'Constructing and De-constructing the ACP Group: Actors, Strategies and Consequences for Development', *Geopolitics*, 12: 635–55.

Taylor, I. (2011) *The International Relations of Sub-Saharan Africa*, New York: Continuum.

Taylor, I. and Williams, P. (2004) *Africa in International Politics: External Involvement on the Continent*, London and New York: Routledge.

Wissenbach, U. (2009) 'The EU's Response to China's Africa Safari: Can Triangular Co–operation Match Needs?', *European Journal of Development Research*, 21: 662–74.

Part II

Actors and contexts

2

The European Union as an actor in Africa: internal coherence and external legitimacy

Fredrik Söderbaum

Historically, the policy of the European Union (EU) towards the African, Caribbean and Pacific (ACP) group of countries has been based on a particular trade–aid relationship with former colonies. With the adoption of the Cotonou Partnership Agreement (CPA) (European Union, 2000), this relationship has been redefined in a variety of ways, and observers have described it as a symmetric 'partnership' between 'equal' partners trying to achieve a common agenda. Since the early 2000s, intercontinental cooperation between the EU and Africa as a whole has become increasingly important. The EU's own Strategy for Africa was a significant step in this direction (European Commission, 2005). However, African leaders criticised it due to the fact that they had not been properly consulted. This led in February 2007 to the initiation of talks on a joint strategy, to be developed and owned by both parties. The resulting Joint Africa–EU Strategy (JAES) (Council of the European Union, 2007) was adopted at the second EU–Africa Summit in Lisbon in December 2007. The JAES now serves as the overarching policy framework for intercontinental relations, complementing rather than replacing other frameworks such as the Cotonou Agreement and the Union for the Mediterranean (former Euro-Mediterranean Partnership, EMP).

Similar to the CPA, the JAES seeks to 'move away from a traditional relationship and forge a real partnership characterised by equality and the pursuit of common objectives' (point 9a). Furthermore, and again similar to the CPA, the JAES is significant in its intention to create a more overtly political relationship between the two continents. The JAES outlines eight thematic strategic partnerships, which reach well beyond the traditional spheres of aid and development, also including partnerships in fields such as peace and security, democratic governance and human rights, trade and infrastructure, energy, climate change and migration. These and other associated strategies reveal that we are facing a major transformation of the EU–Africa relationship. The fundamental question addressed in this chapter is to what extent and under what circumstances the EU should be seen (and is acting) as 'one' in its relations with Africa. More specifically, is the EU best understood as a single and unitary actor, as a dispersed actor or even as no actor at all in its relations with Africa? Is there variation across different policy areas and what explains the outcome? Even if the two aspects are related, the concern is not so much with the content of the EU's policy towards Africa as with the construction of it – and the EU's performance as an actor.

The focus on the internal unity of the EU needs to be understood in the context of the vibrant discussion about the EU's role and influence in world politics. Differing views abound about what type of political animal the EU is, about its nature as an actor and about the impact of its external relations. Today's EU was not designed to become a global actor, and sceptics argue that the EU has ineffective foreign policies due to the diffuse interests of its member states. From such a rather orthodox and realistic perspective, the EU is seen merely as a potential actor in world politics. However, a large number of other scholars argue that today's EU has become a force on the world scene and that it is gradually becoming more unified and coherent, albeit more in some policy areas and counterpart regions and countries than in others.

This chapter compares three specific policy areas: trade, aid and security. These policy areas are particularly interesting because of the varied historical, political, legal and institutional configurations within the EU. Even if the EU mainly speaks with one voice in trade policy, the unity of the EU is debated, particularly whether the Commission, the Council or individual member states are most important for shaping the EU's trade policy with Africa. Compared to the field of trade, the EU's policies in the field of development cooperation and security policy are often regarded as even more ambiguous and pluralistic, not the least because in these policy fields decision-making is either 'shared' between EU institutions and EU member states, or based on inter-governmental decision-making. Key policy-makers, especially from the Commission but sometimes also from individual EU member states, claim that the making of the EU as an efficient and legitimate global actor across different fields of foreign policy depends on a strengthening of the EU's supranational institutions, instruments and policies, where the Commission or the Council must, so the argument goes, play a leading role (Bretherton and Vogler, 2006). Such attempts at centralisation and communitarisation of decision-making and policy are contested, and it is important to analyse the tensions and paradoxes between the EU's central institutions and those of the individual EU member states as these are played out in the different policy areas. Before moving on to this analysis, the next section discusses the concept of the EU as an actor.

The EU as an international actor

Any inquiry of EU's role in Africa depends on our conceptualisation and theorisation of the EU as an actor and as 'one'. Manners (2010) argues that the role of the EU in world politics has been told and untold in a variety of ways over the past five decades, ranging from mythologies of the EU as a bull, a third force, a civilian power, a normative power, a gendered construction of the EU's weakness, and, finally, a pole of power in a multipolar world. Of course, there are more general questions that could be asked about the EU's power and identity and that may influence the EU as an actor and its actorness, but the framework adopted here focuses in particular on the unity of EU's supranational institutions and the relationship between these supranational institutions and the EU member states.

EU actorness is a still under-researched phenomenon, which has come to life due to the transformation of the EU and its increasing outreach in world politics. There have been an increasing number of studies on EU as an international actor and its actorness during the two last decades (Allen and Smith, 1990; Ginsberg, 1999; Jupille and Caporaso, 1998; Bretherton and Vogler, 2006; Söderbaum and van Langenhove, 2006; Orbie, 2008; Söderbaum and Stålgren, 2010a). Actorness for a region is not necessarily the same as for nation-states, although there are of course certain similarities. One unique feature of regional actorness, compared with that of nation-states, is that it must be created by voluntary processes, and be based on dialogue and cooperation rather than coercion and hierarchy. Furthermore, the EU can act as a collective actor in international affairs and be seen as 'one' by both outsiders and its own citizens, for instance when signing a trade agreement or disbursing aid. However, being an international and global actor is more demanding, than simply being a regional organisation or a region. Hence, the fact that the European Commission does 'something' (for example, disbursing aid) is not enough for a claim to possess actorness (purposive capacity to act), at least not for 'substantial' actorness. Similarly, highly stated ambitions and a normative agenda are not automatically translated into actorness, which Christopher Hill elegantly showed in an influential study about the EU's 'capabilities–expectations gap' (Hill, 1993).

An important distinction in the literature is between presence and actorness (Bretherton and Vogler, 2006; Hettne, 2010). The two are closely related. Presence stems from the fact that simply by existing, and due to its relative weight (demographically, economically, militarily and ideologically), the EU has an impact on the rest of the world. Its footprints are seen everywhere. The EU is for example the largest donor in the world, the size of its economy is comparable to that of the US and it is also setting up a military capacity meant to be used outside the region. This provokes reactions and creates expectations from the outside. Presence is a complex and comprehensive material variable, depending on the size of the actor, the scope of its external activities, the relative importance of different issue areas and the relative dependence of various regions upon the European market. A stronger presence means more repercussions and reactions and thereby a pressure to act. In the absence of such action, presence itself would diminish. The crucial question is to what extent the EU's strong international presence is actually transformed into a purposive capacity to shape the external environment by influencing the world, in this case in Africa.

Actorness implies a scope of action and room for manoeuvre, in some cases even a legal personality, which is rare in the case of regions. It suggests a growing capacity to act in a way that follows from the strengthened presence of the regional unit in different contexts. Indeed, literature on EU actorness is very clear that the EU's external policy is closely connected to endogenous and internal conditions, especially coherence and coordination within the Union (Hill and Smith, 2005; Bretherton and Vogler, 2006; Hettne, 2010). This link between the internal and external is evident in the EU's official policy documents and treaties, which repeatedly stress that without a unified, coherent, consistent and coordinated external policy, the legitimacy of the EU as a global actor is called into question.

Horizontal coherence refers to the level of internal coordination of EU policies, such as for instance in the spheres of aid and trade (sometimes referred to as policy coherence), whereas institutional coherence refers to the way the EU's supranational institutions relate to one another (for example, Commission, Council, Parliament and the Court) (Nuttall, 2005). Vertical coherence indicates the degree of congruence between the external policies of the EU's member states and the EU's supranational institutions. Thereby, a variety of coordination mechanisms are at work, such as the community method, the open method of coordination, the inter-governmental method or a strictly national system of foreign policies that takes place outside the EU's structures.

The EU's member states are also engaging with Africa on a bilateral basis, which has to be considered in any analysis of EU actorness. Many member states have developed rather comprehensive Africa policies covering most policy areas. These policies are first and foremost formulated as national policies, although they are often designed to be reinforced through common or shared EU policies/instruments. Two EU member states, the UK and France, stand out since they have the most comprehensive approach to Africa in all three policy areas covered in this chapter. Some member states, such as Sweden and Denmark, are deeply engaged in development cooperation but much less involved in trade and security matters (Söderbaum and Stålgren, 2010a).

Trade

Many scholars agree that the EU is a strong and recognised economic actor. Formally speaking, the EU member states have subordinated themselves to the EU's common trade policy under the community method. Yet, it is misleading to believe that the EU is a unitary actor in this policy field. It is evident that the Union does not always speak with a single voice in international trade negotiations and there is often a complex coordination process leading up to whatever trade policy finally occurs. Perhaps the most crucial issue in the debate about the EU's external trading regime concerns the role and the interplay of the EU member states in the Council and the European Commission (Aggarwal and Fogarty, 2004; Dür and Zimmermann, 2007). Two main groups of scholars dominate this debate. One group emphasises how EU member states control and guide the Commission, whereas another group of scholars claims that the exclusive competence of the Commission provides it with autonomy over the Council and large room for manoeuvre (Elgström and Frennhoff Larsén, 2010).

As far as the EU's trade policy towards Africa is concerned, this policy is very much shaped by the Union's attempt to establish the much talked about Economic Partnership Agreements (EPAs) with geographically more focused subgroups of the ACP, four in Africa, one in the Caribbean and one in the Pacific. The EU claims to be combining trade and aid policies in a new way under the frame of the EPAs. According to the EU's official discourse, 'the idea is to help the ACP countries integrate with their regional neighbours as a step towards global integration, and to help them build institutional capacities and apply principles of good governance. At the same time, the EU will continue to open its markets to products from the ACP group, and other developing countries' (European Commission, 2004: 10). Closer integration of the African countries and regions into the global economy is, according to the EU's official rhetoric, seen as the most promising way to increase the continent's trade volume while simultaneously fostering development. Consequently, this strategy is considered to bring about mutual benefits, both for the EU and for the weaker partner regions (European Commission, 2004: 3).

There is significant criticism in academia, among policy-makers and in civil society against the EU's policy stance in the EPA negotiations. One point of critique is related to the EU's approach to African regionalism, which has been viewed as contradictory. On the one hand, the EU provides economic and political support to regional integration and to the creation of viable regional institutions on the African continent; on the other hand, the EU has been criticised for undermining African regional integration and cooperation, for example through its insistence on negotiating the EPAs with regional groupings that do not match Africa's existing regional economic trading schemes or regional organisations. This strategy appears to be rather difficult to reconcile with the emphasis on an 'equal' partnership so often referred to in the EU's official policies.

This contradiction is closely linked to the underlying purposes of the EU's revised strategy, and especially the individual preferences of certain EU member states. Mary Farrell (2005) argues that contrary to official rhetoric, the EU's Africa policy and the Cotonou Agreement reflect neo-liberal goals as well as the imposition of strong political conditionalities, rather than the normative agenda repeatedly stated in the EU's official discourse. In her view, this represents 'a triumph of realism over idealism'. In another study written by myself and Björn Hettne, this was referred to as 'soft imperialism' in contradistinction to the EU's much-talked about self-identity as a 'civilian' or 'normative power' (Hettne and Söderbaum, 2005; cf. Manners, 2010).

Although the EU is highly committed to free trade in its official rhetoric and also has one 'common' trade policy, there is a rich literature on the diverging preferences of various EU member states. A basic and often referred to distinction is between liberal and protectionist countries. In general, member states such as Denmark, Finland, the Netherlands, Sweden and the UK belong to the former category, whereas the Mediterranean countries tend to belong to the latter. Another line of division goes between the free-trade-oriented position of the Directorate General for Trade (DG Trade) and those EU member states who argue that free trade is the best strategy for growth and development, and the more 'development-friendly' member states who argue that the special needs of the least developed countries (LDCs) must be taken into account (Elgström, 2009: 452). The Nordic countries belong to the 'development-friendly' group, which results in divergences that cannot simply be categorised along the liberal-protectionist axis. However, it appears that the free trade position has been dominant throughout the negotiations.

Previous research on the functioning of the EU suggests that divergences among member states may not only impact negatively on the Council, but may also make the Commission weak, by watering down its proposals to the lowest common denominator. However, as Elgström and Frennhoff Larsén (2010) show, differing preferences among member states in the EPA negotiations provided the Commission with significant autonomy *vis-à-vis* the Council. This autonomy was further reinforced 'due to its informational and procedural advantages given by its institutional position as sole negotiator' (Elgström and Frennhoff Larsén, 2010: 206). Hence, the particular institutional solution within the EU in the field of trade (for example, built around exclusive competence of the EU which strengthened the Commission autonomy) resulted in fairly strong EU actorness.

Elgström and Frennhoff Larsén (2010) emphasise a second factor, namely the importance of Commission unity. According to previous research in the field, internal fragmentation within the Commission may negatively affect its effectiveness and assertiveness *vis-à-vis* member states (Carbone, 2007). It is well-known that there were some tensions between DG Trade and DG Development/DG Agriculture during the early phases of the EPA negotiations. However, gradually the unity within the Commission increased and DG Trade consolidated its leadership (Elgström and Frennhoff Larsén, 2010). Internal coherence of the Commission increased and enabled it to play a dominant role *vis-à-vis* the EU member states and the Council.

Development cooperation[1]

The EU's development policy is rooted in the colonial relations of its member states. From the mid-1960s the then European Community (EC) established special relations with its former colonies beginning with the so-called Yaoundé Conventions. These relations were constantly expanded under the aid and trade agreements of the Lomé Conventions from 1975 on, which were then passed onto the CPA in 2000. With the adoption of the Maastricht Treaty in 1993, development policy was formally introduced as an area with shared or joint competence, where both the EU and the national governments can make decisions. Following nearly a decade of dubious performance in this area, negotiations between the Council and the Commission resulted in a joint policy statement in 2000 stipulating the principles and objectives of the EU's development policy. After the European Parliament was brought into discussions, these principles and objectives were revised in 2005 in the so-called 'European Consensus on Development' (European Union,

2006). This consensus also took into account external events such as the Millennium Development Goals (MDGs) and the terrorist attack in September 2001 (Carbone, 2007).

The EU's objective to become a global development actor has been stimulated by efforts to consolidate the EU internally as well as by global and multilateral development. Indeed, the European Commission's official brief is to systematically and constructively exploit the potential for complementarity and synergy within the EU and to assist the member states in developing their own aid systems and to promote the Union's joint position in the multilateral aid architecture. The debate about a common EU development policy carries a particular emphasis in the delineation of roles between the EU and the member states. The so-called 'value added' of the EU and the European Commission is an important, but also contested, element in the discussion on EU development policy. According to the Commission, 'Community action is more neutral than action by the Member States, which have their own history and are bound by a specific legal system. Community solidarity and the Community's integrated approach to cooperation are undoubtedly major assets' (European Commission, 2000: 4). The Commission also claims to provide 'added value' through its ability to formulate and defend a common European position globally (European Commission, 2004: 7). Against this background, the Commission's priority of promoting a common European position within global and multilateral coordination initiatives, such as the Paris Declaration on Aid Effectiveness and the MDGs, becomes apparent (OECD/DAC 2005a). The EU's official view is that it should strive toward being a single unified actor at all 'levels of governance' in the development community (on the multilateral, inter-regional, regional, as well as country levels) (European Commission, 2004: 6–7).

There is a reasonable and at times relatively sophisticated degree of coordination in the donor community in Africa (OECD/DAC, 2005a, 2005b). This type of coordination is above all taking place within multilateral frameworks, such as the Paris Agenda, the MDGs, Poverty Reduction Strategy Papers (PRSPs) and a variety of budget support mechanisms or sectoral or thematic approaches. There is some success, especially in budget support in several countries in Africa and in sectoral coordination, where we witness a division of labour and specialisation with 'lead donors' as centres of coordination. The European Commission is a large and influential donor in these processes, but the EU does not perform as a coordinating mechanism (Söderbaum and Stålgren, 2010b).

It is fair to say that there is a certain degree of success in EU-based coordination, especially regarding coordination of general policy, both in Brussels and on specific policy issues on the country level in many African countries. In Tanzania, for instance, Heads of Cooperation meetings are taking place on a monthly basis, and in Zambia coordination has been taking place slowly. However, these coordination efforts are mostly limited to agreements on more technical issues or information sharing. The added value of EU coordination appears to lie in procedural matters, monitoring specific EU commitments and in the attempt to promote division of labour (Delputte and Söderbaum, 2012). Hence, the EU cannot be understood as a viable coordination mechanism on the ground. As a senior official of the Delegation of the European Commission in Mozambique points out, describing the Commission's role in the field of HIV/Aids:

> The Commission is almost a Byzantine bureaucracy in certain respects.... All the others harmonize including the Norwegians, the Dutch, the Irish. And they ask me that as we have agreed in principle on so many things, why can't you also take part in this? I try to tell them that it is not because we don't want to, but we have rules. But I know they still think this is an odd position. (Interview in Maputo, Mozambique. February 2005)

The problematic role of the European Commission is shared by other bilateral donors. One donor representative described the Delegation of the European Commission as 'probably the most difficult donor to cooperate with because of its unique and bureaucratic administrative routines and funding mechanisms' (Interview in Maputo, February 2005). Another EU member state representative described the Commission as 'someone who likes to go his own ways and always follows the dictates from Brussels instead of supporting existing coordination efforts' (Interview in Maputo, February 2005). Differently expressed, in contrast to the EU's official rhetoric and the ambitions of the European Consensus on Development, the EU is, by and large, not functioning as a platform for coordination between the EU member states (Söderbaum and Stålgren, 2010b). Since individual EU member states can and do continue to conduct international development policy according to national priorities and preferences, a complete communitarisation of international development cooperation is not politically desirable for many EU member states and would presumably be of questionable value for a number of developing countries (Grimm, 2010). Hence, on the ground in Africa, the European Commission can most of the time be understood as 'just another donor', and within the EU family as 'the 28th member state', conducting its own aid policies, rather than serving as the hub for donor coordination within the EU as a whole. Thus, the

EU demonstrates weak actorness in this policy field, and can hardly be said to be acting as 'one' (Söderbaum and Stålgren, 2010b; Delputte and Söderbaum, 2012).

The Commission's failure to be a coordination mechanism within the EU reflects its inability to present to member states its comparative advantage and a coherent 'value added' proposition relative to other coordination mechanisms. Indeed, it is not clear what the EU can do more effectively than the individual member states, nor in what way it enhances aid effectiveness and aid coordination. The limits of the EU as an actor in this field is closely related to its institutional design, in particular the shared competence between the EU and the member states. It is very clear that many of the biggest EU member states do not want to subordinate their aid policies under the EU umbrella or under the Commission's leadership. Due to shared competence, there is no strong pressure for them to coordinate within the EU. The problems to coordinate are then reinforced by the Commission's bureaucratic weaknesses. There is very clear evidence that the Commission is seen as a technocratic donor burdened with heavy bureaucracy. Bureaucratic obstacles are certainly not unique to the Commission, and many EU donors face legal constraints to use procedures other than their own. Yet, the Commission is 'probably the most difficult donor to cooperate with because of its unique and bureaucratic administrative routines and funding mechanisms' (Interview in Maputo, February 2005).

The failure of the EU to act as 'one' in development cooperation is also related to competing 'identity claims' in the donor community. Development cooperation remains a scene for the manifestation of international identities, not only for the European Commission but also for individual member states. The attempts for a centralised and common European aid policy with the Commission in the driver's seat compete with other identity-driven ambitions of bilateral donors, such as France and the UK, but also most other large EU donors such as Sweden, Denmark and the Netherlands who are generally considered as role models in the field. To the extent that development policy is driven by the ambition to manifest one donor identity, these efforts can be seen as a threat to the identities of the donors. Coordination efforts under the banner of a common donor identity, such as the EU, limit the visibility of the individual donors and member states. As one donor official put it: 'A donor who does not give is not a donor' (Interview in Harare, 2001). The EU's ambition to become a coordination mechanism in the field of aid appears to be closely related to its self-serving ambition to manifest its own identity. A senior policy advisor of a EU member state concurred: 'Development policy is a tool for the Commission to build the

EU as a global actor', which certainly reinforces competition between the EU and the separate member states (Interview in Stockholm, January 2007).

Hence, the identity variable may not be compatible with donor coordination and aid effectiveness. But there are links. To some extent, the identity motive can explain the particular type of policy coordination known as 'lead donor', which is a kind of division of labour where one particular donor is given responsibility for leading a particular sector and the other donors are followers. Being a 'lead donor' enables a donor to manifest its own identity for the sector it is in charge of. Similarly, it appears that donor identities are not challenged as much in the multilateral mechanisms for donor coordination as they appear to be within the EU framework. This may very well be related to the very strong ambitions of the European Commission to control the process, which the national donor agencies, for various reasons, are not accepting.

Security[2]

Literature in this area frequently posits the view of the EU as an economic giant but a political dwarf, with the conclusion that its security policy is weak. The EU has nevertheless begun to demonstrate a considerable amount of activity in the security field, especially since the turn of the century. One reason for this lies in the contemporary conceptualisation of security, which goes well beyond conventional large-scale military presence to include, for example, international terrorism, proliferation of weapons of mass destruction, human security and state fragility. In the face of the multiplicity of new threats, the EU member states have been able to overcome some of their internal differences leading to consolidation of the EU as a global actor, with an evident global ambition.

Following the signing of the Treaty of Amsterdam, the European Security Strategy (ESS) (European Council, 2003), adopted in 2003, has become an important framework for the EU, highlighting both present and future global challenges and key threats to international security. Questions of human security are high on the European agenda for the African continent due to the fact that conflicts frequently either provoke or exacerbate the devastating humanitarian situation in conflict prone areas. In addition, through the discourse on the so-called security–development nexus, the EU stresses the relationship between development and security and emphasises the fact that many conflicts in Africa are tightly linked to state fragility. These two discourses have been developed during the last decade both in Africa-focused strategies and frameworks, as well as on a more general level in the ESS and in the EU's Human Security Doctrine (Council of the European Union, 2004). In effect, this means that the EU can deploy conventional military missions or civilian missions under the Common Security and Defence Policy (CSDP), and at the same time rely on various types of humanitarian aid or long-term development cooperation under the leadership of the European Commission.

What is striking about the Union's activities in large parts of Africa is the high level of intra-EU tension, which has resulted in bureaucratic ineffectiveness at most levels of the organisational structures. The relationship between the Commission and the Council, as well as the relationship between EU actors in Brussels and in the field, all come together in this milieu. The overlapping of responsibilities and rivalry is then compounded by coordination weaknesses springing from structural issues related to the nature of the various mandates and diverse instruments of Commission and Council entities, which sabotages the EU's status as a credible actor in the region (Froitzheim *et al.*, 2011; Vines, 2010).

The 'peace and security partnership' within the JAES provides a framework for cooperation between the EU and Africa in this field. This partnership has three priority actions: (i) to reach common positions and implement common approaches to peace and security in Africa, Europe and globally; (ii) the full operationalisation of the African Peace and Security Architecture (APSA); and (iii) predictable funding for Africa-led peace and security operations (PSOs), especially through the African Peace Facility (APF). From 2004 to 2007, 90 per cent of the APF funds were used for the AU Mission in Sudan (AMIS, 306 million euro), while the rest was allocated to the AU Mission in Somalia (AMISOM, 15 million euro), the CEMAC Mission to the Central African Republic (FOMUC, 33.4 million euro) and the AU Mission in the Comoros (AMISEC, 5 million euro) (Pirozzi, 2009).

Looking at EU's involvement in African conflicts, some of the EU's missions deployed under the CSDP, such as the EUFOR Chad/CAR mission in 2006 and Operation Artemis in the Democratic Republic of Congo (DRC) in 2003, had limited mandates, focusing mainly on the stabilisation of the security conditions and the improvement of the humanitarian situation in geographically confined areas with a short-term perspective. These two missions are often assessed as having achieved their objectives. Yet, it is equally clear that in view of their limited mandates and short time frame, these operations had only marginal impact on the conflicts.

The CSDP missions with a more comprehensive mandate and long-term perspective, such as the ongoing EUPOL and EUSEC RD Congo, can be criticised on a number of accounts. Importantly, their weaknesses are often directly linked to the limits and inefficiencies of the EU as an actor. Indeed, the institutional complexity within the EU is so profound that its role and efficiency are severely constrained by inter-institutional conflict and rivalries within the Council and between different missions, between the responsibilities and activities of the Council and Commission and between the EU as a whole and the particular interests of the member states (especially France and Belgium) (Smis and Kingah, 2010; Lurweg, 2011). Furthermore, although the EU tries to be present on the ground, inadequate exchange of information between the Delegations hampers effective policy design and implementation (Vines, 2010). Dysfunctional EU security governance also arises due to the multitude of actors, an overlap of bilateral and EU policies and top-down approaches from Brussels. All this is then further exacerbated by low staff competence resulting from the overrepresentation of inexperienced and junior employees due to the fact that senior experts are very reluctant to be deployed in the extreme working and living conditions of conflict prone areas in Africa.

In examining the role of the EU in the DRC (which is the conflict with the most comprehensive EU involvement), it is hard to avoid the conclusion that the EU's key political goal is not mainly to solve the myriad of problems – if that is indeed possible in such a difficult environment (Lurweg, 2011). The most important goal appears instead to be to build the EU's presence and identity as a peace-builder and security actor. Starved of resources and beset by institutional in-fighting, bureaucratic turf wars and an inability to deal with the very nature of the entity that passes itself off as the Congolese state, almost without exception EU interviewees in the eastern DRC were dismissive of their own organisation's efforts in conflict management, security sector reform and peace-building. For instance, one EU representative based in eastern DRC felt totally cut off from other EU units, and had no idea what either Brussels or the EU Delegation in Kinshasa did with the information that was provided to them (Interview in Goma, November 2010). Another EU representative in eastern DRC lamented 'I do not know what I am doing here ... the EU's involvement is purely political' (Interview in Goma, November 2010).

Clearly, there is a tendency for all external powers and donors (including individual EU member states) to focus on specific projects in order to get immediate and visible results as a means to justify the expenditure of resources to domestic constituencies. This seems to result from the fact that in complex environments such as the one in DRC, long-term goals are difficult to achieve in the short-term. Yet it is in these complex environments where comprehensive and long-term peace-building strategies are needed the most. When evaluating the EU's policy delivery in the eastern DRC, actual effectiveness is not necessarily as important as projecting a presence and being identified as 'being there', even if being there does not help resolve the ongoing conflict situation. In other words, the EU's activities in the eastern DRC fit Gegout's (2009: 411) dismissive observation that the 'EU missions were carried out first and foremost to promote the EU as a security actor, not to help civilians in conflicts'.

Conclusion

This chapter suggests significant variation in the way the EU can be seen as a unified actor in its relationship with Africa, even if there are also some intriguing similarities across various policy fields. The EU often speaks with one voice in the field of trade. The EU's institutional coherence has increased over time in the EPA negotiations hand in hand with reduced differences within the Commission and the consolidation of DG Trade's leadership position *vis-à-vis* other institutional actors and certain EU member states. Yet, the EU's policy stance in the EPAs is widely contested by civil societies in Europe and in Africa, resulting in a challenge of actorness.

Although the Cotonou Partnership Agreement, the European Consensus on Development as well as the JAES have lead to an improvement in the EU's development policies towards Africa, the Union is not at all a coherent development actor and it is not functioning as a coordination mechanism. Rather, the European Commission is often acting (and perceived) as 'just another donor' or as 'the 28th member state'. Hence, the discussion about the EU as the world's largest donor is to a large extent a play with words since it refers to both common EU aid and the bilateral aid by the individual EU member states (which in effect have relatively little to do with one another). The limits of EU actorness is a consequence of, firstly, the ambiguity surrounding the 'value added' of the EU compared to bilateral and multilateral aid as well as the inability of the Commission to lead the process towards more coherence, and, secondly, the fact that the urge to build the EU's identity as a global aid actor competes with the interests and identities of many individual member states.

The EU's security partnership with Africa has become stronger over time, and the EU is increasingly seen as one security actor on the African continent, not the least through a number of military and civilian peace operations. However, the EU's effectiveness and impact as an actor is severely constrained by failures of inter-institutional synchronisation between the Commission and the Council. This is in turn tightly linked to the fact that the EU is rather selective in its involvement in conflict management and peace-building on the African continent. Several missions were/are very specific and limited (such as Operation Artemis, EUFOR Chad/CAR) and with a marginal impact on conflict management and peace-building. This travels well with the fact that the EU actively tries to reduce/downplay its own role in complex conflicts and humanitarian emergencies through buying in to the politically correct policy of 'African solutions to African problems'. However, the weakness of this policy is that it transfers responsibility and accountability to African institutions without providing them with enough resources and capacities to solve the complex conflicts. Nevertheless, it suits European policy-makers in the sense that the EU's involvement can be selective and limited, and the EU does not have to take the blame for or the shame of failure, while at the same time it can boost the EU's visibility and its identity as a security actor.

Looking at the EU across the three policy fields, there is evidence of the EU as a coherent and unified actor, albeit to different degrees depending on various circumstances and institutional configurations. In many regards, the EU is not capable of pursuing a coherent and coordinated policy. There are too many diverging interests within the Union, and the negative effect on actorness of the EU's complicated and even inefficient institutional machinery should not be underestimated. There is widespread belief that the Lisbon Treaty will help to improve the situation, for example through the abandonment of the pillar structure, the changes to the Council Presidency as well as the establishment of the European External Action Service (EEAS).

The EU's actorness is challenged both by the EU's own identity-building efforts and by its own geostrategic interests, and it remains to be seen whether the Lisbon Treaty will be able to offset such interests. The EU is strongly concerned with building its identity as a global actor and with gaining political power in a realist sense. To the extent that behaviour will be based on geostrategic interests, the EU's policies will be contested both internally within the Union and by its counterparts. For instance, it is quite clear that the EU's trade policies in Africa have failed to instil confidence in its partners, which is seen in the very controversial debate around the EPAs. A long-term stable policy and what is constantly referred to as a 'partnership' can hardly be based on dialogue unless it is a two-way traffic. Importantly, any policy based on narrow self-interest or 'soft imperialism' will meet resistance from within the EU as well as from its partners, which will only undermine the EU as an actor. The EU's quest for identity-building is particularly evident in the field of development cooperation, but also in the security sector. Indeed, there is considerable evidence that too much of what the EU (particularly the Commission) does in these two policy fields is ultimately aimed at building the EU's identity as a global actor. Even if such identity-building can be legitimate for certain purposes, the fundamental problem is that genuine actorness cannot be built mainly around symbolic presence and proud policy declarations that are detached from sound policy and real performance. Under such circumstances, the scope for improvement offered by the Lisbon Treaty and the EEAS are mainly cosmetic and will not change the underlying logic.

Notes

1 This chapter draws on a number of interviews conducted in several African countries and some EU member states in 2001, 2007 and 2010. Interviewees requested full anonymity.
2 This section builds on research undertaken together with Meike Froitzheim within the research project on EU as a Global-Regional Actor in Security and Peace (EU-GRASP) (www.grasp.eu), and has received funding from the European Community's Seventh Framework Programme (FP7/2007-2013) under grant agreement number 225722. Also see Froitzheim *et al.* (2011).

References

Aggarval, V.K. and Fogarty, E. (2004) *EU Trade Strategies: Between Regionalism and Globalisation*, Basingstoke: Palgrave Macmillan.

Allen, D. and Smith, M. (1990) 'Western Europe's Presence in the Contemporary International Arena', *Review of International Studies*, 16 (1): 19–37.

Bretherton, C. and Vogler, J. (2006) *The European Union as a Global Actor*, London and New York: Routledge.

Carbone, M. (2007) *The European Union and International Development: The Politics of Foreign Aid*, London: Routledge.

Council of the European Union (2004) *A Human Security Doctrine for Europe: The Barcelona Report of the Study Group on Europe's Security Capabilities*, Presented to EU High Representative for Common Foreign and Security Policy Javier Solana Barcelona, 15 September 2004, available at: http://www.consilium.europa.eu/uedocs/cms_data/docs/pressdata/solana/040915 CapBar.pdf (accessed 5 March November 2012).

Council of the European Union (2007) *The Africa–EU Strategic Partnership: A Joint Africa–EU Strategy*, Lisbon, 16344/07 (Presse 291), 9 December.

Delputte, S. and Söderbaum, F. (2012) 'European Aid Coordination in Africa: Is the Commission Calling the Tune?', in S. Gänzle, S. Grimm and D. Makhan (eds), *The European Union and Global Development: An 'Enlightened Superpower' in the Making*, Basingstoke: Palgrave Macmillan.

Dur, A. and Zimmermann, H. (2007) 'Introduction: The EU in International Trade Negotiations', *Journal of Common Market Studies*, 45 (4): 771–87.

Elgström, O. (2009) 'Trade and Aid? The Negotiated Construction of EU Policy on Economic Partnership Agreements', *International Politics*, 46 (4): 451–68.

Elgström, O. and Frennhoff Larsén, M. (2010) 'Free to Trade? Commission Autonomy in the Economic Partnership Agreement Negotiations', *Journal of European Public Policy*, 17 (2): 205–23.

European Commission (2000) *The European Community's Development Policy*, COM (2000) 212, 26 April.

European Commission (2004) *A World Player: The European Union's External Relations*, Luxembourg: Office for Official Publications of the European Communities.

European Commission (2005) *EU Strategy for Africa: Towards a Euro-African Pact to Accelerate Africa's Development*, COM (2005) 489, 12 October.

European Council (2003) *A Secure Europe in a Better World: European Security Strategy*, Brussels, 12 December.

European Union (2000) 'Partnership Agreement between the Members of the African, Caribbean and Pacific Group of States of the One Part, and the European Community and its Member States, of the Other Part, Signed in Cotonou on 23 June 2000', *Official Journal of the European Communities*, L 317/3, 15 December.

European Union (2006) 'Joint Statement by the Council and the Representatives of the Governments of the Member States Meeting within the Council, the European Parliament and the Commission on European Union Development Policy: "The European Consensus"', *Official Journal of the European Union*, C 46/1, 24 February.

Farrell, M. (2005) 'A Triumph of Realism over Idealism? Cooperation between the European Union and Africa', *Journal of European Integration*, 27 (3): 263–83.

Froitzheim, M., Söderbaum, F. and Taylor, I. (2011) 'The Limits of the EU as a Peace and Security Actor in the Democratic Republic of the Congo', *Africa Spectrum*, 46 (3): 45–70.

Gegout, C. (2009) 'EU Conflict Management in Africa: The Limits of an International Actor', *Ethnopolitics*, 8 (3): 403–15.

Ginsberg, R.H. (1999) 'Conceptualizing the European Union as an International Actor: Narrowing the Theoretical Capability–Expectations Gap', *Journal of Common Market Studies*, 37 (3): 429–54.

Grimm, S. (2010) 'EU Policies Toward the Global South', in F. Söderbaum and P. Stålgren (eds), *The European Union and the Global South*, Boulder, CO/London: Lynne Rienner Publishers.

Hettne, B. (2010) 'EU Foreign Policy: The Interregional Model', in F. Söderbaum and P. Stålgren (eds), *The European Union and the Global South*, Boulder, CO/London: Lynne Rienner Publishers.

Hettne, B. and Söderbaum, F. (2005) 'Civilian Power or Soft Imperialism: The EU as a Global Actor and the Role of Interregionalism', *European Foreign Affairs Review*, 10 (4): 535–52.

Hill, C. (1993) 'The Capability–Expectations Gap, or Conceptualising Europe's International Role', *Journal of Common Market Studies*, 31 (3): 305–25.

Hill, C. and Smith, M. (eds) (2005) *International Relations and the European Union*, Oxford: Oxford University Press.

Jupille, J. and Caporaso, J.A. (1998) 'States, Agency, and Rules: The European Union in Global Environmental Politics', in C. Rhodes (ed.), *The European Union in the World Community*, Boulder, CO: Westview Press.

Lurweg, M. (2011) 'Coherent Actor or Institution Wrangler? The EU as a Development and Security Actor in Eastern DR Congo', *African Security*, 4 (2): 100–26.

Manners, I. (2010) 'Global Europa: Mythology of the European Union in World Politics', *Journal of Common Market Studies*, 48 (1): 67–87.

Nuttall, S. (2005) 'Coherence and Consistency', in C. Hill and M. Smith (eds), *International Relations and the European Union*, Oxford: Oxford University Press.

OECD/DAC (2005a) Aid Harmonization Web-site, available at: www.aidhar monization.org/ah-cla/ah-browser/indexabridged?area_list=DonorCoop &master=master (accessed 25 January 2005).

OECD/DAC (2005b) 'Survey on Harmonisation and Alignment – Progress in Implementing Harmonisation and Alignment in 14 Partner Countries', Fourth Draft: OECD/DAC.

Orbie, J. (2008) *Europe's Global Role: External Policies of the European Union*, Aldershot: Ashgate.

Pirozzi, N. (2009) 'EU support to African Security Architecture: Funding and Training Components', European Union Institute for Security Studies Occasional Paper 76.

Smis, S. and Kingah, S. (2010) 'Unassertive Interregionalism in the Great Lakes Region', in F. Söderbaum and P. Stålgren (eds), *The European Union and the Global South*, Boulder, CO/London: Lynne Rienner Publishers.

Söderbaum, F. and van Langenhove, L. (eds) (2006) *The EU as a Global Player: The Politics of Interregionalism*, London: Routledge.

Söderbaum, F. and Stålgren, P. (eds) (2010a) *The European Union and the Global South*, Boulder, CO/London: Lynne Rienner Publishers.

Söderbaum, F. and Stålgren, P. (2010b) 'The Limits to Interregional Development Cooperation in Africa', in F. Söderbaum and P. Stålgren (eds), *The European Union and the Global South*, Boulder, CO/London: Lynne Rienner Publishers.

Vines, A. (2010) 'Rhetoric from Brussels and Reality on the Ground: The EU and Security in Africa', *International Affairs*, 86 (5): 1091–108.

3

The EU's Africa policy between the US and China: interests, altruism and cooperation

Gorm Rye Olsen

Africa's international position has changed significantly since the beginning of the twenty-first century. This has very much to do with the rise of China as a global power but it also has to do with the strongly increased American interest in Africa. For some, these changes have challenged the prominent position that Europe has had on the continent for decades. The official rhetoric of the Chinese government is that the Chinese–African relationship is not a threat to anyone (Alden and Hughes, 2009). Nevertheless, non-African actors, not least the European Union (EU) and the United States (US), have been sceptical and openly critical. For China, Africa is a crucial supplier of crude oil, an important supplier of raw material and an attractive market for its manufactured products. For the US, Africa has become more prominent in its foreign policy since the September 2001 terrorist attacks, and Washington has clearly regarded it as an important component in the global war on terror. For the EU, Africa has been considered one of the best geographical areas where it can test its potential as a global actor.

Against this background, this chapter explores the evolution of EU–Africa relations against the rise in interest of both China and the US in the African continent. In particular, the assumption (or the official discourse) is that the EU still plays a prominent role, mainly because it is a different type of actor in respect to the other two 'powers'. In this view, the EU is not motivated by narrow, selfish or 'national' interests – as is the case for Beijing and Washington – but more by some altruistic reasons, and to a certain extent is more attentive and sensitive towards African needs and perspectives. The chapter scrutinises the role of the three different actors in Africa, concentrating on two important policy fields, namely security and development aid. However, before the empirical analysis, it is necessary to introduce the theoretical reflections which inform the following sections, paying special attention to the interests which drive the behaviour of the three actors in question.

Theoretical reflections

This chapter proposes a theoretical framework which is mainly inspired by neo-classical realism, complemented with some elements of small state theory. Further, it is inspired by the debates on the nature of the current world order, and here it draws on liberalism.

Neo-classical realism, like realism, stresses the importance of interests, but most importantly for our analysis is the fact that it seeks to take into account the possible impact of ideas and values on foreign policymaking. Values, ideas and norms become embedded in government institutions and in the broader 'culture' of the state, thus affecting policy decisions (Rose, 1998; Taliaferro *et al.*, 2009). Kitchen (2010), in particular, distinguishes between different types of values and ideas, with so-called 'intentional ideas' being of particular relevance for this paper as they are normative suggestions seeking to establish goals for foreign policy. It is possible to argue, therefore, that the notion that ideas and values may influence international relations conforms to the concept of 'soft power' found in the liberal tradition in international relations. Furthermore, small state theory suggests that issues such as reputation, prestige and esteem are important to many states because they are a means of exercising power in the international arena (Thorhallsson and Wivel, 2006; Petersen, 2000; Lawler, 2007). Foreign policy actors, thus, base policy decisions on their interpretation of what is in the national interest of their own countries. Also, their decisions are influenced by the norms and values embedded in the national institutions, as well as other issues such as the international esteem and reputation of their own countries.

This chapter proposes to place the neo-classical realist approach within a broader theoretical framework, linking the concept of 'intentional ideas' and soft power to the significance of values and norms in international affairs. The liberal approach puts a great deal of emphasis on the development of common norms and values as guidelines for policy initiatives launched by states. In recent years, scholars in this tradition have been preoccupied with the emergence of new global powers such as China and India and the implications for the existing world order. Ikenberry (2008, 2009), for instance, argues that the Western order has a remarkable capacity to accommodate rising powers and in fact has already started facilitating Chinese integration (see also Koivisto and Dunne, 2010). The increasing interdependence between countries (regions and continents) is supposed to push the development of common norms and values. It is also expected to exert strong pressure for international cooperation which, some years ago, led Krasner to reflect on the possible emergence of regimes – defined as 'principles, norms, rules and decision-making procedures around which actor expectations converge within a given issue area' (Krasner, 1983: 1) – as an indication of increased international collaboration within specific policy fields, such as security or development policy (see also Levy *et al.*, 1995).

In line with liberalism, this paper assumes that the increasing interdependence between states (regions and continents) puts pressure on the US, China and the EU to act in relation to Africa. It does not assume, however, that this pressure leads to the establishment of a regime aimed at addressing the challenges of Africa. Nor does it assume that a regime focusing on a given issue area such as security or development assistance is emerging. It is instead expected that these three external actors intervene in Africa on the basis of two (more or less) interlinked interests. Firstly, they may react on the basis of what they consider their 'national' interests, which is basically in agreement with the realist tradition. Secondly, they may also react, to a greater or lesser extent, on the basis of soft power concerns, such as intentional ideas, norms and values, but also prestige and reputation.

Different actors, different interests

There is a general consensus that China is in Africa mainly for economic reasons. The increasing demand for raw materials and natural resources is closely linked to rampant industrial growth in China. China has become the world's second largest consumer of oil, with Africa playing a very important role in servicing this demand (Jiang, 2009). It also imports significant proportions of its consumption of iron ore, copper, cobalt, platinum, cobalt, timber and other raw materials. With food security also becoming a growing concern for Beijing, the Chinese government has encouraged investments in African agriculture and fisheries (Naidu and Mbazima, 2008). Africa is particularly appealing for Chinese companies as they increasingly seek new markets for their goods, but more generally, the Chinese government has promoted a 'going global' strategy with the aim of making Chinese corporations 'more competitive in the international economy and to benefit from global commercial learning' (Naidu and Mbazima, 2008). In addition to these strong economic interests, there is also a political-strategic interest to take into account: the traditional Chinese approach to cooperation – emphasising mutual benefits and state-to-state interactions – gives Beijing significant influence on their African partners. The overall aim is that of making new allies, and, by doing so, Beijing expects to strengthen the African voting bloc within the UN and have a group of countries challenging the global dominance of the US and the West (Alden, 2007; Chong, 2008).

Historically, American interests in Africa have been marginal. While in previous decades Africa was of secondary importance compared to other regions, there is no doubt that since the beginning of the 2000s it has acquired a much more prominent position in overall US foreign policy strategy (Olsen, 2011). In the post-9/11 period, which radically changed the American approach to, and perception of Africa, US policy on the continent has been driven mainly by two concerns: security and access to oil. The global war on terror made the US realise that Africa was an important component within the international security architecture. The growing American dependence on oil from Africa, however, is a long-standing issue and therefore is not linked to the events of September 2001 (Olsen, 2008). In fact, as part of the overall American strategy to reduce its dependence on Middle Eastern oil, US companies have for a number of years been heavily involved in oil exploration in Africa. In the mid-2000s, for instance, the US relied on as much as 15 per cent of its total oil imports from Africa, and that level is expected to increase to 25 per cent by 2015 (Pham, 2005).

In the case of the EU, it can be difficult to clearly identify a coherent approach to Africa. The major development aid programme is, to a certain extent and with some reservations, an illustration of the EU's commitment to economic and social development. But the new global context places these traditional aims into a new setting, linking development very closely to security. Adding an actor dimension to this discussion, it can be argued that one of the main interests for the EU in Africa is the explicitly formulated ambition to play a greater role on the international scene, based on the self-awareness of being a unique international actor. This may be the result of the activities of the European Commission, which has primary responsibility in important policy areas, such as trade and (partially) foreign aid. At the same time, the Council secretariat has developed a strong interest in Africa, by promoting a number of ESDP missions. Finally, the potential role of individual EU member states adds further complications. Some member states, in particular France, may have an interest in Europeanising their bilateral Africa policies, with the EU becoming a 'tool' to pursue their national interests (Bagoyoko and Gibert, 2009).

In sum, the three actors have diverse interests in relation to Africa. Both the US and China are clearly preoccupied with their own national interests. However, China is not driven only by commercial interests, but also by some political-strategic interests, which, of course, are also selfish as far as they contribute to increasing Beijing's international influence. When it comes to the European Union, the picture is more ambiguous. It can be argued that the EU's interest in respect to the continent is both altruistic and selfish. Moreover, it should be noted that to, more than the other two actors, the EU has engaged in dialogue with the African Union (AU). At the same time, while the EU and the AU may have a number of interests in common, they certainly have different interests: whereas the European Union is increasingly preoccupied with security and stability, the African Union (and of course its members) is more interested in issues related to trade and economic development (Sheriff and Ferreira, 2010; Mair and Peretto, 2010).

Security interventions

Since the turn of the century, China has been increasingly involved in UN peace-keeping operations. Almost 75 per cent of its peace-keepers are deployed in Africa, most notably in Darfur, the Democratic Republic of the Congo (DRC), Cote d'Ivoire, the Eritrea-Ethiopia border. Moreover, Chinese troops only participate in multilateral operations which have an explicit UN mandate (Shelton, 2008; Gill and Huang, 2009). This involvement in UN peace-keeping operations is the result of political choices. Chinese leaders have attempted to build an image of the country as 'a responsible global power', which accepts its role in advancing global peace and stability (Gill *et al.*, 2007; Shelton, 2008). For Taylor (2008: 8), 'China's emerging role in peace operations is a part of a pragmatic reorientation and reassessment of Beijing's political interests by policy-makers, who are now more concerned with looking like a responsible great power and less of a developing country bent on protecting state sovereignty at all costs.' An important example is its role in the fight against piracy off the coast of Somalia. China contributed a limited number of patrol vessels, which were not only deployed to assist Chinese cargo ships and tankers transporting oil from Port Sudan (Pham, 2009), but also sent a clear signal 'that China wishes to participate more actively in international security' (Ji and Kia, 2009: 1). It has to be mentioned, however, that the Chinese vessels were not part of a coordinated, international operation, even though they engaged in intelligence sharing with other countries operating in the region (Ji and Kia, 2009).

The US has been heavily engaged in African security since September 2001. The idea, clearly expressed in the 2002 American National Security Strategy (White House, 2002), was to focus on those countries which could represent a threat to international stability. US military spending has substantially increased compared to the 1990s (Chong, 2008: 21). Moreover, the strong influence from the Pentagon and the CIA contributes to explaining the inclination towards choosing military means and instruments (Schraeder, 2006; McFate, 2008; Olsen, 2008; Ploch, 2009). At the same time, while there has been a strong reluctance to deploy troops on the ground, there have also been increased offers to provide training to African armies. Nevertheless, the American approach to conflict management in Africa has been influenced by traditional great-power thinking. National interests have been in the forefront, so much so that the US has tended to forget its African partners (Burgess, 2009; Berman, 2009). For Nathan (2009: 60), the failure of the US administration to consult the African Union and African leaders when it launched Africom cannot be seen simply as a 'communications lapse but as indicative of the superpower's arrogance, ignorance of African politics and disregard for the efforts of Africans to enhance their own security'.

The EU has been particularly active in promoting security in Africa, both through military and civilian missions, and by supporting the African Peace and Security Architecture (APSA). Throughout the 2000s it launched several military operations aimed at managing violent conflicts on the continent. Some have been particularly important for the EU as a global security actor, so much so that some observers have talked of Africa as a field of experimentation for the evolving ESDP (Bagoyoko and Gibert, 2009) In particular, Operation Artemis, in the crisis-ridden Ituri province of the DRC in the summer of 2003, was the first military mission to deploy a pure EU military force outside of Europe, and was implemented without using NATO facilities. Operation Atalanta, launched in the Gulf of Aden and in the Indian Ocean in December 2008, was the first EU naval operation. On the one hand, it was meant to deter piracy and support the vessels of the World Food Programme deliver food aid to Somalia. On the other hand, it was meant to protect European vessels whilst strengthening the Union's international ambitions by deploying a naval force (Chafer and Cumming, 2010). All the African ESDP missions illustrate the challenge of interpreting the nature of European interests. There is no doubt that France has played a decisive role in bringing about the decisions which have led to the deployment of EU troops in most crisis situations. At the same time, the other member states in the Council voted in favour of the deployment of the troops knowing that France had specific interests in most cases, but also that in most cases it supplied the majority of the soldiers (Olsen, 2009; Chafer and Cumming, 2010). More significantly, the overall aim to establish the EU as an influential international actor by means of its involvement in conflict prevention and management in Africa should never be forgotten.

Parallel to these 'unilateral' missions, the EU has been engaged in 'multilateral' operations in close cooperation with the AU. Through the African Peace Facility (APF), the EU has handsomely supported the AU with financial and capacity-building resources. This was most clearly the case with the two AU operations in Darfu/Somalia (AMIS I and AMIS II), but also in the AU missions in Burundi, the Central African Republic and the Comoros. For Cilliers (2008: 12), 'without the African Peace Facility, it is unlikely that the AU would have been able to undertake any of these missions. Since implementation of the African Peace Facility, the relationship between the AU and the EU has developed quite strongly, resulting in the EU strategy for Africa'. The establishment of the APF can be explained by the wish to take care of European interests. Firstly, it was in the EU's interests to manage, and if possible, contain violent conflicts in Africa, not least because of their potential to spillover into Europe. Secondly, the EU member states wanted to avoid sending European soldiers into some remotes areas of Africa (Olsen, 2009). Thirdly, strengthening the APF, and more generally the APSA, was a way for the two key EU member states, the UK and France, not only to increase their influence over European security, but also to maintain their permanent seats in the UN Security Council (Chafer and Cumming, 2010). The EU–AU collaboration is based on a clear division of labour: the African Union deploys troops and the EU supplies economic resources and advice in support of the troops, disarmament and general elections. No doubt the APF is a unique instrument, and actually one of the few means the AU has at its disposal to intervene in the field of conflict management. However, the situation where the EU is relegated to the role of mere funder of the AU's initiatives does not seem to supply the final answers to the complex security situations in Africa. In fact, the EU has only been partially successful in reinforcing the capacity and impact of the AU on African and international politics (Pirozzi, 2009; Sherriff and Ferreira, 2010).

Summing up, the American interventions in Africa are based on more or less narrow national security concerns, whereas the Chinese and European interventions are driven by 'national' interests mixed up with their ambition to become important international players in their own right. However, it still needs to be emphasised that the EU, despite a number of limitations, has somehow been active in supporting the efforts of the AU in the implementation of the African Peace and Security Architecture. Neither is it to be neglected that the EU has been willing to deploy troops on the ground in order to maintain peace and stability until the UN could take over. It is also important to note that China has gradually become more willing to engage in conflict management in Africa, in particular by contributing with troops to UN missions. This ultimately suggests that China is slowly adjusting to 'Western' norms and values within the need for increased international cooperation.

Development aid interventions

In recent years, China has made significant steps into Africa by economic means. Aid, in particular, has caused the most worries and criticism among traditional donors. There seems to be general agreement that Chinese development assistance does not qualify as foreign aid according to DAC criteria. In fact, it is difficult to separate aid from other economic development instruments, since Beijing uses a whole range of different financial instruments in combination. China bundles aid and state-subsidised loans into a non-transparent system, prefers bilateral arrangements, ties aid to its goods and services and uses project aid in sharp contrast to the DAC consensus of transparency, untied aid, donor coordination, programme aid and budget support (Mohan and Power, 2008). In sum, it seems that China 'operates outside the global aid regime' (Brautigam, 2008: 212). In spite of, or rather because of, these characteristics, Chinese aid is very attractive to many African countries: it is seen as an additional source of funding and it is not tied to any political or economic conditions (Brautigam, 2009). In fact, China may be criticised for having pursued its own political and strategic interests when giving aid to Africa, but for some observers it is 'much like the traditional donors did' (Samy, 2010: 82–3). Moreover, it should be added that China has increasingly started sharing interests with Western countries in a number of respects such as promoting good governance and stability, which are increasingly perceived as preconditions in reaping the benefits from its involvement in Africa (Ferreira, 2008: 4; Gu *et al.*, 2008: 289).

US development assistance to Africa has increased rapidly since the early 2000s (Radelet and Bazzi, 2008), but at least one-third can be classified as 'military' (Dagne, 2006). US foreign aid policy, in fact, has been strongly influenced by its national security interests. The 2002 American National Security Strategy recognised that poverty and weak and failing states may represent a threat to international stability and ultimately to the United States. Moreover, not only did it express reservations about how aid had been used in Africa, but it also emphasised the fact that it was legitimate to use aid as a tool to enhance US national security interests rather than as a means to promote economic and social development in recipient countries (Hills, 2006). Therefore, it is hardly a coincidence that USAID under George W. Bush was placed under direct control of the Department of State (Copson, 2009). Neither, is it a coincidence that USAID on its official internet homepage has highlighted that the American core priorities in sub-Saharan Africa are related to issues of security and good governance. Unsurprisingly, since 2000, the patterns of aid allocations to Africa have changed: a decreasing share has gone to the poorest countries, while some 'strategic' countries (for example, Angola, Ethiopia, Sudan, Uganda) have experienced significant increases. Also, it appears that since debt relief, humanitarian assistance and HIV/AIDS programmes accounted for a considerable proportion of the increased aid, it would be difficult to categorise all the new available resources as 'pure' development assistance (Radelet and Bazzi, 2008). More significantly, the strong focus on security interests makes it difficult to accommodate the development priorities of the recipient governments and to realise the norms on harmonisation, alignment and ownership as enshrined in the 2005 Paris Declaration on Aid Effectiveness (OECD, 2008).

Also EU aid to Africa has significantly increased since the early 2000s. At the EU Summit in Barcelona, the EU member states promised to make an efforts to set aside at least 0.39 per cent of their collective GNI by 2006. The goal was revised upwards in 2005 as the EU countries agreed to reach 0.51 per cent in 2010 and 0.7 per cent by 2015 – it was also decided that half of those increases would be allocated to African countries (Olsen, 2008). Nevertheless, official figures for 2010 suggest that the EU member countries have fallen short of their commitments and failed to reach the declared target. There is no doubt that the EU is committed to promoting economic and social development in Africa and to the achievement of the Millennium Development Goals (MDGs). This is also shown by its contribution to the global agenda on aid effectiveness. But there is also no doubt that the EU has become increasingly preoccupied with security issues (Olsen, 2009). Orbie and Versluys (2009: 77), for instance, argue that a 'securitisation' of European development policy has taken place since the early 2000s. The significant amount of money allocated for security-related programmes ultimately seem to suggest that selfish concerns have become increasingly more important in comparison to altruistic concerns.

Conclusion

This chapter has investigated the evolution of the relationship between the European Union and Africa *vis-à-vis* the US and China, drawing on a theoretical framework which has emphasised the role of interests, together with that of norms and values in shaping policy decisions. The starting point was the view that the EU would be able to maintain its traditionally strong position on the continent, in spite of the Chinese and American advance. This would be because the EU's approach is allegedly different in that it is not based solely on selfish and (more or less narrow) national interests, but is also motivated by elements of altruism, including a degree of responsiveness to African concerns and demands. The empirical analysis, however, does not support this argument. Rather, it points towards a different conclusion, namely that it is not very easy to detect significant differences between the reasons that drive the EU and China in their approach to Africa. All this means that the traditional prominent position of the European Union in Africa is under significant pressure, not so much from the US but more from China.

More specifically, in the field of security, China and the European Union do not base their policies solely on narrow national and material interests. The only significant difference between Beijing and Brussels appears to be the stronger EU commitment to cooperate with the AU, as witnessed by the establishment of the African Peace Facility. In the case of the US, the strongest concern has been the fight against terrorism. The decision to establish Africom also points to how differently the US and EU have responded to the demands of African leaders. In the case of development aid, the fact that the US gives large amounts of resources to Africa to promote its national security agenda is hardly surprising. The official Chinese position is that its development assistance is of mutual benefit to both parties: in return for its assistance, especially in the infrastructure sector, China gets access to markets. At the same time, China has strived to strengthen its role in the international arena and project the image of a responsible global power. Similarly, in the case of the EU, the large aid programme is a reflection of the impact of norms and values, but it is also a reflection of the continuing ambition to become a significant international actor.

References

Alden, C. (2007) *China in Africa*, London: Zed Books.

Alden, C. and Hughes, C.R. (2009) 'Harmony and Discord in China's Africa Strategy: Some Implications for Foreign Policy', *The China Quarterly*, 199: 563–84.

Bagoyoko, N. and Gibert, M.V. (2009) 'The Linkage between Security, Governance and Development: The European Union in Africa', *Journal of Development Studies*, 45 (5): 789–814.

Berman, E.G. (2009) 'Déjà-vu all over again: US Missteps in Promoting Peace in Africa', *Contemporary Security Policy*, 30 (1): 32–4.

Brautigam, D. (2008) 'China's Foreign Aid in Africa: What do We Know?', in R.I. Rotberg (ed.), *China into Africa*, Washington, DC: Brookings Institution Press, pp. 197–216.

Brautigam, D. (2009) *The Dragon's Gift: The Real Story of China in Africa*, Oxford: Oxford University Press.

Burgess, S. (2009) 'In the National Interest? Authoritarian Decision-Making and the Problematic Creation of the US Africa Command', *Contemporary Security Policy*, 30 (1): 79–99.

Chafer, T. and Cumming, G. (2010) 'Beyond Fashoda: Anglo-French Security Cooperation in Africa since Saint-Malo', *International Affairs*, 86 (5): 1129–48.

Chong, X.-Y. (2008) 'China and the United States in Africa: Coming Conflict or Commercial Coexistence?', *Australian Journal of International Affairs*, 62 (1): 16–37.

Cilliers, J. (2008) 'The African Standby Force: An Update on Progress', ISS Paper 160, March.

Copson, R.W. (2009) 'Out of the Kilter: US Assistance to Sub-Saharan Africa', available at: www.royalafricansociety.org/index.php?id=414&option=com_contnet&task=vi (accessed 12 October 2009).

Council of the European Union (2007) *The Africa–EU Strategic Partnership: A Joint Africa–EU Strategy*, adopted at the Second EU–Africa Summit, Lisbon, 9 December.

Dagne, T. (2006) *Africa: US Foreign Assistance Issues*, Washington, DC: Congressional Research Service.

European Union (2008) *European Union Military Coordination of Action Against Piracy in Somalia (EU NAVCO)*, Brussels: Council Secretariat.

Ferreira, P.M. (2008) 'Global Players in Africa: Is There Scope for an EU–China–Africa Partnership?', *Estratégia*, available at: www.ieei.pt/files/6PMFerreira.pdf (accessed 30 June 2012).

Gill, B. and Huang, C.-H. (2009) 'China's Expanding Peacekeeping Role: Its Significance and the Policy Implications', SIPRI Policy Brief, February.

Gill, B., Huang, C. and Morrison, J.S. (2007) 'Assessing China's Growing Influence in Africa', *China Security*, 3 (3): 3–21.

Gu, J., Humphrey, J. and Messner, D. (2008) 'Global Governance and Developing Countries: The Implications of the Rise of China', *World Development*, 36 (2): 274–92.

Hills, A. (2006) 'Trojan Horses? USAID, Counterterrorism and Africa's police', *Third World Quarterly*, 27 (4): 629–43.

Ikenberry, G.J. (2008) 'The Rise of China and the Future of the West', *Foreign Affairs*, January/February.

Ikenberry, G.J. (2009) 'Liberal Internationalism 3.0: America and the Dilemmas of Liberal World Order', *Perspectives on Politics*, 7 (1): 71–87.

Ji, Y. and Kia, L.C. (2009) 'China's Naval Deployment to Somalia and Its Implications', *EAI Background Brief No. 454*, 29 May 2009.

Jiang, W. (2009) 'Fuelling the Dragon: China's Rise and Its Energy and Resources Extractions in Africa', *The China Quarterly*, 199: 585–609.

Krasner, S.D. (1983) 'Structural Causes and Regime Consequences: Regimes as Intervening Variables', in S.D. Krasner (ed.), *International Regimes*, Ithaca, NY: Cornell University Press, pp. 1–21.

Krasner, S. (1995) 'Sovereignty, Regimes and Human Rights', in V. Rittberger (ed.), *Regime Theory and International Relations*, Oxford: Oxford University Press, pp. 391–430.

Kitchen, N. (2010) 'Systemic Pressures and Domestic Ideas: A Neoclassical Realist Model of Grand Strategy Formation', *Review of International Studies*, 36: 117–43.

Koivisto, M. and Dunne, T. (2010) 'Crisis, What Crisis? Liberal Order Plus Building World Order Conventions', *Millennium: Journal of International Studies*, 38 (3): 615–40.

Lawler, P. (2007) 'Janus-Faced Solidarity: Danish Internationalism Reconsidered', *Cooperation and Conflict*, 42 (1): 1010–26.

Levy, M., Young, O.R. and Zürn, M. (1995) 'The Study of International Regimes', *European Journal of International Relations*, 1 (3): 267–330.

McFate, S. (2008) 'US Africa Command: Next Step or Stumble?', *African Affairs*, 107 (426): 111–20.

Mair, S. and Petretto, K. (2010) 'Peace and Security: Unremitting Challenges for African–European Relations', in *Beyond Development Aid: EU–Africa Political Dialogue on Global Issues of Common Concern*, Portugal: Europe Africa Policy Research Network.

Mohan, G. and Power, M. (2008) 'New African Choices? The Politics of Chinese Engagement', *Review of African Political Economy*, 115: 23–42.

Naidu, S. and Mbazima, D. (2008) 'China–African Relations: A New Impulse in a Changing Continental Landscape', *Futures*, 40: 748–61.

Nathan, L. (2009) 'Africom: A Threat to Africa's Security?', *Contemporary Security Policy*, 30 (1): 58–61.

OECD (2008) *The Paris Declaration on Aid Effectiveness and the Accra Agenda for Action*, Paris: OECD, available at: www.oecd.org/development/aideffectiveness/34428351.pdf (accessed 5 March 2012).

Olsen, G.R. (2008) 'The Post September 11 Global Security Agenda: A Comparative Analysis of United States and European Union Policies towards Africa', *International Politics*, 45 (4): 457–74.

Olsen, G.R. (2009) 'The EU and Military Conflict Management in Africa: For the Good of Africa or Europe?', *International Peacekeeping*, 16 (2): 245–60.

Olsen, G.R. (2011) 'Civil–Military Cooperation in Crisis Management in Africa: American and European Union Policies Compared', *Journal of International Relations and Development*, 14 (3): 333–53.

Omotola, J.S. (2010) 'Globalization, New Regionalism and the Challenge of Development in Africa', *Africana*, June, 103–36.

Orbie, J. and Verslys, H. (2009) 'The European Union's International Development Policy: Leading or Benevolent?', in J. Orbie (ed.), *Europe's Global Role: External Policies of the European Union*, Farnham and Burlington: Ashgate, pp. 67–90.

Petersen, N. (2000) 'National Strategies in the Integration Dilemma: The Promises of Adaptation Theory', in H. Branner and M. Kelstrup (eds), *Denmark's Policy towards Europe after 1945: History, Theory and Options*, Odense: Odense University Press, pp. 72–99.

Pham, J.P. (2005) 'US national interests and Africa's strategic significance', *American Foreign Policy Interests*, 26: 19–29.

Pham, J.P. (2007) 'Panda in the Heart of Darkness: Chinese Peacekeepers in Africa', *World Defence Review*, available at: http://worlddefensereview.com/pham102507.shtml (accessed 18 September 2009).

Pham, J.P. (2009) 'Return of the Somali Pirates', *World Defence Review*, available at: www.defenddemocracy.org/index.php?option=com_content&task=view&id=11 (accessed 25 November 2009).

Pirozzi, N. (2009) *EU Support to African Security Architecture: Funding and Training Components*, Occasional Paper, February, No. 76, Paris: European Institute for Security Studies.

Ploch, L. (2009) *Africa Command: US Strategic Interests and the Role of the US Military in Africa*, CRS report for Congress, 2 October.

Radelet, S. and Bazzi, S. (2008) 'US Development Assistance to Africa and the World: What Do the Latest Numbers Say?', CCG Notes, Centre for Global Development, 15 February, Washington, DC.

Rose, G. (1998) 'Neoclassical Realism and Theories of Foreign Policy', *World Politics*, 51 (1): 144–72.

Samy, Y. (2010) 'China's Aid Policies in Africa: Opportunities and Challenges', *The Round Table*, 99 (406): 75–90.

Schraeder, P. (2006) 'The Africa Dimension in US Foreign Policy in the Post-9/11 Era', Paper prepared for IV Flad-IPRI International Conference on 'Strategy and Security in Southern Africa', Lisbon, 12–13 October.

Shelton, G. (2008) 'China: Africa's New Peacekeeper', *The China Monitor*, 33: 4–5.

Sheriff, A. and Ferreira, P.M. (2010) 'Between the Summits: Background Paper', in *Beyond Development Aid: EU–Africa Political Dialogue on Global Issues of Common Concern*, Portugal: Europe Africa Policy Research Network.

Taliaferro, J.W., Lobell, S.E. and Ripsman, N.M. (2009) 'Introduction: Neoclassical Realism, the State and Foreign Policy', in S.E. Lobell, N.M. Ripsman and J.W. Taliaferro (eds), *Neoclassical Realism, the State and Foreign Policy*, Cambridge: Cambridge University Press, pp. 1–41.

Thorhallsson, B. and Wivel, A. (2006) 'Small States in the European Union: What Do We Know and What Would We Like to Know?', *Cambridge Review of International Affairs*, 19 (4): 651–68.

Taylor, I. (2008) 'China's Role in Peacekeeping in Africa', *The China Monitor*, 33: 6–8.

White House (2002) *The National Security Strategy of the United States of America*, Washington, DC, September.

4

Locating the EU's strategic behaviour in sub-Saharan Africa: an emerging strategic culture?

Richard G. Whitman and Toni Haastrup

There is now a need for a new phase in the Africa-EU relationship, a new strategic partnership and a Joint Africa–EU Strategy as a political vision and roadmap for the future cooperation between the two continents in existing and new areas and arenas. (Council of the European Union, 2007)

Conflict is often linked to state fragility. Countries like Somalia are caught in a vicious cycle of weak governance and recurring conflict. We have sought to break this, both through development assistance and measures to ensure better security. Security Sector Reform and Disarmament, Demobilisation and Reintegration are a key part of postconflict stabilisation and reconstruction, and have been a focus of our missions in Guinea-Bissau or DR Congo. This is most successful when done in partnership with the international community and local stakeholders. (Council of the European Union, 2008)

The EU's Common Foreign and Security Policy (CFSP) and Common Security and Defence Policy (CSDP) – and formerly the European Security and Defence Policy (ESDP) – have featured sub-Saharan Africa as a central strand of activities since their foundation (Nuttal, 2000; Holland, 1997, 2005; Howorth, 2007). However, to date there has been relatively little analysis of the importance of sub-Saharan Africa to refining the EU's understanding of what constitutes security and of the security threats that inform its foreign and security policies in the region. Where sub-Saharan Africa has featured in analysis of the CFSP/CSDP it has been to explore policy towards third countries, and individual CSDP operations and activities (see Pirozzi, 2009; Elowson, 2009).

This chapter is interested in how the EU's strategic preferences – as evidenced by its *behaviour* in Africa – inform the notion of a *strategic culture* in the Union's external relations. Its focus is not on the full array of the EU's CFSP towards sub-Saharan Africa but on an examination of its CSDP operations. Sub-Saharan Africa is of considerable interest as a 'test-bed' of CSDP activity. Since 2003, the continent has experienced the full range of both civilian and military types of CSDP activities (Whitman and Wolff, 2012). Indeed, some distinctive operations have only been carried out in sub-Saharan Africa: the first operation carried out outside Europe without any NATO support (Artemis), the first naval operation (EUNAVFOR) and the EU support to the African Union's Mission in Somalia (AMIS) are just three examples. With ten of the EU's total of 24 CSDP operations to date taking place in sub-Saharan Africa, the continent has been an important theatre of operations for the development of the operational practices of the CSDP.

The existing literature has examined the decision-making processes that resulted in some of these CSDP operations. In addition, case studies have been used to analyse EU decision-making processes as a basis for accounting for particular policy outcomes (Krause, 2003; Sicurelli, 2008; Bagoyoko and Gibert, 2009). Where there is currently a gap in the literature is the extent to which the EU's CSDP activities in sub-Saharan Africa have contributed to the development of an EU strategic culture. In this chapter, we argue that there are three indicators that motivate EU activities in Africa; that those indicators frame the EU's strategic behaviour; and that they contribute to the development of the EU's own strategic culture.

The chapter proceeds by introducing the notion of strategic behaviour and its relation to the concept of strategic culture as applied to the EU. Subsequently, it identifies three frames of EU external relations, derived from an analysis of strategic declarations: the security–development nexus, the human security imperative and preference for local enforcement. The aim is to assess the extent to which these frames are found in the EU's behaviour in Africa. Hence, the chapter examines the EU's Africa CSDP missions through five distinctive operational types. We argue that these frames, which motivate the EU's CSDP activities, constitute a basis for understanding the EU's strategic culture. By assessing CSDP missions in Africa based on these types, the chapter gives context to the EU's foreign and security policy in the region.

Shaping an EU strategic culture?

Debate around whether the EU possesses a strategic culture has been on-going since the foundation of the CSDP in the late 1990s and is currently the subject of extensive analysis (Schmidt and Zyla, 2011). Contemporary debate poses questions such as: What are the characteristics of strategic culture? Is it important for the EU to possess a strategic culture? Crucially, is the EU developing a strategic culture? If so, what are its characteristics? Why is the notion of the EU's possession, or not, of a strategic culture of such importance? The answers to these questions lie in part in international relations literature on strategic cultures, and in the more recent literature that focuses on the possibility of an EU strategic culture. This literature is concerned with the assertion that there is a relationship between the strategies pursued by individual international actors. It assumes that these actors 'have different predominant *strategic preferences* that are rooted in the early or formative experiences of the state, and are influenced to some degree by the philosophical, political, cultural, cognitive characteristics of the state and its elites' (Johnston, 1995: 34, emphasis added). Thus, the strategic culture of an international actor constitutes the essence of that state's identity.

Analysis of strategic culture originally developed during the Cold War, with a predominant focus on the two superpowers. It made generalisations about the superpowers' appetites for risk and propensity in the use of force used to inform strategies for the conduct of nuclear war in the US (Johnston, 1995). From there, a burgeoning literature developed which examined the strategic culture of a variety of states, including those of individual EU member states (Meyer, 2006). Applying the original concept to the EU represents a particular set of empirical and theoretical challenges. In the first instance, there is the existence of 27 distinctive security cultures, alongside a putative EU strategic culture. The interrelationship between those individual security cultures and the EU's emergent one raises the question as to whether the processes of member states' preferences are symbiotic with those of the EU as a whole.

More generally, and as with all concepts in the social sciences, there is considerable debate about the use of the concept of strategic culture. One key area of debate concerns the relationship between strategic culture and strategic behaviour. The distinction between these two concepts is of crucial relevance for the study of the EU. For Gray (1999: 50), strategic behaviour 'means behaviour relevant to the threat or use of force for political purposes'. The relationship between strategic culture and strategic behaviour can be further conceptually distinguished: '*strategic culture* can be conceived as a context out there that surrounds, and gives meaning to, *strategic behaviour*, as the total warp and woof of matters strategic that are thoroughly woven together, or as both' (Gray, 1999: 51). In other words, strategic behaviour is found in those preferences utilised by the actor to respond to specific security situations. Although the two are often conflated, there is an obvious distinction between strategic culture and strategic behaviour. Yet to date, the debate on the existence of an EU strategic culture has largely glossed over definitions in a rush to judge whether the EU has an embryonic strategic culture. The nature of what constitutes strategic culture has therefore been less contested than whether the EU is acquiring one.

In 2003, Rynning identified a dividing line within the literature on the EU and this line still holds (Rynning, 2003). He distinguished between optimistic and pessimistic assessments on the EU's possession of a strategic culture. What divides these assessments is the conclusion as to whether the EU is gaining both the ability and the confidence to use military force to address perceived threats to EU security. Cornish and Edwards (2001) sought to evaluate whether the EU has acquired a strategic culture by examining four areas: (1) military capabilities; (2) whether CSDP experiences are engendering a sense of reliability and legitimacy for autonomous EU action; (3) whether the policy-making processes of the EU now ensure a political culture with the appropriate level and depth of civil-military integration; and (4) the evolving relationship between the EU and NATO. Cornish and Edwards entwine elements of strategic culture and behaviour, which we are seeking to distinguish between here. They conclude in their 2005 work that the EU's strategic culture is work in progress (Cornish and Edwards, 2005: 820).

Subsequent work on general stocktaking of CSDP exercises concludes that the policy domain, and by implication the EU's strategic culture, is something of a 'curate's egg' (Menon, 2009). As Schmidt and Zyla (2011) explain, the scholarship on strategic culture demonstrates different approaches to epistemology, ontology and methodology. And while it is clear that the concept has much to offer as the basis for empirical work, it is also apparent that authors are defining and using the concept in very different ways when applying it to the EU. Consequently, there is no single notion of strategic culture underpinning the most recent body of empirical work explicitly examining the civilian and military components of the CSDP for evidence of its existence (Haine, 2011; Kammel, 2011; Pentland, 2011; Rummel, 2011; Schmidt, 2011).

In the same vein as the most recent scholarship, this chapter uses the notion of strategic culture to confer insight into the notion of an EU strategic culture. And more specifically, it concurs with Norheim-Martinsen (2011) that the EU's strategic culture is not about common interests but about preferred means. If we accept that the EU is developing a strategic culture in terms of preferred means, then how and where can this be identified? In the next section, by examining declaratory statements and the EU's CSDP operations in Africa, we argue that the whole of EU strategic *behaviour* in this region can inform the debate about the emergence of an EU strategic *culture*.

The EU in sub-Saharan Africa: making the case for strategic culture

Examining the emergence of an EU strategic culture with reference to sub-Saharan Africa may not seem the obvious choice, given the EU's sustained foreign and security policy engagement in the Western Balkans. Indeed, the EU has been in the Balkans since the foundation of the CFSP in 1993, through the development of the European Security and Defence Identity (ESDI) and on to the eventual creation of the CSDP. However, two distinct disadvantages accrue to such an examination of the Western Balkans. The first is that the EU has defined an end point to its engagement with states in this region, in the context of a route map to an eventual accession to the EU. The second is that there have been more CSDP operations in sub-Saharan Africa than in the Western Balkans, where there have been six CSDP operations to date. Therefore, the EU's CSDP operations, which are implementations of the EU's strategic preferences or strategic behaviour in sub-Saharan Africa, provide an important case through which we can begin to understand the emergence and evolution of the EU's strategic culture. Additionally, recent studies confirm Africa as a relevant testing ground for the EU's strategic culture (Schmidt, 2011; Haine, 2011; Rummel, 2011).

In this examination of the EU's strategic behaviour in sub-Saharan Africa, the chapter will draw some conclusions on the wider EU strategic culture. We will examine two elements of the EU's engagement with the continent. The first element seeks to clarify the terms through which the EU has defined sub-Saharan Africa as a theatre for its foreign security and defence operations. Thus, we examine the context of EU–Africa relations through key strategic declaratory instruments. These strategic declarations refer to key policy statements which guide the EU's relationship with third parties (Whitman, 1998; Whitman and Manners, 1998). The second element examines the EU's actual behaviour through the CSDP operations in Africa where the EU has used either the threat or the use of force as a policy instrument.

One of the new opportunities introduced by the EU's engagement in Africa is the increased interest in conflict prevention and crisis management (Olsen, 2002, 2009; Vines, 2010). Preventing conflict has been a preoccupation for the EU since the conflicts that erupted during the split up of former Yugoslavia and in the post Soviet Union. The EU's approach to conflict prevention (and management when necessary) is based on the identification of the root causes of these conflicts. It has been argued that the EU's very own existence is a good example of conflict prevention (Cameron, 2001) and so the desire to replicate this success internationally is ingrained in its identity. Additionally, given its limited capabilities to manage conflict properly in the immediate post-Cold War era, the EU has sought to carve a niche in conflict prevention. Pursuing a foreign policy based on this niche area of conflict prevention also bolsters the EU's role in global politics. Thus, in the continued quest for international status, the EU has emerged as a viable actor in the area of conflict prevention in Africa.

The EU's approach and its intended responses to international security challenges are set out in multiple documents or strategic declarations. In the period following the Cold War especially, the EU used a set of strategic declarations which provide the framework within which its policies towards Africa are being defined and organised. In this analysis, these strategic declarations are used as markers that establish a component that guides the EU's strategic behaviour, inform the EU's African policies and shape the EU's strategic culture (Manners, 2002). These strategic declarations reveal that the EU has established three inter-related frames for its foreign and security policies: the security–development nexus, the human security imperative and a preference for local enforcement.

The *security–development nexus* describes the links made in foreign policy practice between the usually divergent areas of development and security. The security–development nexus is holistic and addresses the interconnectedness of security issues with traditional poverty reduction strategies, the result of which has been an enhancement of EU foreign policy (Hadfield, 2007). According to Kerr (2007: 92), 'empirical observations and several data-collection studies reveal the significance of that nexus'. Bagayoko and Gibert (2009) further contend that the reality of Africa's developmental challenges and the competition among EU institutions to drive the EU's Africa policy determines this nexus.

The security–development nexus also complements the *human security imperative*. Human security in particular shifts the focus from the protection of states and regimes to the protection of the individual and their communities. Both of these strands of approach are accepted by the EU's African partners (Bagayoko and Gibert, 2009). This is evidenced in Africa's own declaratory instruments such as the 'Protocol Establishing the Peace and Security Council' (of the African Union) (African Union, 2002) and the 'Conference on Security, Stability, Development and Cooperation in Africa' (African Union, 2000).

Finally, the EU supports the sub-contracting of operational activity to African third parties, thus highlighting a *preference for local enforcement*. This frame, like the other two, is a mutually constituted one between the EU and its Africa partners. Although local enforcement relieves Europeans of bearing the cost of interventions, it also gives the space for African governments and institutions to determine the appropriate types of interventions to a crisis situation.

To understand how these three frames have emerged, it is best to start by examining the relevant declarations to the EU's strategic preferences or behaviour. A key starting point is the European Security Strategy (ESS) signed by the member states' heads of state and government in December 2003 (European Council, 2003). In the words of the heads of state and government:

> The European security strategy reaffirms our common determination to face our responsibility for guaranteeing a secure Europe in a better world. It will enable the European Union to deal better with the threats and global challenges and realise the opportunities facing us. An active, capable and more coherent European Union would make an impact on a global scale. In doing so, it would contribute to an effective multilateral system leading to a fairer, safer and more united world. (European Council, 2003)

Furthermore, as the European Council conclusions also noted, the appropriate consequence of the 'strategic orientation[s]' contained in the document was that they had to 'mainstream them into all relevant European policies' (Whitman, 2004). Consequently, the ESS is supposed to provide the EU and its member states with the road map towards greater global impact (Whitman, 2006). The ESS defines Europe's security interests and priorities across three parts of the document: global challenges and key threats – this priority identifies 'the security environment'; strategic objectives – how to address these threats; and policy implications for Europe. Sub-Saharan Africa features in each of these three sections of the document.

In the identification of global challenges and key threats, the immediate historical context within the practice of international relations influences the ESS. As a result, the document touches on international security issues such as terrorism, proliferation of weapons of mass destruction, regional conflicts, state failure and organised crime. Where the ESS identifies these threats, sub-Saharan Africa is an illustrative example that links the lack of socio-economic development to political instability and conflict. Consequently, the ESS asserts that in Africa security is a precondition for development (European Council, 2003: 2). This is an embryonic expression of how the security–development nexus frames EU external relations policy broadly – and, specifically, the EU's policy towards Africa.

The ESS identifies three strategic objectives as a second priority area in the EU's strategic behaviour. These objectives include, 'addressing the threats', 'building security in our neighbourhood' and an 'international order based on effective multilateralism'. All three objectives apply to Africa as a whole in the sense that the EU commits to address all threats wherever they emerge. The building of security in the neighbourhood also applies to Africa inasmuch as five (North) African countries are part of the neighbourhood. Nevertheless, the most relevant objective to this analysis of sub-Saharan Africa is the objective of 'effective multilateralism'.

'Effective multilateralism' has become an overarching objective of the EU's foreign policy. Arguably, it is the EU's equivalent of the US Cold War policy of containment, meaning it is a key driver of the EU's international activities. Multilateralism is structured around getting as many participants in the project of peace as possible. Thus, the EU takes many of its security cues from the UN Security Council. Critics of the EU propensity for multilateralism decry the insistence on this method as an excuse for inaction (Lindberg, 2005; Chamorel, 2006). Nevertheless, it informs many of the EU's external activities and indeed forms part of the culture of EU foreign policy. When the security strategy addresses the practice of multilateralism, it identifies the AU as a primary interlocutor for Africa. This is framed as a part of the EU's collaboration with regional organisations and strengthening of global governance structures.

The third section of the ESS concentrates on support for capabilities development, which has been a collective concern in international relations since the early 1990s. By contributing to institutional development, the EU hopes to build structures that foster some if not all of its own values, especially democracy, the rule of law, respect for human rights and peace. The EU has especially contributed to the development of skills in 'security' and 'justice' institutions like the police and army forces in post conflict countries in Africa with the aim of fostering lasting peace. Although the EU did not identify an African candidate for the 'strategic partnerships' in the security strategy of 2003, subsequent development of EU–Africa relations, which indicates the preference for a regional approach to security, has led to the African Union being elected as the default representative of Africa states. While this has not usurped the EU relationship with individual African countries or subregions, there has been a redirection of EU strategic engagement to the continental level. This has been the motivation for EU–AU dialogues since 2007, and it further underscores the EU's preference for effective multilateralism.

In its December 2008 five-year review of the implementation of the ESS, the EU has summarised the evolution of its foreign and security policy. The review also shows how sub-Saharan Africa fits within the EU's wider grand strategy (Council of the European Union, 2008). Firstly, this review document explicitly notes the concept of human security. The refinement of this concept as a guiding principle of EU external relations highlighted its importance to the implementation of the ESS following the initial publication in 2003 (Whitman, 2006). Secondly, this review document focuses on Somalia as illustrative of the inter-linkage between security and, development, and thus, the security–development nexus. EU engagements in Guinea-Bissau and in the DRC are also cited as instances where the EU's intervention has been driven by the desire for post-conflict stabilisation and reconstruction, which reinforces both the human security imperative and the security–development nexus. From the point of view of the EU, the 2008 review represents a good snapshot of the EU's evolving policy towards Africa over the last half decade. Both the ESS and the five-year review point to the developing strategic culture of the EU. These observations are mirrored in those strategic declarations that deal specifically with sub-Saharan Africa.

From 2003 onwards, strategic declaratory statements of the EU on sub-Saharan Africa have contained significant reference to both the security–development nexus and the human security imperative, indicating that the EU has 'uploaded' these two key foreign policy frames into its strategic objectives for the continent.

The first such reference is the Common Position adopted in January 2004 concerning conflict prevention, management and resolution in Africa (Council of the European Union, 2004). This document establishes a number of principles that have guided EU policy. Firstly, the EU seeks to 'contribute to the prevention, management and resolution of violent conflicts in Africa by strengthening African capacity and means of action in this field' (Council of the European Union, 2004). The EU sees its roles as working between the nexus of security and development to help build post conflict societies that promote democratic values, including respect for the rule of law and human rights. Secondly, the EU aims to implement the policy of close cooperation with the UN and regional and sub-regional organisations. The move from region-to-region cooperation coincided with the development of the EU's own capabilities to pursue political and security engagements internationally. This move, which has now led to the EU–AU relationship, begun in the late 1990s to support the AU predecessor, the Organisation of African Unity, in building its conflict management capabilities so as to manage the post-Cold War flux on the continent. Third, the Common Position established that conflict prevention, management and resolution needed to be tackled through capacity building at the international, regional and country levels, thus reinforcing the views established in general EU strategic documents.

This initial Common Position has thus been the platform, on which the EU has developed several dimensions to its external security relations in Africa. This approach has focused on capacity building, and the disarmament, demobilisation and reintegration of combatants, and on combating the destabilising accumulation and spread of small arms and light weapons.

In 2005, the EU launched the EU Strategy for Africa (Council of the European Union, 2005), as the framework document on EU foreign policy towards Africa. The document dealt with all aspects of EU foreign policies towards Africa, taking input from member states' as well as Commission experiences. It emphasised the security–development nexus through its central objective of giving EU support to African states in their attempt to achieve the Millennium Development Goals (MDGs). Specifically, the EU Strategy for Africa aimed to 'strengthen its support in the areas considered prerequisites for attaining the MDGs (peace, security, good governance), areas that create a favourable economic environment for growth, trade and interconnection and areas targeting social cohesion and environment'. Furthermore, it noted that:

> The EU will step up its efforts to foster peace and security by means of a wide range of actions, ranging from the support for African peace operations to a comprehensive approach to conflict prevention addressing the root causes of violent conflict. These actions also target cooperation in the fight against terrorism and the non-proliferation of weapons of mass destruction, as well as support for regional and national strategies for disarmament, demobilisation, reintegration and reinsertion in order to contribute to the reintegration of excombatants – including child soldiers – and stabilisation of post-conflict situations. (Council of the European Union).

Importantly, the EU Strategy for Africa also emphasised a new approach to EU–Africa relations as based on ownership, equality and partnership. Commitment to these three principles encouraged greater coordination of EU external relations with African institutions.

The themes raised in the EU Strategy for Africa were reinforced at the second Africa–EU Summit, which was held from 8 to 9 December 2007 in Lisbon. The summit was conducted at the level of heads of state and government from Africa and the EU. Running through the key declarations and documents agreed at the summit – the Lisbon Declaration and the Joint Africa–EU Strategy *(JAES)* – the two sides characterised the relationship as a 'strategic partnership', thus signifying a new approach to EU–Africa relations. This partnership is structured through eight partnership or cluster areas which have their objectives set out in two-year action plans. Unlike previous strategic declarations, the JAES presents a shared vision of both EU and African states, with their regional organisations acting on their behalf. Further, it goes beyond the previous focus on sub-Saharan Africa to incorporate the five North African countries for an EU engagement with the whole of Africa. Importantly, this shared vision prioritises peace and security in its own right, not simply as an extension of previous development focused engagement. Indeed, the JAES is replete with references to human security and the security–development nexus is presented as a shared understanding that underpins the objectives for the partnership. This is important in that within the confines of new EU engagement in Africa, the strategic behaviour of the EU *vis-à-vis* the continent is in part constituted by its African partners. Thus, the JAES notes, 'Africa and Europe understand the importance of peace and security as preconditions for political, economic and social development' (Council of the European Union, 2007). In other words, the potential for a strategic culture for the EU is being defined by the joint priorities of Africa and the EU (Council of the European Union, 2007)

As one of eight cooperation areas, the 'Peace and Security' section of the JAES and its attendant First and Second Action Plans are primarily concerned with the EU facilitating African ownership of conflict prevention and conflict management mechanisms, and building capabilities of African institutions through the certification of the African Peace and Security Architecture (APSA). In this, the EU's role is as mentor and supporter (Council of the European Union, 2007). The First Action Plan especially highlights two elements of the EU's strategic behaviour that is region-specific to sub-Saharan Africa. The EU's preference for local enforcement has been systematically codified in its recent dealings with Africa. This idea of local ownership motivates EU capability building projects for the APSA. In addition to broad institutional development, the EU supports specific aspects of the Architecture, including the operationalisation of the Continental Early Warning System and the implementation of the African Stand-by Force (ASF) through an investment of up to 1 billion euro.

Another area of EU support is through the African Peace Facility (APF), which relies on EU funds. By building African capabilities through the APF, the EU hopes to diminish its direct military interventions on the continent. The APF finances AU security missions to ensure that the African Union is able to undertake its new responsibilities to provide peace and security in Africa. The Facility provides the financial support to facilitate capacity building by African states and the AU, particularly for the training of African troops to perform peace and security operations. This arrangement thus provides for the third frame, which allows the EU to delegate activities to local actors. The turn towards local ownership in African security suggests that at least in part, the implementation of the EU's strategic preferences on the continent is being shaped by activities involving non-EU actors (see Pirozzi, 2009 for further details of EU support to the APSA). In examining the EU's strategic declaratory instruments as statements of the EU's preferences, it has become clear that in addition to the human security imperative and security-development nexus, the preference for local enforcement or local ownership has emerged as an overarching frame of EU foreign policy engagement in Africa.

Operationalising strategic preferences: the EU's strategic behaviour

Sub-Saharan Africa has seen the most significant cluster of CSDP activity since the initiation of such operations in 2003. These CSDP operations provide an important basis from which to assess the operational activity aspects of the EU's strategic behaviour. Each of the individual CSDP operations has been the subject of academic and policy analysis (Pirrozi, 2009; Wolff and Whitman, 2012). Previous analysis focused on the motivations behind the deployment of each of the operations; the difficulties with converting the mandate of the General Affairs and External Relations Council (GAERC) – Foreign Affairs Council since the entry into force of the Lisbon Treaty – into a CSDP operation; and whether the operation constituted a successful realisation of its objectives. Unlike previous works, this chapter considers the CSDP operations in their totality with the aim of discerning the relevant patterns that characterise the EU's strategic behaviour.

Thus far, there have been ten sub-Saharan African CSDP operations. These will first be considered against a typology of operational types and then against a set of rationales providing an indication as to where each fits with the three frames of strategic behaviour outlined above.

The sub-Saharan CSDP operations have been categorised based on the mandate criteria outlined by EU member states authoring the operation. All the EU's CSDP operations can also be placed on both a civilian–military spectrum and defined in terms of their operation type. These CSDP operations are found within five operational types or categories: (1) reform-focused operations, (2) logistical assistance, (3) military deployments, (4) police support and (5) border assistance and monitoring. While the EU attempts to incorporate civilian and military dimensions into its operations, the majority of its deployments in Africa have usually been militarised.

Two out of Africa's EU CSDP operations to date can be characterised as *police support* operational types: EUPOL Kinshasa and EUPOL DR Congo. EUPOL Kinshasa (2005–07) was the first EU civilian mission deployed in Africa. Its objective was to help integrate the different factions of the post-conflict Democratic Republic of Congo (DRC) into the Congolese National Police Integrated Police Unit (IPU) in the capital Kinshasa. This new Unit was created from members of warring factions within the DRC society to help guard and guide the transitional institutions. The IPU consisted of 1,008 staff who played a key role in building confidence among former opponents (Whitman, 2009). Further, it was effective enough to assist with security during the 2006 elections in the DRC. The EU provided 4.37 million euro and 30 experts over a two-year period to provide support for local police training. EUPOL DR Congo succeeded EUPOL Kinshasa between 2007 and 2009 to develop the justice institutions in post-conflict DRC further. Specifically, it was tasked with creating an organised crime unit and developing linkages between the new police and justice system (Whitman, 2009).

Three operations have been conducted under the category of *reform-focused operations*. This operational type focuses on the promotion of the rule of law or democratic norms and security sector reform. The first of this type of operation (2008–10) was the completed security sector reform mission in Guinea-Bissau (EU SSR Guinea-Bissau). The second mission (2005–12), EUSEC DR Congo, established at the request of the DRC government, was a capacity-building mission to reform the Congolese army. The EU had nine officials assigned to key posts within the Congo's security sector to train and mentor key officials in key positions within the Ministry of Defence and the army. The third operation is the EU Somalia Training Mission that was initiated in April 2010. This is a military training operation for the development of the Somali security sector with the aim of strengthening the Transitional Federal Government and building new institutions. The operation provides specific military training as part of the EU's support to the African Union Mission to Somalia (AMISOM). Provision of training capability involves various partners in addition to the AU and EU, such as Uganda, the United States and the United Nations. This example constitutes a division of labour in international security and highlights the implementation of effective multilateralism as stated in the ESS.

Only one CSDP operation to date has fallen into the category of *logistical assistance*. This operation is the EU's multidimensional support to the African Union Mission in Sudan (AMIS). It involved EU provision of technical support to the AU, to assist it in the mounting of the AU's first-ever large-scale peace support operation (AMIS II) in the Darfur Region of Sudan. The EU also provided military equipment, assets and planning advice to the AU. Under this remit, the EU has also provided financial and political support. This arrangement between the EU and AU ended on 31 December 2007 when AMIS was succeeded by the hybrid mission of UN and AU, UNAMID.

Concerning the *military deployments*, these are specific activities in support of Petersberg tasks. The Petersberg tasks are the EU's military and security priorities including humanitarian aid and peace-keeping missions. Four operations fall within this category of military deployment. The first of this is the Artemis operation in the Congo (June–September 2003). Artemis was a stop-gap mission provided by the EU in Bunia, Congo, pending the deployment of a substantive amount of UN troops. Artemis was the first out-of Europe mission conducted by the EU and it consisted of mostly troops from France, but included personnel from 15 other EU countries and five non-EU countries.

This mission was followed by the EUFOR DRC (April–November 2006) in the same region. The aim was to provide security for the general election process in the DRC. The UN had insisted that security was a key priority for elections as the DRC made the transition towards democracy. Consequently, the EU provided civilian protection support to the existing UN mission. This mission illustrated the tandem approach of German, Spanish and French forces within an EU framework.

The third operation military deployment operation is the EUFOR Chad–Central African Republic operation. Its purpose was to protect the camps of refugees and displaced persons in the east of Chad and the north east of the Central African Republic. The mandate of this operation was one year long (15 March 2008–15 March 2009). The EU operation facilitated the delivery and free movement of humanitarian aid.

The fourth operation of this type is the ongoing EU NAVFOR Somalia/Operation Atalanta – the first mission of the EU NAVFOR. The primary mission of this operation is devoted to anti-piracy and anti-robbery operations off the coast of Somalia, although it has also conducted rescue missions for fishermen. EU NAVFOR also assists in the safe delivery of food from the World Food Programme, other humanitarian aid and the protection of AMISOM shipping. The United Kingdom hosts the operational headquarters of NAVFOR while several other member states have contributed personnel to the mission. In addition, the EU NAVFOR has established the Maritime Security Centre – Horn of Africa which provides 24 hour patrol in the Gulf of Aden to protect shipping interests from criminality in the area. EU NAVAFOR has shown leadership and encouraged a consortium of patrol on the seas including NATO, China, Japan, India and Russia. The budget of this operation was 8.05 million euro in 2011.

The assessment of each CSDP operation has been through a combination of examination of the mission mandate, the activities undertaken during the missions duration and the actors involved in the implementation. The material used as the basis for the assessment is with the use of the IISS Strategic Surveys (2005–08) (IISS, 2005–08) and cross referred to development indices (UNDP, 2004–07), EU documentation and secondary source analysis.

The CSDP operations can be categorised according to their correspondence to different aspects of the EU's strategic behaviour identified in the strategic declarations above. The majority of the EU's sub-Saharan African CSDP operations demonstrate evidence of the security–development nexus as providing a rationale for intervention. For example, the locations of EU intervention in the DRC, the CAR and Guinea-Bissau are all countries that are placed low on the development indices. These countries – and indeed the region – reflect low levels of socio-economic development, marred by political instability and violent conflict. Thus, the design and implementation processes of the operational types are cognisant of the linkages between security and development. The CSDP missions aim to create or support the creation of the conditions, that is, a non-violent space, wherein socio-economic development and political stability through democratic rule can flourish.

There is a clear sense that the human security imperative is a key rationale for all EU intervention. All ten operations had civil protection dimensions, either through direct engagement, as was the EU mission in Chad and CAR in the DRC, or through the protection of humanitarian aid, as off the coast of Somalia.

The scope of the CSDP operations in Africa is also indicative of the preference for local enforcement. Firstly, the CSDP military operations often have a short, set duration and often with a narrow mandate. Secondly, the EU preference has been to engage in activities to supply technical and logistical knowledge with the aim of increasing indigenous capacity. This period of training is often longer. This has often been the case within the civilian dimension of operations and is evident in the EUPOL and SSR operations. The importance of local enforcement is also evident in at least one military mission where the situation demanded African peace-keeping capacity, the AMIS operation. In this case, the conflict demanded local African knowledge as well as long-term commitment, which the EU was incapable of offering. Thus, its best offer of assistance was providing equipment and assets, planning, technical assistance, and tactical and strategic transportation.

We see how the majority of the EU's sub-Saharan African CSDP operations confirm the three frames through which the EU demonstrates its strategic preferences. If we take strategic culture to be the context within which strategic preferences are being deployed (Gray, 1999), we then assume that the EU will always engage in international security whenever it can find justification in the three frames. Biscop and Coelmont (2010) argue that the EU lacks consistency in its deployments and that this inhibits the development of a strategic culture for the EU. Citing the engagement in the DRC, ostensibly to ensure human security through civilian protection, they also argue that the EU lacks any underlying strategy towards Africa. Indeed, the lack of consistent engagement might speak to the weak actorness of the EU in Africa. Yet, we would maintain that given the consistent identification of the three components of strategic behaviour at those times when the EU does engage in Africa, there is evidence of strategic culture in EU–Africa relations.

In short, identifying components of EU strategic behaviour through the analysis of the EU's Africa CSDP operations provides an understanding of the EU's strategic culture. However, an interesting dimension to this story of the EU's strategic behaviour is that of the preference for local enforcement. Although local empowerment serves as a cornerstone of the EU's emerging strategic culture, it might also pose a challenge in that the opportunity to establish strategic culture through CSDP operations becomes lessened when the EU ceases to participate. Further, the continued bi-lateral engagement in Africa as we have seen in French, British and American intervention of Libya jeopardises an 'EU' strategic culture, especially one supported by a joint vision and understanding of local ownership between Europeans and Africans.

Conclusion

This chapter has sought to evaluate the potential for an EU strategic culture by assessing its strategic behaviour in Africa. An evaluation of strategic declarations and relevant CSDP operations indicates that the EU's strategic behaviour is framed by three indicators: the *security–development nexus*, the *human security imperative* and the *preference for local enforcement*. These frames provide the context for most of the EU's external (security) relations by giving meaning to its strategic preferences. While this chapter allows us to reflect on issues concerning contemporary EU–Africa relations, as well as the EU engagement in international security, it also raises questions that can only be answered through further empirical work to refine the analysis on the EU's *strategic culture*.

The three frames identified here give us the requisite tools to begin making further assessments of the EU's strategic behaviour in the attempt to develop a more definitive framework for its strategic culture. Examination of the EU's foreign security and defence policy engagement with sub-Saharan Africa has allowed for the identification of components of the EU's strategic behaviour. The chapter suggests that the EU policy pursued towards sub-Saharan African demonstrates three characteristics of the EU's strategic behaviour, highlighting the potential this then has for understanding a broader EU strategic culture. By highlighting the contexts of the EU's strategic behaviour as part of an emerging culture, the importance of strategic culture is found as being closely linked to the EU's international identity.

Sub-Saharan Africa has proved to be an invaluable case study through which to examine the potential for the EU's strategic culture. Over recent years, the continent has become an increasingly important venue for the EU's foreign, security and defence policy. Nevertheless, the dynamics of recent EU–Africa relations, which prioritise local ownership, partnership and equality, as well as inter-regionalism could affect how we view strategic culture with regards to the EU. The increased emphasis on local ownership to a certain extent means that its third partners could define the emergence of a strategic culture to a small extent. Additionally, the resulting preference for local enforcement could lead to the contraction of EU operational activities as we see them now on the continent. This will then limit the opportunity for continued evaluation of the emerging strategic culture through CSDP operations.

References

African Union (2000) 'Conference on Security, Stability, Development and Cooperation in Africa', available at: http://www.africa-union.org/Special_Programs/CSSDCA/cssdca-solemndeclaration.pdf (accessed 5 March 2012).

African Union (2002) 'Protocol Relating to the Establishment of the Peace and Security Council of the African Union', available at: www.au.int/en/sites/default/files/Protocol_peace_and_security.pdf (accessed 5 March 2012).

Bagoyoko, N. and Gibert, M. (2009) 'The Linkage between Security, Governance and Development: The European Union in Africa', *Journal of Development Studies*, 45 (5): 789–814.

Biscop, S. and Coelmont, J. (2010) *A Strategy for CSDP: Europe's Ambition as a Global Security Provider*, Brussels: Egmont/Academia Press.

Cameron, F. (2001) 'The European Union and Conflict Prevention', UNIDR

Background Papers, available at: www.unidir.org/pdf/EU_background_papers/ EU_BGP_01.pdf (accessed 11 October 2011).

Chamorel, P. (2006) 'Anti-Europeanism and Euroskepticism in the United States', in T.L. Ilgen (ed.), *Hard Power, Soft Power and the Future of Transatlantic Relations*, Aldershot: Ashgate.

Cornish, P. and Edwards, G. (2001) 'Beyond the EU/NATO Dichotomy: The Beginning of a European Strategic Culture', *International Affairs*, 77 (3): 587–603.

Cornish, P. and Edwards, G. (2005) 'The Strategic Culture of the European Union: A Progress Report', *International Affairs*, 81 (4): 801–20.

Council of the European Union (2004), *European Council 12 and 13 December 2003 Presidency Conclusions*, 5381/04, Brussels, 5 February.

Council of European Union (2005) *The EU and Africa: Towards a Strategic Partnership*, 15961/05 (Presse 367), 19 December.

Council of the European Union (2007) *The Africa–EU Strategic Partnership: A Joint Africa–EU Strategy*, 16344/07 (Presse 291), 9 December.

Council of the European Union (2008) *Report on the Implementation of the European Security Strategy – Providing Security in a Changing World*, Brussels, S407/08, 11 December.

Elowson, C. (2009) 'The Joint Africa–EU Strategy: A Study of the Peace and Security Partnership', FOI Report, March, available at: www2.foi.se/rapp/foir2736.pdf (accessed 11 October 2011).

European Council (2003) *A Secure Europe in a Better World: European Security Strategy*, Brussels: European Union.

Gray, C.S. (1999) 'Strategic Culture as Context: The First Generation of Theory Strikes Back', *Review of International Studies*, 25: 49–69.

Hadfield, A. (2007) 'Janus Advances? An Analysis of EC Development Policy and the 2005 Amended Cotonou Partnership Agreement', *European Foreign Affairs Review*, 12 (1): 39–66.

Haine, J.Y. (2011) 'The Failure of a European Strategic Culture – EUFOR CHAD: The Last of its Kind?', *Contemporary Security Policy*, 32 (3): 582–603.

Holland, M. (ed.) (1997) *Common Foreign and Security Policy: The Record and Reforms*, London: Pinter.

Holland, M. (ed.) (2005) *Common Foreign and Security Policy: The First Ten Years*, 2nd edition, London: Continuum.

Howorth, J. (2007) *Security and Defence Policy in the European Union*, Basingstoke: Palgrave.

International Institute for Strategic Studies (2005) *Strategic Survey 2005*, available at: http://www.iiss.org/publications/strategic-survey/strategic-survey-archive/ (accessed 5 March 2012).

International Institute for Strategic Studies (2006) *Strategic Survey 2006*, available at: http://www.iiss.org/publications/strategic-survey/strategic-survey-archive/ (accessed 5 March 2012).

International Institute for Strategic Studies (2007) *Strategic Survey 2007*, available at: http://www.iiss.org/publications/strategic-survey/strategic-survey-archive/ (accessed 5 March 2012).

International Institute for Strategic Studies (2008) *Strategic Survey 2008*, available at: http://www.iiss.org/publications/strategic-survey/strategic-survey-archive/ (accessed 5 March 2012).

Johnston, A.I. (1995) 'Thinking about Strategic Culture', *International Security*, 19 (4): 33–64.

Kammel, A.H. (2011) 'Putting Ideas into Action: EU Civilian Crisis Management in the Western Balkans', *Contemporary Security Policy*, 32 (3): 625–43.

Kerr, P. (2007) 'Human Security', in A. Collins (ed.), *Contemporary Security Studies*, Oxford: Oxford University Press, pp. 92–108.

Krause, A. (2003) 'The European Union's Africa Policy: The Commission as Policy Entrepreneur in the CFSP', *European Foreign Affairs Review*, 8 (2): 221–37.

Manners, I. (2002) 'Normative Power Europe: A Contradiction in Terms', *Journal of Common Market Studies*, 40 (2): 235–58.

Menon, A. (2009) 'Empowering Paradise? The CSDP at Ten', *International Affairs*, 85 (2): 227–46.

Meyer, C.O. (2006) *The Quest for a European Strategic Culture: Changing Norms on Security and Defence in the European Union*, Houndmills: Palgrave Macmillan.

Norheim-Martinsen, M. (2011) 'EU Strategic Culture: When the Means Becomes the End', *Contemporary Security Policy*, 32 (3): 517–34.

Nuttall, S. (2000) *European Foreign Policy*, Oxford: Oxford University Press.

Olsen, G.R. (2002) 'The EU and Conflict Management in African Emergencies', *International Peacekeeping*, 9 (3): 87–102.

Olsen, G.R. (2009) 'The EU and Military Conflict Management in Africa: For the Good of Africa or Europe', *International Peacekeeping*, 16 (2): 245–60.

Pentland, C.C. (2011) 'From Words to Deeds: Strategic Culture and the European Union's Balkan Military Operations', *Contemporary Security Policy*, 32 (3): 551–66.

Pirozzi, N. (2009) 'EU Support to African Security Architecture: Funding and Training Components', Occasional Paper 76, EUISS Paris.

Rummel, R. (2011) 'In Search of a Trademark: EU Civilian Operations in Africa', *Contemporary Security Policy*, 32 (3): 604–24.

Rynning, S. (2003) 'The European Union: Towards a Strategic Culture?', *Security Dialogue*, 34 (4): 479–96.

Schmidt, P. (2011) 'The EU's Military Involvement in the Democratic Republic of Congo: Security Culture, Interests and Games', *Contemporary Security Policy*, 32 (3): 567–81.

Schmidt, P. and Zyla, B. (2011) 'European Security Policy: Strategic Culture in Operation?', special issue of *Contemporary Security Policy*, 32 (3).

Sicurelli, D. (2008) 'Framing Security and Development in the EU Pillar

Structure: How the Views of the European Commission Affect EU Africa Policy', *Journal of European Integration*, 30 (2): 217–34.

United Nations Development Programme (2004) *Human Development Report 2004*, New York: UNDP.

United Nations Development Programme (2005) *Human Development Report 2005*, New York: UNDP.

United Nations Development Programme (2006) *Human Development Report 2006*, New York: UNDP.

United Nations Development Programme (2007) *Human Development Report 2007*, New York: UNDP.

Vines, A. (2010) 'Rhetoric from Brussels and the Reality on the Ground: The EU and Security and Africa', *International Affairs*, 86 (5): 1091–108.

Whitman, R.G. (1998) *From Civilian Power to Superpower? The International Identity of the EU*, Basingstoke: Macmillan.

Whitman, R.G. (2004) 'NATO, the EU and ESDP: An Emerging Division of Labour?', *Contemporary Security Policy*, 25 (3): 430–51.

Whitman, R.G. (2006) 'Road Map for a Route March? (De-)civilianizing through the EU's Security Strategy', *European Foreign Affairs Review*, 11 (1): 1–15.

Whitman, R.G. (2009) 'The EU and Sub-Saharan Africa: Developing the Strategic Culture of the Union's Foreign, Security and Defence Policy', paper presented at European Union Studies Association 11th Biennial Conference, 23–25 April, Marina Del Rey, California.

Whitman, R. and Wolff, S. (eds) (2012) *The European Union as a Global Conflict Manager*, London: Routledge.

Whitman, R.G. and Manners, I. (1998) 'Towards Identifying the International Identity of the European Union: A Framework for the Analysis of the EU's Network of Relationships', *Journal of European Integration*, 21 (3): 231–49.

Clientelism and patronage

In order to understand Africa's place in the inte[r]
absolutely vital to grasp the state–society co[n]
across many parts of SSA. These clearly ha[ve]
a fundamental – possibly decisive – influ[ence]
continent's international relations. Y[et]
looked, despite Falola (2006: 181)
the patron–client system in a c[ountry]
behaviour and activities of m[any]
both domestically and int[ernationally]
ties of governance in [many]
external processes is[...]
practices, global
Africa's elites [and]
the contine[nt]
Obvio[usly]
gener[...]
fra[...]

Afr[ica ...] ... [a] passive bystander, devoid of agency and acted upon. Yet it is surely crucial to reject the idea that Africa is a victim (Taylor, 2004). As Jean-François Bayart has asserted, 'the discourse on Africa's marginality is a nonsense' (2000: 267). Africa has never existed separate from the world, but rather has been inextricably entwined in world politics and has continually exercised its agency. In substantial terms, SSA cannot be seen to pursue a relationship with the world as 'Africa is in no sense extraneous to the world' (Bayart, 2000: 234). The continent is globally dialectically connected and both determines and is determined by myriad developments, actors and structures, both internal and external, if such an artificial separation is to be cited. This is so much more the case now in the early twenty-first century, when the place of Africa in the international system increases in importance and stature year-on-year as a multitude of actors jostle for economic and political influence and as changes within Africa reconfigure the continent. Before discussing how these new actors are increasing the prominence of Africa at the global level, analysing the domestic milieu is vitally important.

rnational system, it is
nplexes that are evident
ve critical implications and
ence upon many aspects of the
t such essentials are often over-
noting that grasping the 'nature of
ountry is necessary to understand the
embers of the political class and warlords'
rnationally. Critically analysing the modali-
rge parts of SSA and how they combine with
essential if we wish to comprehend the diplomatic
interactions and broad international relations of
nd its ordinary citizens and in order to fully comprehend
t's dynamic place in the world.
sly, in discussing something as broad as 'the African state',
lisations are necessary, and the applicability of general conceptual
neworks to each individual SSA country is contingent and dependent
upon myriad factors. Having said that, it cannot be denied that many
post-colonial African countries, bounded by formal frontiers and with
an international presence at various international institutions, function
quite differently from conventional understandings of what a formal
essentialised Western state. In fact, many – possibly most – African
states are not institutionally functional.

That Africa states should not conform to Western models is, of
course, not surprising. In order for Africanists to understand the politics
of the state on the continent, the concept of neo-patrimonialism has
largely become the standard tool of analysis (LeVine, 1980; Jackson
and Rosberg, 1982; Callaghy, 1984; Sandbrook, 1985; Bayart, 1993;
Bratton and van de Walle, 1994, 1997; Chabal, 1994; Tangri, 1999).
This feature of politics in many parts of SSA has profound implications for any attempt to situate and appreciate the proper context
and behavioural patterns of Africa's international relations. Under a
neo-patrimonial system the separation of the public from the private is
recognised (even if in practice only on paper) and is certainly publicly
displayed through outward manifestations of the rational-bureaucratic
state – a flag, borders, a government and bureaucracy, etc. However,
in practical terms the private and public spheres are habitually not
detached and the outward manifestations of statehood are often facades
hiding the real workings of the system. In many African countries,
particularly those marked by enclave economies, the official state

bureaucracies inherited from the colonial period, however weak and ineffective, have become dysfunctional and severely constrained in their official, stated, duties. Many post-colonial African leaders have rather relied on effected control and patronage through capturing power over the economy, rather than through the state in the form of a functioning administration (Chabal, 2009).

Of course, clientelism and patronage are not unique to Africa (see Lemarchand and Eisenstadt, 1980; Clapham, 1982; Fatton, 2002), nor are neo-patrimonial regimes. Categorising SSA states as neo-patrimonial is certainly not to exoticise them, or even to place a normative value on their dynamics. Yet it is vital to understand that many African bureaucracies are 'patrimonial-type administrations in which staff [are] less agents of state policy (civil servants) than proprietors, distributors and even major consumers of the authority and resources of the government' (Jackson and Rosberg, 1994: 300). Handing out bureaucratic posts has become an important way in which leaders can secure support. This support stems from the fact that 'being appointed into the government is tantamount to being given the opportunity to fill one's pocket with state-owned wealth and also the opportunity to develop one's part of the country if the appointee has any sense of loyalty to his people' (Fru Doh, 2008: 40). And so the patronage networks spread from the capital outwards.

The composition of Africa's diplomatic corps is a graphic example of this, with fierce competition and pressure on the patrons to deliver the best jobs (at the United Nations, in the developed world's capitals, etc.). In many African countries, those who hold the highest diplomatic ranks are not the best qualified – they are just the best connected. This of course has often grave implications for the competency of Africa's representation overseas. Such clientelism is central to neo-patrimonialism, with widespread networks of clients receiving services and resources in return for support. This is well understood and even expected in many African countries, reflecting the mutual benefits that neo-patrimonialism confers to both patron and client. Indeed, the exercise of personalised exchange, clientelism and corruption is internalised and constitutes an 'essential operating codes for politics' in Africa (Bratton and van de Walle, 1997: 63). This is 'accepted as normal behaviour, condemned only in so far as it benefits someone else rather than oneself' (Clapham, 1985: 49). Those not in the loop may resort to insurrection as a means of accessing the resources and power that accrues to those who have captured the state. In some countries, corruption, working the system and subverting the state is 'a firmly established practice ... to the point where citizens no longer realize it is a wrongdoing. Because corruption is now the norm, anyone trying to do the correct thing by serving [the people] without stealing or asking for bribes can lose his job ... as his colleagues accuse him of snatching the morsel from their mouths' (Fruh Doh, 2008: 141).

Sovereignty and 'development'

As outlined above, control of the state serves the twin purposes of lubricating patronage networks and satisfies the selfish desire of elites to self-enrich themselves, in many cases in a quite spectacular fashion; that is, what lies at the heart of the profound reluctance by African presidents to hand over power voluntarily and why very many African regimes end messily, often in coups (Decalo, 1990). In most cases, the democratic option is either absent or is not respected by the loser – 'no party will accept to be unsuccessful and so form the opposition, which according to Africa's political trend, would be deprived of power and the means to those favours for their camp' (Fruh Doh, 2008: 51). Politics in Africa thus tends to be a zero-sum game.

National development and a broad-based productive economy is far less a concern (in fact, might stimulate opposition) to elites within many such systems than the continuation of the gainful utilisation of resources for the individual advantage of the ruler and his clientelistic networks. This has serious implications for the efficacy of international developmental assistance – something that most donors refuse to confront or even acknowledge, however unpalatable it may be. Furthermore, even where elites may not actively block progress, such international notions of 'development' have to be mediated through African expectations. In short, the nature of the state–society complex in Africa has critical implications for the continent's local political and economic structures, development, international relations, its interactions with external players and Africa's general place in the world.

It might be argued that acknowledgment of the sovereign status of many African state formations, however dysfunctional and fictitious, has allowed and even encouraged the current situation whereby many African citizens are materially worse off than they were under colonialism. Gaining control of an African state immediately supplies recognition and prestige from the outside world and provides external diplomatic backing and access to aid, which then further lubricates the patronage networks on which the state is predicated. In addition, assuming office automatically leads to membership in an elite club of African rulers who, as has been repeatedly demonstrated, band together for mutual support and protection against both external threats and, regrettably, domestic opposition to their rule. Such recognition, be it external or intra-African, is based on a concept of sovereignty that grants opportunities to rulers of even the most dysfunctional and weakest states. The use and abuse of the notion of sovereignty also allows an assortment of non-African actors to successfully construct commercial and military alliances with state leaders and their courtiers as well as with private corporations.

In short, many state elites in Africa use the mantle of sovereignty not to promote the collective good but to bolster their own patronage networks and to weaken those of potential challengers. The international system is complicit in such a charade (see Taylor and Williams, 2001). Malgovernance is aided, even perpetuated, on the continent by the doctrines of sovereignty and non-interference, and it is no coincidence that Africa's elites are among the most enthusiastic defenders of these principles. This remains the case, despite the African Union's ostensible claim to provide an increased scope for intervention.

The persistence of bad leadership has often been explained through the argument that 'dependent' elites in Africa are agents or compradors working for 'the West', whose notionally collective interests are somehow served by a perpetually crippled Africa. However, apart from their tendency to jealously cling to power, a great many African presidents are astonishingly inefficient and probably unable to create the conditions that might facilitate the West's supposed mission to exploit the continent for all its worth. One struggles to see how the collapse of Africa's infrastructure and productive capabilities aid global capitalism, or how widespread hyperinflation, war, and pestilence benefit the West.

Indeed, contra to the notion that Africa is a passive bystander to global processes, African elites have generally proven themselves excellent arch-manipulators of the international system. The continued flow of resources in the form of development assistance, even to countries where the ruling elites manifestly do not care about development, bears testament to this. Even allegedly omnipotent international agencies such as the International Monetary Fund and the World Bank have failed in achieving meaningful results in most SSA countries *vis-à-vis* their reform projects, as African governments have fought tooth and nail to protect their sinecures and prebends.

In fact, subversion has led to partial reform where there are considerable gaps between stated and actual commitments to reform. This is because donor-supported reforms have within them measures that would cut considerably the opportunities for informal manipulation over economic resources, rent-seeking and the ability to show favour to clients by state actors. Thus what occurs is the partial reform syndrome where aid-recipient administrations manipulate the reform process in order to protect their patron–client bases (van de Walle, 2001). Although Western governments 'have not renounced their self-proclaimed right to influence the course of events' on the continent (Bayart, 2000: 239), the politics of resistance to attempted neo-liberal reconfiguration is very much alive in Africa (Harrison, 2002).

Indeed, partial reform allows African elites to cast themselves as 'responsible partners' and in doing so has stimulated increased flows of aid. Studies have shown however that though donor funding may improve access to education and health, a moral hazard emerges whereby undesirable behaviour by state elites is in danger of being stimulated – however unintentionally – because elites know that their mistakes or inappropriate behaviour, such as corruption, excessive military spending (see Tangri and Mwenda, 2001), will be covered by the ambiguous efforts of international organisations and NGOs. Equally so, the liberal project to transform Africa's authoritarian politics into workable democracies has largely stalled – and in many cases the donors do not seem to mind too much.

Implications for Africa's international relations

As the processes outlined above have unfolded since independence, many African states have increasingly succumbed to modes of governance where the elites (invariably in alliance with non-African partners) have effectively undermined the formal and institutionalised structures of their own states. This process has involved both internal and international elements. The informalisation of politics and institutional processes has resulted in the multiplying of informal markets, popular survival strategies (increasingly operationalised through emigration), forms of privatisation that depend on the patronage and largesse of diverse global actors and, in some extreme cases, the criminalisation of the very state itself. Often, such a 'rolling back of the state' has gone hand in hand with the privatisation strictures of the international financial institutions (IFIs), though such outcomes are no doubt quite different from what donor community had envisioned when it promoted liberalisation as the way to 'set the market free'.

An international relations of questionable statehoods across SSA is of profound importance for any discussion of the continent's dialectical interactions with Europe and the wider world. Yet many analyses of Africa's place in world politics suffer from an inability to conceptualise processes, events and structures that fall within the realm of what is usually considered private, illegal or, worse, mundane and apolitical. Rectifying these inadequacies would require, according to Bayart (2000: 246), paying close attention not only to what transpires within government structures, but also at 'the trading-post, the business-place, the plantation, the mine, the school, the hospital, and the Christian mission-station'.

In fact, whilst the Western-derived (and approved) state model has increasingly foundered, Africans, through a dialectic of structural pressures and their own political agency, have continually interacted with the world in ways that accommodate ideas of personal and communal progress and order. Although these concepts are defined in ways that do not necessarily resonate with dominant liberal approaches (Chabal and Daloz, 1999), they nonetheless represent African agency and are rational and careful responses to the irresponsibility of the continent's elites and the stress placed on SSA by global pressures.

Problematically, many previous studies ignore such dynamics. Depending upon frameworks that are exclusively state-centric in both their ontology and approach, such analyses fail to pay due attention to the critical roles played by non-state actors in the continent's international relations, particularly the international financial institutions, development and humanitarian NGOs and multinational/transnational corporations. Private (and occasionally public) corporations, diasporic communities, sportsmen and women, musical collaborators and criminal networks all flourish next to, together with and 'beneath' the more readily observable state-to-state interactions that generally makes up most studies of SSA's international relations.

Obviously, the society of states and the international organisations that its members have established and are members of remain important contexts for appreciating Africa's global place. There can be no serious argument advancing the notion that the African state should be dismissed as a foundational element in studying the international relations of SSA. However, the society of states itself exists beside and in a mutual relationship with a global political economy which increasingly demonstrates alternative, non-state sites of authority and where actors engaged in business increasingly bypass formal political boundaries (either legally or illegally). Mature analyses of Africa's place in the world necessitates an understanding how these two contexts, the society of states *and* the non-state world interacts with the global political economy and influences the affairs of SSA's peoples, communities and governments. With this in mind, examining the institutional efforts to re-fashion Africa and how this may affect the continent's international relations is now embarked upon.

The African Union and NEPAD

The African Union (AU) was launched in July 2002, effectively replacing the Organisation of African Unity (OAU), which had been the premier continental organisation in Africa. For much of its existence, the OAU had, in effect, acted as the trade union of the African heads of state and had, unfortunately, largely lost much of its credibility. One of the main criticisms of the OAU was that its Charter was, in the post-Cold War era when democratisation and human rights (largely, admittedly, Western-defined) were in the fore, out-of-date. In particular, its narrowly defined concept of sovereignty was seen to, in practice, protect dictators hiding behind the principle of non-interference. This principle, drawn up in the 1960s at a time of great power machinations and continued imperialist adventures in Africa, was understandable from its particular historical origins. However, it was used by both African leaders and their various extra-African allies, be they capitalist or socialist, to bolster the position of incumbent elites, often against international sanction. The net effect of this for the average African citizen was largely negative as it was widely observed that this principle, originally designed to prevent outside interference in the era of decolonisation and the Cold War, was exploited to defend autocrats, often against their own people. It was perhaps this, more than anything else, that made the OAU suspect in the eyes of many, particularly as 'non-interference' meant that the OAU only played a very limited role in quelling Africa's conflicts.

Like the OAU, the AU aims to unify the 54 African member states politically, socially and economically, and is loosely modelled on the European Union (EU). But, like its predecessor, the AU is an ambitious project fraught with all sorts of difficulties. It is difficult, for example, to see how such vitally needed unity will be achieved, given the current tensions that continue to wrack Africa and the aforementioned state–society complexes across the continent, which militate against serious cooperation in terms of development. Furthermore, ideas for the AU include creating an African parliament and a court of justice, but presently it is quite difficult to see how the AU can be democratic and have a parliament if many of its constituent member states are not. Fundamentally, in Africa there is an urgent task to achieve greater economic unity and integration. Such regionalisation will, if past experience of regional initiatives elsewhere are anything to go by, eventually entail the creation of a common currency, a unified central bank and ultimately actual economic union. As a first stepping stone, the current regional economic blocs, such as ECOWAS, SADC and COMESA need to be rationalised and made more efficient. Yet at present many African countries still conduct more trade with their former colonial masters than with each other, a situation that continues to foster dependence on outsiders and which hampers African integration and unity. Overlap between regional blocs remains a major problem within Africa: the case of COMESA and SADC probably being the first case to come to mind.

At the same time, challenges posed by ongoing processes of globalisation pose severe questions for the AU as the premier organisation to manage Africa's international relations. The AU was launched at a time when there had been growing questioning of the basic neo-liberal philosophy that underpins contemporary capitalism, frequently cast within the catch-all term 'globalisation'. Some observers may proffer the view that this juncture opens up space for Africa and that perhaps the AU may be the vehicle to advance this. However, there remain limitations, both externally through the workings of the powerful global market and its capitalist actors, and internally through the actions and attitudes of African elites themselves. The essential acceptance of the basic tenets of the ongoing world order reflects the actuality that many elites from Africa are, in the main, just as interested in maintaining the global system as their colleagues in the North. Certainly, African elites are not passive victims and at least some fractions of such elites have been among the strongest supporters of the so-called Washington Consensus. In this sense, such leaders have been among the social forces that have facilitated liberalisation, despite veritable downsides for the average African.

Central in this process has been the New Partnership for Africa's Development (NEPAD). NEPAD was launched in Abuja, Nigeria, in October 2001; it arose from the mandate granted to five African heads of state (Algeria, Egypt, Nigeria, Senegal, South Africa) by the then Organization of African Unity (OAU) to work out a development programme to spearhead Africa's renewal. It might be said that NEPAD temporarily succeeded in placing the question of Africa's development onto the international table, and for a time managed to obtain a fairly high profile and awareness. In doing this, NEPAD's promoters claimed it was a political and economic programme aimed at promoting democracy, stability, good governance, human rights and economic development on the continent. NEPAD was essentially sold as a bargain: African countries would set up and police standards of good government across the continent – whilst respecting human rights and advancing democracy – in return for increased aid flows, private investment and a lowering of obstacles to trade by the West. An extra inflow of $64 billion from the developed world was touted as the 'reward' for following approved policies on governance and economics.

A great deal of expectations were raised about the possibilities opened up by NEPAD, particularly with regard to the promise to develop a credible peer review process to advance democracy and good government in Africa. Much of this was wholly unrealistic. The logic and modus operandi of neo-patrimonial rule and the dominance and nature of extractive economies in Africa – and their relationships with the international system – meant that NEPAD's strictures on good governance and democracy could never be implemented without eroding the very nature of the post-colonial African state and undermining the positions of incumbent elites – an unlikely possibility. In fact, NEPAD systematically ignored the reality that power in African politics must be understood as the utilisation of patronage and clientelism and operates within neo-patrimonial modes of governance, which was the antithesis of NEPAD's own vision for Africa. Indeed, in spite of the façade of the modern state, which was taken as NEPAD's starting point, power in most African polities progresses informally, between patron and client along the lines of political reciprocity, is intensely personalised and is not exercised on behalf of the general public good. 'The state itself remains the major vortex of political conflict precisely because it presides over the allocation of strategic resources and opportunities for profit making' (Othman, 1989: 114).

The irony was that the type of solutions advanced by NEPAD would have deprived rulers of the means to maintain their patronage networks. In short, to have an Africa based on the enunciated principles of NEPAD would have actually eroded the material base upon which the neo-patrimonial state was predicated. And yet NEPAD seemed to advance the idea that the very same African elites who benefitted from the neo-patrimonial state would now commit a form of class suicide. The possibility always seemed improbable. With very few exceptions, the majority of heads of state involved in NEPAD were quintessentially heads of neo-patrimonial regimes and certainly did not regard their rule as 'temporary' or that institutional law should constrain their preeminence. In other words, most African presidents behaved in ways that were the exact opposite of what NEPAD said regarding good governance. What this meant was that the commitment shown by state elites had to be taken with a pinch of salt. After all, countries that signed up to NEPAD included such models of good governances as Angola, Burkina Faso, Cameroon, Congo (Brazzaville), Egypt, Gabon, Kenya, Malawi, Nigeria, Sierra Leone, Uganda and Zambia. None would have passed NEPAD's own strictures on clean government.

Of course, returning to the AU, it may be said that imagining that resistance to neo-liberalism might be located in the elites of Africa is, to put it mildly, naïve, and the potentiality of the AU as a site for any such project is highly curtailed. Defending globalisation and the advancement of specific externally oriented interests and values, whilst ameliorating the excessively negative aspects of this project, is the new message; yet this is obviously problematic as there is a quite definite contradiction between, on the one hand, supporting global free trade and, on the other, committing oneself to somehow changing the rules of the system to ensure greater equity. The point is that many African elites, having bought into the globalisation discourse, actively encourage an unquestioning stance towards foreign direct investment (FDI). Investment itself may not be problematic, as all developing states need capital in order to finance development, setting aside for one moment the fact that there is actually a net outflow of capital from Africa. However, it is the type of FDI that is welcomed and the manner in which this is managed and guided (or not) by the host country that is crucial. The question is how to make those corporations engaging in FDI in Africa development oriented. How the AU resolves this issue and the continued push for further liberalisation will be a crucial test for the new body. Neo-liberal reforms in Africa have generated considerable social conflict and there is a profound inconsistency in advancing both highly welcome democratic reform and continued economic adjustment and austerity.

Ultimately, two points need to be made about the AU. Firstly, if it did not exist, it would have to be invented and so the AU is the vehicle thus chosen; it must be worked with and engaged and cannot be dismissed. Secondly, any institution can only be as strong as its members. By this I include both states and citizens. It is true that at its most basic level the AU is a compact between states. But, this does not mean that it can be left alone by activists, intellectuals and 'ordinary' people; after all, it is the people that demand unity and development and it is the people – the 'African' part of the AU – that need to engaged if the project is to succeed. Certainly, any strategy for Africa's renewal needs to be grounded not in the elites but in the ordinary citizens, based on basic human needs. Otherwise such a project remains subject to a wide variety of destabilising forces, not least if elites seek to duck out from the commitments they themselves have made. This is especially the case as Africa's place in the global system becomes ever more important as new actors scramble for the continent alongside the 'traditional' European and American actors, primarily spurred on by Africa's importance as a source of energy.

The 'new' Africa and oil

As the twenty-first century progresses, Africa's place in the global political economy is increasing in importance. In particular, there has been growing international interest in Africa as a source for energy. So much so, that there is now talk of a 'scramble for Africa's oil', redolent of the nineteenth century's scramble for Africa (Klare and Volman, 2006b; Watts, 2006; Ghazvinian, 2007; Clarke, 2008). It is important to remember that although this unprecedented attention is relatively new, the presence of oil in Africa is not; oil extraction on the continent began in the 1950s, whilst exploration was started much earlier (Soares de Oliveira, 2007). Yet Africa has now emerged as a hugely important source of oil in the global economy. This is largely due to new discoveries and the instability of oil markets in the Middle East, which compels the search for alternative supply locations.

These and other factors have resulted in major oil corporations from around the world increasingly focusing their attention towards diversifying oil supplies and looking towards sub-Saharan Africa. American oil corporations largely control the oil fields of those economies that have recently discovered oil reserves, such as Equatorial Guinea and São Tomé and Príncipe, whilst British and American oil interests dominate Nigeria and French companies lead in Gabon and Congo-Brazzaville. Chinese corporations dominate the oil sector in Sudan. Meanwhile, actors from Brazil, India, Japan, Malaysia, South Korea, are aggressively competing and seeking access across the continent.

With a new energy policy in the United States bent on diversifying Washington's sources of oil and with the increased activities of new actors, a perceived rush for Africa's oil is now a feature of the continent's international relations. The burgeoning oil fields in sub-Saharan Africa, particularly in the Gulf of Guinea, have become of major geo-strategic importance to the oil-dependent industrialised economies. In fact, it might now be stated that the United States does not just buy oil from Africa, 'in many ways it is *dependent* on African oil' (emphasis added) (Barnes, 2005: 236). Thus, African oil has become a 'matter of US national strategic interest' (Obi, 2005: 38), granting the Gulf of Guinea 'major strategic relevance in global energy politics' (Alao, 2007: 168). Concomitantly, Africa has now emerged as a major site for competition between various oil corporations from diverse nations. Within the next few years, perhaps the largest investment in the continent's history will be seen as billions of dollars are poured into exploration and oil production in Africa. For instance, three of the world's largest oil corporations (Shell, Total and Chevron) are targeting 15 per cent, 30 per cent and 35 per cent respectively of their global exploration and production budgets on Africa (Ghazvinian, 2007: 7).

The characteristics of the new scramble stimulate some of the dynamics that underpin this new rush into Africa's oil fields by external actors. Firstly, at the global level there is mounting anxiety that future oil supplies will not meet global demand, particularly within a wider context where emergent economies such as Brazil, China and India are rapidly increasing oil consumption to feed their growing economies. The gap between global supply and demand may be reached as early as 2025, according to some analyses (Klare and Volman, 2006a: 609). Thus, although the actual quantity of African oil reserves are low in comparison to those presently found in the Middle East, in a context marked by deep anxiety over future supplies Africa's reserves (c. 9 per cent of the world's total) are extremely significant (Ghazvinian, 2007).

An important characteristic of the scramble is that whilst it is not solely a race between Chinese and US corporations, the scramble's dynamics are heavily influenced by the roles and activities of actors from these two states (Frynas and Paulo, 2007: 230). Policy-makers in both nations have identified African oil as vital to their respective nations' national interest, albeit for different motives. It is apparent that policy analysts in Beijing see the broader global political milieu as being intrinsically linked to Chinese energy security and feel that in the current environment China is vulnerable until, and unless, it can diversify its oil sourcing and secure greater access to the worlds oil supplies (Taylor, 2006: 937). Between 2002 and 2025 it is estimated that Chinese energy consumption will rise by 153 per cent (Klare and Volman, 2006b) and China is now the second largest consumer of oil globally (Shinn, 2008). In order to fuel such a growing demand, Chinese oil corporations have entered into the competition for African oil. In fact, China's economic and diplomatic strategy has for the most part focused on oil-rich states lacking American investment; the majority of Chinese African oil imports come from the oil-rich states of Angola, Sudan, Equatorial Guinea, DRC and Nigeria (Taylor, 2009).

From the American perspective, the 'war on terrorism and preparations for war against Iraq ... enormously increased the strategic value of West African oil reserves' (Ellis, 2003: 135). The high level of interest from such major importers has certainly raised the level of competition over Africa's oil. Whilst corporations headquartered in other states – Britain, Brazil, France, India and Malaysia, for example – are playing important roles in the ongoing scramble and equally striving to build up their oil portfolios in Africa, it is the ostensible Sino-American competition for oil on the continent that has grabbed the most headlines: 'There is also little doubt that the interest in Africa's oil and gas resources has spurned a rivalry between international actors in Africa, notably the American and Chinese governments' (Frynas and Paulo, 2007: 230).

Arguably, unlike the colonial scramble for Africa, African agency is far more present in the contemporary rush for oil. Many African governments are quite proactive in their roles within today's context. Whilst we might aver that the nineteenth-century scramble 'was driven and dictated by European colonial interests', in the current scramble for oil, 'African leaders act in the role of decision-makers' (Frynas and Paulo, 2007: 235). Whilst it is true that many African states are rich in oil but lack sufficient capital to exploit these resources and thus create formative conditions whereby African elites might be seen as dependent upon external actors to facilitate exploitation (Goel, 2004: 482), the ability of the government to negotiate favourable contracts should not be discounted. In fact, many oil-rich African governments are quite skilful in playing the oil game – albeit for the benefit of the incumbent elites, rather than the broad masses.

New actors

It is well-known that both China and India have gone into Africa in a big way and that this has propelled the continent's importance in international relations. Yet other than these two old 'new' actors, numerous other states and corporations are endeavouring to secure contracts and a presence on the continent. Brazil for instance is increasingly active in Africa. This is perhaps natural given that Brazil has the second largest black population of any country in the world, after Nigeria. Trade and investment are central to Brazil's new interest in Africa. Solidarity has also been presented as central to Brazil's efforts towards intensifying its links with Africa as well as with the rest of the developing world. Indeed, it was with the change of government and the assumption of president Luiz Inácio 'Lula' da Silva in January 2003 that kick started the deepening of relations with Africa, not only at the rhetorical level but also with practical policies. Importantly, Brazil has one of the strongest economies among developing countries, and, according to Soares de Lima and Hirst (2006), the key foreign policy aspiration of Brazil is to achieve international recognition a major power in world affairs. As part of this, Brasilia seeks to assert an independent voice within international affairs (see de Arimatiia da Cruz, 2005). Developing close ties with other developing nations is central to this – as the India–Brazil–South Africa Dialogue Forum demonstrates. Of course, Brazil is deeply involved in the *Comunidade de Paises de Lingua Portuguesa* (Community of Portuguese-Speaking Countries), an eight-member organisation of which the majority (five) are African. Africa is now seen 'as the most important experiment of South–South cooperation' by Brasilia, 'accompanied by an increase in bilateral trade' (Visentini, 2009: 5).

Russia is another actor that is increasingly important on the continent. During the Soviet era, Africa was an important element in Moscow's foreign policy calculations. This changed under Gorbachev (see Matusevich, 1999), and, although Russia never left SSA, the post-Soviet period was marked by a certain disengagement, particularly under Boris Yeltsin's chaotic regime (see Shubin, 2004). This seems to be now changing (Matusevich, 2006). In 2009, President Dmitry Medvedev paid his first official visit to Africa (and the first by a Russian head of state for more than three years), visiting Egypt, Nigeria, Angola and Namibia. During Medvedev's tour, Russia's energy giant Gazprom signed a 2.5 billion dollar deal with Nigeria's state-operated oil company to invest in a new joint venture. Although the commercial attractions of SSA are clear for Russian corporations, the political dimension cannot be ignored. A newly assertive Moscow is now actively seeking to project itself as a global player, increasingly in regions far from its traditional spheres of influence. Indeed, Moscow's foreign policy is dominated by efforts to reverse the decline of the 1980s and 1990s. This entails promoting international conditions conducive to facilitating Russia's reconstruction as a major power (MacFarlane, 2006). Moscow favours a multipolar world, with several strong regional centres. The notion of the BRICs (Brazil, Russia, India and China) is part of this and Moscow has been enthusiastically pushing such a configuration (Armijo, 2007). Africa will fit into such considerations as a site where Russian corporations can seek out contracts, particularly in the energy realms, and as a place where more and more Russian investment will likely be located. Of course, competition with Chinese commercial interests is part of this.

There are other new actors other than Brazil and Russia. Iran for instance is stepping up its contacts in Africa, to the extent of offering to host an Iran–Africa Summit. This is part of Iranian president Mahmoud Ahmadinejad's assertive foreign policy. While Tehran is probably most interested in expanding an Islamic bloc to embrace African states – especially one that might displace the supremacy of Sunni Arab states – African elites are more likely more interested in economic advantages. When nations such as Turkey host a Turkey–Africa Cooperation Summit, as they did in August 2008, the scale of this upsurge of interest is clear. Demonstrating its commitment, Ankara covered the full expenses of each attending Head of State or Government and of the 53 members of the AU, only Mozambique, Swaziland and Lesotho were

not represented. Turkey has in fact increased its trade with Africa from
5.4 billion to 13 billion dollars in under three years, with a target to
reach 30 billion dollars by 2010 (*Mail and Guardian* (Johannesburg)
18 August 2008). Similarly, Mexico's 'foreign policy strategy towards
Africa [now] seeks to build a new and much closer relation with this
region. Through a wider and more efficient diplomatic presence, the
promotion of mutual understanding, the enhancement of co-operation,
trade and investments, Mexico wishes to strengthen its links with the
African nations' (Mexican Foreign Secretary Patricia Espinosa, quoted
by *Reuters* (Cape Town), 19 February 2009).

These new actors bring new ways of doing things, new markets in
which to sell their goods, alternative sources of financing and assistance
and an increased ability to lower Africa's dependence on traditional
partners, most notably Europe and the US. It is a fact that Africa's partnerships are diversifying, with significant increases in trade, FDI, aid,
from these emerging partners, but in African studies, the extent of these
new developments is often obscured by a focus on the 'old' relations of
ex-colonial powers and the United States. Until recently, Africa was seen
by the West as marginal and of little political interest. However, over the
last 15 years or so, emerging powers have made significant inroads into
Western political and economic dominance in Africa and this has caused
a degree of reflection in the West regarding attitudes towards 'the hope-
less continent'. Today, any discussion of Africa's international relations
needs to note the changing economic status of emerging partners and
their growing political influence. This new diversity of partners is a tre-
mendous opportunity for Africa. Each wave of countries engaging with
Africa brings with it new products, capital goods, technology, know-
how and expertise and development experience. Each also brings new
ways of doing things which question previous assumptions. The emerging partners also often have a comparative advantage in their outward
engagement with Africa. For instance, they are able to access large pools
of finance and capital reserves (mostly through state incentives and subsidised support). They also uphold a version of the developmental state
model that encourages a particular approach to business that enables
private enterprise and mercantile commerce. Clearly, this state capitalist
model is powerfully competitive.

In summary, the technological transfer and cooperation inherent in
these ties will arguably alter Africa's economic and governance landscape. Africa is moving from its traditional 'post-colonial' North–South
relations towards a more diverse, more business-centric set of partnerships with a wider array of actors. Already this has prompted renewed
competitive interest from the West in Africa – potentially providing

African governments with greater leverage. This wider spectrum of relations is potentially good news for Africa – but some longstanding challenges remain, centred around some of the issues of governance and state–society relations that have been mentioned above.

Conclusion

As Africa's international relations diversify and intensify to include new actors as well old players and to cover increasingly varied processes and dynamics involving both state and non-state actors, the study of the continent's extraversion is more and more apposite. Despite its historic neglect by international relations, Africa is likely to become more and more central to the discipline. The resurgence of politico-economic interest in Africa poses a complicated array of negative pressures on the continent and its international relations that very much outweigh any limited opportunities for progressive change. Hazards spring from both the behaviour of external actors, be they corporate or state, as well as the actions of comprador elites in control of rentier economies. Furthermore, with the upsurge of interest in oil by industrialised and emerging economies, there is the very real possibility that African elites may chose to ignore blandishments about the necessity to practice good government (however defined).

From the outside in – and contradicting the self-proclaimed discourse that foreign actors promote democracy and good governance in Africa – kleptocratic and authoritarian trends have been remarkably tolerated by external players, be they autocracies such as China, self-proclaimed democracies as found in the EU or institutions ostensibly grounded in self-perceived 'sensible' economic policies, such as the IFIs. In fact, the political make-up of the external actor makes very little concrete difference in the process by international actors of engagement with African spaces and particularly the need for oil has accelerated a long-existent trend of forgetting the rhetoric about good governance in favour of naked geopolitics (Schraeder, 2001). Of prime concern is that this new interest in Africa may reify in Africa what one analysis has termed the phenomena of the successful failed state (Soares de Oliveira, 2007). These entities would, by any normal measure of a state's capabilities and performance, be considered as failed, in that there is chronic leadership, weak and undiversified economies, fragile institutions and low levels of human development. Yet whilst being marginalised by the rest of the world, such states are inherently engaged with it. And whilst possessing the attributes of a failed/failing state, there persists a paradoxical sustainability where the presence of resources maintains the interest and attention of the international community who uphold relationships with such states, granting them (or rather, their elites) legitimacy (Soares de Oliveira, 2007). This 'legitimacy' not only can serve to play out in domestic terms, but also at the global level. And in such circumstances, if a state's 'success' is determined by their international legitimacy, recognised sovereignty and the ability to interact at the global level (rather than how they serve their citizenries), then the presence of resources has granted success to some fundamentally dysfunctional African states. Angola, Equatorial Guinea, Nigeria and Sudan all spring to mind at this juncture.

Of course, there are possibilities – however remote – that Africans may benefit from this increased interest in the continent. The plethora of externally delivered infrastructure built by, for example, Chinese construction companies as part of broader packages to secure contracts for Chinese actors would be one such instance. And it is not *impossible* that accrued revenues may help implement constructive change, perhaps in the direction of the welfare reforms pursued by some Middle Eastern oil producers. It is just unlikely. Regrettably, the pursuit of energy and profit is likely to fit a familiar pattern – and is intimately entwined with consumption and lifestyle patterns outside of the African continent. As Obi (2007: 17) notes, what is needed is the 're-organisation of production in the continent in ways that lift it out of its marginal position in the globalised division of labour which since the days of the 'old scramble' has defined it as an object of domination and exploitation by forces from "outside"'. Yet in the present climate, this situation is not likely to change any time soon. Though Africa's place in international relations has certainly increased in recent years, how the continent will negotiate its position with external actors, particularly given the general dysfunctional state of many African states and its continental body, the AU, remains an issue of great concern.

References

Alao, A. (2007) *Natural Resources and Conflict in Africa: The Tragedy of Endowment*, Rochester, NY: Rochester University Press.

Armijo, L. (2007) 'The BRICs Countries (Brazil, Russia, India, and China) in the Global System', special issue of *Asian Perspective*, 31 (4).

Barnes, S. (2005) 'Global Flows: Terror, Oil and Strategic Philanthropy', *Review of African Political Economy*, 34 (104/5): 1–22.

Bayart, J.-F. (1993) *The State in Africa: The Politics of the Belly*, London: Longman.

Bayart, J.-F. (2000) 'Africa in the World: A History of Extraversion', *African Affairs*, 99: 217–67.

Bratton, M. and van de Walle, N. (1994) 'Neopatrimonial Regimes and Political Transitions in Africa', *World Politics*, 46 (4): 453–89.

Bratton, M. and van de Walle, N. (1997) *Democratic Experiments in Africa: Regime Transitions in Comparative Perspective*, Cambridge: Cambridge University Press.

Callaghy, T. (1984) *The State–Society Struggle: Zaire in Comparative Perspective*, New York: Columbia University Press.

Chabal, P. (1994) *Power in Africa: An Essay in Political Interpretation*, New York: St Martin's Press.

Chabal, P. (2009) *Africa: The Politics of Suffering and Smiling*, London: Zed Books.

Chabal, P. and Daloz, J.-P. (1999) *Africa Works: Disorder as Political Instrument*, Oxford: James Currey.

Clapham, C. (ed.) (1982) *Private Patronage and Public Power: Political Clientelism in Modern States*, London: Pinter.

Clapham, C. (1985) *Third World Politics: An Introduction*, London: Croom Helm.

Clapham, C. (1996) *Africa and the International System*, Cambridge: Cambridge University Press.

Clarke, D. (2008) *Crude Continent: The Struggle for Africa's Oil Prize*, London: Profile Books.

de Arimatiia da Cruz, J. (2005) 'Brazil's Foreign Policy Under Luis Inacio "Lula" da Silva: An Early Assessment of a Leftist President', *Politics and Policy*, 33 (1): 13–35.

Decalo, S. (1990) *Coups and Army Rule in Africa*, Newhaven: Yale University Press.

Ellis, S. (2003) 'Briefing: West Africa and its Oil', *African Affairs*, 102 (406): 135–38.

Falola, T. (2006) 'Writing and Teaching National History in Africa in the Era of Global History', in P. Zeleza (ed.), *The Study of Africa: Disciplinary and Interdisciplinary Encounters*, Dakar: Codesria.

Fatton, R. (2002) *Haiti's Predatory Republic: The Unending Transition to Democracy*, Boulder, CO: Lynne Rienner.

Fru Doh, E. (2008) *Africa's Political Wastelands: The Bastardization of Cameroon*, Bamenda: Langaa Research and Publishing.

Frynas, J. (2004) 'The Oil Boom in Equatorial Guinea', *African Affairs*, 103 (413): 527–46.

Frynas, J. and Paulo, M. (2007) 'A New Scramble for African Oil? Historical, Political, and Business Perspectives', *African Affairs*, 106 (423): 229–51.

Ghazvinian, J. (2007) *Untapped: The Scramble for Africa's Oil*, London: Harcourt.

Goel, R. (2004) 'A Bargain Born of Paradox: The Oil Industry's Role in American Domestic and Foreign Policy', *New Political Economy*, 9 (4): 467–92.

Harrison, G. (2002) *Issues in the Contemporary Politics of Sub-Saharan Africa: The Dynamics of Struggle and Resistance*, Basingstoke: Palgrave.

Jackson, R. and Rosberg, C. (1982) *Personal Rule in Black Africa: Prince, Autocrat, Prophet, Tyrant*, Los Angeles, CA: University of California Press.

Jackson, R. and Rosberg, C. (1994) 'The Political Economy of African Personal Rule', in D. Apter and C. Rosberg (eds), *Political Development and the New Realism in Sub-Saharan Africa*, Charlottesville: University of Virginia Press.

Klare, M. and Volman, D. (2006a) 'The African "Oil Rush" and US National Security', *Third World Quarterly*, 27 (4): 609–28.

Klare, M. and Volman, D. (2006b) 'America, China and the Scramble for Africa's Oil', *Review of African Political Economy*, 33 (108): 297–309.

Lemarchand, R. and Eisenstadt, S. (eds) (1980) *Political Clientelism, Patronage and Development*, New York: Sage.

LeVine, V. (1980) 'African Patrimonial Regimes in Comparative Perspective', *Journal of Modern African Studies*, 18 (4): 657–73.

MacFarlane, N. (2006) 'The "R" in BRICs: is Russia an Emerging Power?', *International Affairs*, 82 (1): 41–57.

Matusevich, M. (1999) 'Perestroika and Soviet Policy Shift in Africa', *Nigerian Forum*, 20 (7–8).

Matusevich, M. (ed.) (2006) *Africa in Russia, Russia in Africa: Three Centuries of Encounters*, Trenton, NJ: Africa World Press.

Montague, D. (2002) 'Stolen Goods: Coltan and Conflict in the Democratic Republic of Congo', *SAIS Review*, 22 (1): 103–18.

Obi, C. (2005) 'Globalization and Local Resistance: The Case of Shell Versus the Ogoni', in L. Amoore (ed.), *The Global Resistance Reader*, London: Routledge.

Obi, C. (2009) 'Scrambling for Oil in West Africa?', in H. Melber and R. Southall (eds), *A New Scramble for Africa*, Scottsville: University of Kwazulu-Natal Press..

Othman, S. (1989) 'Nigeria: Power for Profit – Class, Corporatism, and Factionalism in the Military', in D.B. Cruise O'Brien, J. Dunn and R. Rathbone (eds), *Contemporary West African States*, Cambridge: Cambridge University Press, pp. 113–44.

Prior, C. (2007) 'Writing Another Continent's History: The British and Pre-colonial Africa, 1880–1939', *eSharp*, 10: 1–16.

Sandbrook, R. (1985) *The Politics of Africa's Economic Stagnation*, Cambridge: Cambridge University Press.

Schraeder, P. (2001) 'Forget the Rhetoric and Boost the Geopolitics: Emerging Trends in the Bush Administration's Policy Towards Africa, 2001', *African Affairs*, 100 (400): 387–404.

Shaxson, N. (2007) *Poisoned Wells: The Dirty Politics of African Oil*, Basingstoke: Palgrave.

Shinn, D. (2008) 'Military and Security Relations: China, Africa and the Rest of the World', in R.I. Rotberg (ed.), *China into Africa: Trade, Aid and Influence*, Washington, DC: Brookings Institution Press, pp. 155–96.

Shubin, V. (2004) 'Russia and Africa: Moving in the Right Direction?', in I. Taylor and P. Williams (eds), *Africa in International Politics: External Involvement on the Continent*, London: Routledge.

Soares de Lima, M. and Hirst, M. (2006) 'Brazil as an Intermediate State and Regional Power: Action, Choice and Responsibilities', *International Affairs*, 82 (1): 21–40.

Soares de Oliveira, R. (2007) *Oil and Politics in the Gulf of Guinea*, London: Hurst & Company.

Tangri, R. (1999) *The Politics of Patronage in Africa: Parastatals, Privatization and Private Enterprise*, Trenton: Africa World Press.

Tangri, R. and Mwenda, A. (2001) 'Corruption and Cronyism in Uganda's Privatization in the 1990s', *African Affairs*, 100 (398): 117–33.

Taylor, I. (2004) 'Blind Spots in Analyzing Africa's Place in World Politics', *Global Governance*, 10 (4): 411–17.

Taylor, I. (2006) 'China's Oil Diplomacy in Africa', *International Affairs*, 82 (5): 937–59.

Taylor, I. (2009) *China's New Role in Africa*, Boulder, CO: Lynne Rienner.

Taylor, I. and Williams, P. (2001) 'South African Foreign Policy and the Great Lakes Crisis: African Renaissance Meets Vagabondage Politique?', *African Affairs*, 100 (399): 265–86.

Taylor, I. and Williams, P. (2004) 'Understanding Africa's Place in World Politics', in I. Taylor and P. Williams (eds), *Africa in International Politics: External Involvement on the Continent*, London: Routledge.

van de Walle, N. (2001) *African Economies and the Politics of Permanent Crisis, 1979–1999*, Cambridge: Cambridge University Press.

Visentini, P. (2009) 'Prestige Diplomacy, Southern Solidarity or "Soft Imperialism"? Lula's Brazil–Africa Relations (2003 onwards)', Working paper, Federal University of Rio Grande do Sul, Porto Alegre.

Watts, M. (2006) 'Empire of Oil: Capitalist Dispossession and the Scramble for Africa', *Monthly Review*, 58 (4): 1–17.

6

African regionalism: external influences and continental shaping forces

Mary Farrell

The Joint Africa–EU Strategy (JAES) (Council of the European Union, 2007) marked another phase in the cooperation between the two continents that had its origins in the post-colonial era for the African countries, and for the new European community founded under the Treaty of Rome (1957). Presented as a strategic partnership among the 27 countries of the European Union (EU) and the 53 countries of Africa, it was framed with the intention to redefine the relations between the two parties so as to address global challenges through common action. Ambitious in its scope and long-term perspectives, the JAES proposed an action plan around eight thematic partnerships to address four main areas: peace and security, governance and human rights, trade and regional integration and development. The strategy coincided with renewed regional dynamism in both Africa and Europe: the African continent had taken a number of steps in the direction of closer cooperation at continental and sub-regional levels, while the EU had enlarged and deepened its own integration processes.

Among the EU's external relations portfolio, promotion of regional integration had become a focus of European policy particularly towards the African countries, though also with other parts of the world. African regionalism itself rests on complex, multilayered and diverse processes embracing a number of political actors, including national governments, business, civil society and, increasingly, external actors such as the EU. However, though the African Union's (AU)'s own structures appear to be closely modelled on the EU's institutional set-up, there are significant pressures and forces emanating from domestic political and social structures on the continent that compel political action at supranational level. How does the JAES reflect this diverse mix of actors, and how far does it respond to the different pressures for integration? To what extent can the strategy manage the disintegrative forces that are evident in the African continent, reflected in such realities as the limited regional economic integration among the existing regional economic communities (RECs), the development gap and the recurring ethnic conflicts that continue to impact upon peace and security in several parts of the continent?

This chapter addresses these questions through an examination of the EU's promotion of regional integration, and inter-regionalism in the context of the Joint Africa–EU Strategic Partnership. The next section considers the rationale behind the EU's adoption of regional integration as external policy, and how or whether this defines the EU as a normative actor. The third section considers the JAES in some detail, identifying the historical origins and the objectives of this strategic partnership, the institutional architecture and the respective principal regional actors, and the challenges and limitations of the partnership. The fourth section identifies specific considerations from the African perspective, and considers the synergies between the inter-regionalism of the JAES and the existing sub-regional integration arrangements. The next section focuses upon the case of joint cooperation to develop the provision of infrastructure on the African continent, followed by a review of the Economic Partnership Agreements (EPAs) as intra- and inter-regional integration agreements. The chapter concludes with a summary of the prospects and limitations of the JAES as a case of inter-regionalism, and as a concrete initiative to promote security and development in Africa.

Promotion of regionalism

Since the 1990s, the EU has been scrutinised for its external relations, and the manner in which the polity has sought to exercise influence in the international arena. Much of the academic research has taken the approach that the EU is some kind of international actor and a global economic power, distinct from the sovereign state actor that takes centre stage in the realist international relations literature yet at the same time posing an analytical challenge by virtue of its status as 'more than an inter-governmental organisation but less than a fully-fledged European state' (Hill and Smith, 2005: 13). The absence of hard power and military capability is compensated for by a willingness to engage in external relations at a global level through the multilateral arena in the pursuit of peace-keeping operations, trade and global environmental negotiations, and a growing portfolio of strategic partnership agreements (bilateral and inter-regional), all of which demonstrate the increasing actorness of the EU in the international arena. The literature captures the complexity and diversity of this actorness even as the absence of consensus on how to characterise it continues to underscore the very vitality of the EU's international relations (Smith, 2010).

The EU's self-referential identity as a normative international actor, where external relations policy is shaped by normative concerns, such as respect for human rights, the rule of law, and multilateralism, can be traced to the Maastricht Treaty and the subsequent commitment of the EU member states to coordinate actions and positions in those policy areas where shared competence defined the mode of national and supranational decision-making. As the promotion of norms became more explicit in EU external policy, the academic literature sought to capture the multilevel governance model cloaked in the normative concerns of the political community, a community that was singularly unable to agree on just how far normative concerns out-weighed more rationalist motivations in external actions (Lucarelli and Manners, 2006; Scheipers and Sicurelli, 2008; Bickerton, 2009). More often than not, the politics of external relations policy reflected a compromise – between multilateralism and regionalism, or between conditionality and partnership (Nicolaïdis, 2010). This compromise is clearly evident in EU–Africa inter-regionalism, and in the promotion of regional integration.

Research on the EU's promotion of regional integration has tended to contextualise it from three different standpoints: the export of the EU model of regional integration, the expansion of regional governance and the spread of European norms (Farrell, 2009; Fawcett and Hurrell, 1995; Hurrell, 2007; Gamble and Payne, 1996). European regionalism is regarded around the world as a model of cooperation that might be emulated or, equally, regarded as a form of regionalism to be rejected as an explicit manifestation of the European experience (Börzel and Risse, 2009; Fawcett and Gandois, 2010). However, making case study comparisons can be difficult because of the different terminology to be found in the literature, and the different empirical processes that may be discernible in any given region.[1] Among the many regionalism processes that can be identified in practice, a distinction can be made between formal inter-state regional cooperation for the purposes of establishing inter-state regional regimes and policies in certain issue areas; regional consolidation where the 'region' takes on 'actorness' to manage relations among the constituent member states, and with the rest of the world; and the informal regionalisation associated with increased levels of economic and social interaction, creating ultimately greater interdependence among the members of the wider regional society. Both formal and informal processes may over time result in the emergence of regional awareness and a sense of regional identity, though such outcomes are never certain even in the most 'regionalised' communities.

The EU's promotion of regional integration with African partners constitutes the formal inter-state regional cooperation, and in practice much of this cooperation under the JAES takes place as part of the high-level, multilayered institutional architecture that has been created.[2] However, in practice the promotion of regional integration is based on complex and dynamic processes that co-exist amidst often competing logics – the logics of economic and societal integration, of power politics and security (Hurrell, 2007: 130). The European Commission's own definition of regional integration would seem to recognise this complexity: 'regional integration is the process of overcoming, by common accord, political, physical, economic and social barriers that divide countries from their neighbours, and of collaborating in the management of shared resources and regional commons' (European Commission, 2008: 7).

In the case of the JAES, inter-regionalism is presented as a way to promote socio-economic development and good governance, closely intertwined with democracy promotion and human rights, in an agenda that relies strongly on dialogue and mutual cooperation. The Cotonou Agreement had explicitly highlighted a dual-track approach in the context of EU–Africa relations, signalling a departure from historical European policy towards the African region that was based largely on aid and development assistance, and on the grant of non-reciprocal access to European markets for primary products originating from the African, Caribbean and Pacific (ACP) group. The new dual-track approach called for regional integration into the global economy, and for regional intra-integration among groups of African states, on the basis that 'cooperation shall provide effective assistance to achieve the objectives and priorities which the ACP states have set themselves in the context of regional and sub-regional cooperation and integration, including inter-regional and intra-ACP cooperation' (Cotonou Agreement, Article 28) (European Union, 2000).

Broadly speaking, the promotion of regional integration by the EU has taken diverse forms, exemplified by a variety of policy instruments and by a qualitative distinction in strategic intent and in desired outcomes. The enlargement process is one form of regional integration, extending the regulatory system and legal order to new member states – this is the most comprehensive and permanent way of 'exporting' the European governance system, relying heavily on the use of conditionality instruments and a mix of coercive and persuasive mechanisms to initiate domestic political change in the applicant countries. Second, the EU has been able to influence regional integration in a broad, general way through normative suasion, prompting other regional communities to adopt certain practices, institutional arrangements, or other forms of governance modelled on the European regional governance system (Farrell, 2009: 8). The role of norms in shaping outcomes is widely acknowledged in the academic literature, and, even in the absence of a specific policy action, an agreement or international treaty, it is possible for a political actor to exercise influence and induce behavioural change (Finnemore and Sikkink, 1998). The EU's effectiveness as a norm exporter and a normative actor has been much discussed, and though there is no general consensus on the substantive impact – since the outcomes will be determined by the nature of the individual agreements made by the EU with other states/actors and by how the target region responds – there remains strong recognition for the socialisation of actors through normative rationality or the logic of appropriateness, where actors are motivated to do the right thing, rather than simply being motivated by self-interest (Börzel and Risse, 2009). The social constructivist argument that actors persuade each other towards the acceptance of certain normative ideas, based on the use of reasoning and 'argument', or dialogue, is also relevant in this regard, emphasising as it does the importance of persuasion as a mechanism for the diffusion of ideas. The EU's promotion of regional integration, thus, resonates with normative statements, not least in the claims that regional integration is the most effective way to guarantee regional security, to promote economic and social development and to guarantee long-term prosperity and growth (Börzel and Risse, 2009: 8).

In addition to internal expansion through enlargement, and the diffusion of ideas through normative rationality, the EU's promotion of regional integration has taken place through the extensive array of inter-regional agreements, including the Cotonou Agreement and the JAES. Inter-regionalism takes a variety of political/institutional forms and policy instruments, covering diverse issue areas (trade, environment, technical assistance, development, infrastructure, political reform) and agreements on aid and trade, to support regional integration, and more comprehensive regional strategies. In the JAES, this is a case of pure inter-regionalism, where bloc-to-bloc cooperation underpins a strategic partnership with multiple goals and objectives. In the case of all three approaches to the promotion of regional integration identified here (enlargement, normative suasion and inter-regionalism), the EU's strategic priorities are clearly identifiable – in the particular case of the JAES, the EU has the chance to exercise international influence through the diffusion of European values, and in some sense to export the EU model of governance.

In both the Cotonou Agreement and the JAES, the European Commission view is based on the traditional model of regional economic integration: regionally integrated markets, business development, regional infrastructure networks, regional governance institutions and policies for sustainable development. The EU instruments adopted to support regional integration in Africa rely mainly on political dialogue, trade policy and financial assistance (through the European Development Fund). Given the leadership role of the European Commission in this area of EU external policy, the possibility for bureaucratic and technocratic influence over African regional integration is all the greater, though this does not diminish the prospects for norm diffusion. However, there is no reason to reject the political dynamics and nature of state interests in the African regional integration processes, since these forces are likely to become stronger rather than weaker as the political dialogue, so prominent to the EU approach, continues in the efforts to implement the strategy.

The Joint Africa–EU Strategy

Following the first Africa–EU Summit of 2000 in Cairo, the stage was set for greater political dialogue on cooperation, extending beyond the existing economic/aid/trade agenda, when the AU and EU agreed on the inter-linkages between security and development, committing themselves to work together towards improving African stability. The Joint Africa–EU Summit in Lisbon (2007) produced the Joint Africa-EU Strategy (JAES) as a comprehensive 'roadmap' for future cooperation. The Action Plan 2008–10, drawn up to implement the goals agreed at Lisbon, bore all the hallmarks of the European approach to governance, as did the arrangements for the multilayered, institutional architecture with multiple actors. Strong on rhetoric, and ambitious in the scope of the proposals for cooperation between the two regions, the JAES promised a 'new political vision' based on 'a Euro-African consensus on values, common interests and common strategic objectives'. Normative values were given emphasis in the promise that the partnership would strengthen 'economic cooperation and the promotion of sustainable development in both continents, living side by side in peace, security, prosperity, solidarity and human dignity' (Council of the European Union, 2007: 2).

The JAES was presented as a strategic partnership of equals, one that went beyond development to include political dialogue that was far-reaching, a partnership that was global and people-centred (beyond institutions), and that treated Africa as a single entity. Pan-African in its membership and its comprehensive coverage of issues that were relevant to both regional partners, the attempts to involve all levels of society (and not just the supranational institutional actors on both sides) marked the concern to ensure 'ownership' of the strategy. In this regard, the strategy aimed at the socialisation of all actors in a very normatively framed discourse, and the adoption of constructivist communication processes where common interests could be determined.

In practice, the inter-regionalism that characterises the JAES is a weak blend of normative values and pragmatism, a compromise of conditionality and partnership and of regionalism and multilateralism. Ownership of the strategy remains limited to the institutional actors and the technocrats in both regions; civil society organisations and business interests are at best minimally involved; member states (European and African) and the African RECs have limited engagement with the strategy. Politics, in the sense of the promotion and negotiation of interests, seems strangely absent from what is often regarded as a massive technocratic exercise, engaging high-level institutions while distancing national and sub-regional society.

The failure to agree on common interests in areas and topics relating to global governance (climate change, migration, global trade talks) would suggest that substantive political dialogue is still some way off, and negotiations over EPAs have remained outside the framework established by the JAES. There is also the question of differences in priorities, with European partners seen to favour peace and security, conflict management and human rights, whereas African partners prioritise trade, economic integration and investment. There remains the challenge of becoming equal partners, of moving away from the traditional donor–recipient relationship and from the 'aid and development' agenda towards negotiations about strategic decision-making in a global world with shifting geopolitical relations. However, in seeking to pursue an agenda constructed around common interests, which can then form the basis for dialogue with the rest of the world, to what extent can the interests of the continent be retained? Regional integration at the level of the RECs continues as work-in-progress, and the prospects for a finely tuned balance between (sub)-regional and global priorities are unclear.

The distance in the aspirations of the European states and the priorities of the African countries widened as financial crisis hit the eurozone in 2010, and the subsequent political instability in North African throughout 2011 highlighted both the inadequacy of existing EU–Africa partnership arrangements and the limitations on actorness, normatively defined or otherwise. The third Africa–EU Summit held in Libya in 2010 under the theme of investment, job creation, and economic growth agreed on the Action Plan 2011–13 (Council of the European Union, 2010a), but there was very little public interest in the event in either Africa or Europe. The Tripoli Declaration (Council of the European Union, 2010b) issued at the end of the Summit reaffirmed commitments made three years previously at the Lisbon Summit, endorsed continued efforts to meet the Millennium Development Goals (MDGs), to make the African Peace and Security Architecture fully operational and to explore new areas of common interest including climate change, energy and science, and the information society and space, but there was little evidence of converging global positions among the key actors.

African perspectives

Notwithstanding the significant political investment in the JAES, there is a mismatch between the European and African expectations, capacities and interests – a mismatch that certainly determines how the partners approach the political dialogue and how agreements are implemented. The broader implications for EU–Africa relations will be considered in the concluding section, so for now we focus on the contemporary context of the 'partnership'. Ironically, the EU complains that Africa lacks a coherent 'face', though in reality who/what is Africa remains an unanswered question (Kotsopoulos and Sidiropoulos, 2007). For the EU, it is the AU that speaks and acts for this 'one Africa'. However, it is questionable whether the AU has the implementation capacity (the AU Commission has 500 staff, compared to the EU's 25,000) or the political mandate from the 53 member states, all of whom are fierce in their defence of national sovereignty. There are also the RECs, seeking deeper economic integration among their members and often derailed in these efforts by recurring security crises.

When the JAES was first agreed, Africa was in the midst of a period of strong economic growth, with stable democracies spreading throughout the continent led by new leaders and political elites keen to maintain these democratic institutions and promote the integration of their economies into the global economy (African Development Bank/ OECD, 2010). The decision to create the AU in 2002, and the adoption of the New Economic Programme for African Development (NEPAD) reflected the changing political and social climate on the continent. Despite the global financial crisis that began in 2008, African leaders remained committed to the goal of global integration, and to the goals and values enshrined in the JAES. However, the rhetoric of partnership in the strategy has not always been matched by agreement on the priorities for action, and differences in interpretation as well as contradictory motivations appear to undermine the path towards common interests.

African priorities rest primarily with trade, economic development and poverty reduction. Support for good governance remains strong, endorsed in a variety of continental fora and instruments including the AU, NEPAD, the work of the United Nations Economic Commission for Africa (UNECA) and the African Peer Review Mechanism (APRM). The EU's promotion of good governance should accord with the African interest, however European motivations are perceived by some African countries as neo-colonial and paternalistic (Kotsopoulos and Sidiropoulos, 2007). Taking just three of the partnership areas agreed under the JAES – migration, trade and regional integration and human rights – a clear perceptions gap can be identified between the two partners. Migration has become a prominent political issue for the European countries, and individually and collectively the EU member states are moving towards a more restrictive policy that will severely curtail the inward movement of people. For the African countries, migration is both a safety valve and a 'brain drain', but ultimately regarded as a manifestation of an unequal exchange between north and south based on historical inequalities in the global system.

The human rights area embodies the European values most strongly, and it is a cross-cutting issue in EU external relations generally. Africa has its own Human Rights Court, and sub-regional arrangements for the implementation of human rights, yet Africans consider that the European focus on human rights fails to capture the distinctiveness of local conditions, and the urgency of specific priorities such as poverty reduction and economic development. With regard to trade and regional integration, the dual strategy in the EPAs (regional integration between African regional groups and the EU *and* intra-regional integration among the African countries) fits African priorities. However, the negotiations between the European Commission and the African countries were protracted, and resort to bilateral negotiations and coercive attempts by the European Commission to forge new 'regions' called into question the identity of the EU as a normative actor. The failure of the EU to secure progress on the World Trade Organization negotiations for the Doha Development Agenda furthermore undermines the credibility of the former's commitment to trade and development, certainly in the minds of the African interlocutors.

The peace and security partnership presents one of the greatest challenges to cooperation, though it has also witnessed some successes since the Summit in Cairo (2000) which emphasised the linkage between security and development. As the EU has little military capacity, its role in peace and security management rests primarily on other instruments – one exception being the 2003 EU peace-keeping mission to the Democratic Republic of Congo (Operation Artemis), undertaken without the involvement of the AU or the RECs. However, the EU's contribution to peace and security relations is primarily based upon the African Peace Facility (APF), a funding mechanism created in 2003 under the 9th European Development Fund in response to a request by African leaders and to be used to support peace-keeping operations as well as institutional capacity-building programmes for the African Peace and Security Architecture. Thus, the APF has been crucial in meeting the financial resource constraints of the AU, particularly in providing support for AU-led peace operations, as well as those of the RECs (Aning and Danso, 2010). An additional financial instrument was created in 2007, the Instrument for Stability, to provide rapid assistance in crises of a non-military nature.

The evolving AU–EU security relations briefly outlined above measure up to the ambitions outlined in the JAES, in so far as values of partnership, ownership and common interests are concerned, though perhaps not fully in the ways envisaged by those European and African leaders. Certainly, the arrangement whereby the EU gives financial assistance and the AU carries out the operations is pragmatic and practical as the AU force has geographical proximity while the EU (without military capability) has the financial resources. Both regional actors have explicitly declared their commitment to peace and stability, in the AU case for the interests of economic development and in the EU case the interest is to promote regional security in Africa by way of ensuring European security. The principle of ownership is observed in the division of labour whereby the AU takes responsibility for the peace-keeping operations on the continent (with the approval of the United Nations Security Council), and all of the AU's security policy actions are conducted within the cooperative institutional setting of the African Peace and Security Architecture (APSA). But an important limitation on the effectiveness of the APSA is the budgetary resources, with AU member states unable or even unwilling to make the necessary contributions to support its operation. So, continent-wide integration in respect of peace and security depends very much on the financial support received from external actors, notably the EU, individual European states (Britain, France and Germany), the United States and other global actors such as China and India.

Infrastructure

This section considers the focus on infrastructure provision as a central component of the JAES, and the implications for EU development policy. Though a relatively new component of EU policy towards Africa, the EU member states were long accustomed to a public policy that supported investment in European transport infrastructure to facilitate the creation of the single market. Investment in infrastructure was regarded by the European states, individually and collectively, as fundamental to regional growth and to regional economic integration (Bafoil and Ruiwen, 2010). In simple terms, the rationale behind the EU's emphasis on infrastructure under the JAES was to support regionally integrated markets at the sub-continental level, and to emulate the processes that had been tried and tested in the European context since the 1980s.

The significance of appropriate and adequate infrastructure to support regional economic integration, and ultimately international integration, was highlighted in the World Bank's 2009 World Development Report, 'Reshaping Economic Geography' (World Bank, 2009a). Prior to this, the Bank had already highlighted the central role of infrastructure in promoting development in the World Development Report 1994, 'Infrastructure for Development' (World Bank, 1994). In the World Bank's view, appropriate infrastructure ensured the spatial connectivity of markets, particularly in the case of economies located at some distance from markets, and it noted that sub-Saharan Africa was affected by unreliable infrastructure (World Bank, 2009b: 23). Addressing the infrastructure gap required a three-pronged approach: institutions that are *spatially blind* and universal, such as regulations on land, labour, education, health, water and sanitation; infrastructure, *spatially connective* and including roads, railways, airports, harbours and communication systems that facilitate the movement of goods, services, people and ideas; interventions, *spatially targeted* to provide preferential trade access and incentives for manufacturing.

On the African continent, actors such as the AU, the African Development Bank and the United Nations Economic Commission for Africa also advocated the support for infrastructure in facilitating regional integration (UNECA, 2010). The United Nations Economic Commission for Africa, in its fourth report on regional integration, expressed the viewpoint succinctly: 'inadequate infrastructure remains one of the chief obstacles to intra-African trade, investment, and private sector development. Programmes to cultivate transport and communications networks, energy resources, and information technology would accelerate trade progress and transform Africa into a haven for investment' (UNECA, 2010: 4).[3]

The JAES highlighted the 'promotion of interconnectivity of Africa infrastructure' in line with AU/NEPAD priorities as the rationale for an infrastructure partnership, targeting transport, water and energy as three areas for action, and the Action Plan 2008–10 proposed to implement the EU–Africa Infrastructure Partnership. Though lacking the high political salience of issue areas such as security or human rights, the need to enhance cooperation on the provision of regionally integrated infrastructure systems is crucial to broader goals of regional and global integration, and to socio-economic development generally. And, while the individual thematic partnerships and the AU–EU strategy as a whole rest on continuing political dialogue, the integrated infrastructure is essential to effective regionalisation. The World Bank estimated that Africa requires 93 billion dollars annually to address the continent's infrastructure deficit (15 per cent of the continent's GDP), and that the current infrastructure services were twice as expensive as elsewhere due to diseconomies of scale and high profit margins in uncompetitive markets.

The Bank's recommendations to pursue regional integration in order to reduce infrastructure costs was made with the explicit recognition of the need to harness political will and build trust among the countries involved, so that regional infrastructure projects could proceed in tandem with regulatory cooperation. The third Africa–EU Summit in Tripoli made a promising start in agreeing the Action Plan 2011–2013, which stated 'recognising the crucial role of infrastructure in regional integration, focus will be on areas such as energy, transport, agriculture, health, water and ICT infrastructure development in Africa, reinforcing the necessary interconnections within Africa and between Africa and Europe'. However, the activities set out in the Action Plan emphasised mainly capacity building, regulatory cooperation and harmonisation of standards and other governance measures. There was very little capital investment likely to support the modernisation and increase in the stock of infrastructure, and while financial support amounting to 3 billion euro was programmed under the European Development Fund, the infrastructure partnership is likely to face significant financial constraints for the foreseeable future. Implementation of the partnership so as to deliver initiatives in the field of trans-African networks is difficult to foresee, while the likelihood of securing the substantial yet necessary financial resources from international donors and from the private sector is much reduced in a global environment characterised by economic uncertainty and financial austerity.

From the preceding discussion, it is clear that the central importance of infrastructure in promoting regional integration is recognised by the EU and by international actors on the African continent, as well as the institutions of global governance, notably the World Bank. However, this coincidence of interests and beliefs does not suggest a convergence in development policies among the international development community. There is no convincing evidence to suggest that the EU regards infrastructure as a development issue. Indeed, a Green Paper on development policy published in 2010 (European Commission, 2010) was regarded by some commentators as focusing too much on the coordination of member states development policies, failing to take adequate notice of the European Commission's role as a global development actor or to initiate a discussion of how a modernised development policy could integrate policies in trade, finance and migration for a new consensus on development (ODI, 2011).

Among the international actors setting out their support for regional infrastructure to support African trade, China has emerged as an active investor on the continent. Chinese investors have already provided major financing for infrastructure projects in West Africa and along the east coast, notably transport projects such as road and rail links connecting up different parts of the continent to provide a hub for new business investment. While it is too early to calculate the longer-term impact on infrastructure and regional growth, the presence of China as an investor and promoter of regional infrastructure means the EU no longer has an influential role on the continent, and its capacity as a normative actor is likely to be undermined. At the very least, the existence of alternative donors offers the African states the option of choosing from a range of external financiers and even enhanced bargaining opportunities.

Economic Partnership Agreements

The negotiations between the European Commission and the African countries to establish EPAs began in 2002, and the agreements were due to come into effect at the beginning of 2008. Progress was slow as the bilateral negotiations came up against growing opposition from individual African states, though some countries signed interim agreements for fear of losing access to the European market. The EPAs were proposed in the Cotonou Agreement (European Union, 2000), which was in essence a departure in the traditional relations between the European and African, Caribbean and Pacific (ACP) countries, introducing the new European approach of development through trade and trade as a substitute for aid. The EPAs were intended to support regional integration, between the EU and African regional groups, and between the African countries within the regional group. However, the EPA negotiations were conducted outside the framework of the JAES, undermining the claim by the EU of a partnership of equals, where 'one Africa' would be the focal point for political dialogue around strategic cooperation in areas of common interest.

The difference between the approach towards the EPA negotiations and the emphasis given to political dialogue and strategic partnerships in the JAES is striking. In the former, the European Commission maintained pressure on countries, negotiating bilaterally while intending to create a regional partnership agreement (four on the African continent), often not engaging in dialogue with existing RECs. The general approach taken by the European side reflected that used in earlier phases of European integration, such as the Single Market Programme, the move to European monetary integration and the accession negotiations with applicant countries – specific conditions to be met in a phased approach and with a target deadline. Using well-tested programme planning, and the political and negotiation skills honed in tough negotiations with European member states and within the WTO, the European Commission proceeded along similar lines in order to deliver the EPAs, but it was forced to concede the pace of negotiations due to the growing concerns of the African countries.

The EPAs would ultimately require economic liberalisation by the participating countries, and it was the anticipated losses in import duty revenue, the transition period for tariff liberalisation and the scope of tariff liberalisation that proved contentious issues for the African countries in the negotiations with the European Commission. With African states unwilling to bear the anticipated financial losses associated with liberalisation and the European side unwilling to make concessions, the negotiations faltered on the differing expectations and perceptions of the two sides. For the African countries, the issues to be negotiated were political, and the removal of non-tariff barriers or requirements to impose (or remove) regulatory policies to support economic liberalisation towards other member countries in the EPA effectively impacted on the state's sovereignty to manage its own policy portfolio. In addition, the perceived marginalisation or down-grading of development by the European Commission in the EPA negotiations gave concern to the African countries, many of whom argued for greater focus on development, and for a relaxation of the interpretation of the WTO requirement that liberalisation should cover 'substantially all trade and transition periods'.

The ACP secretariat was not directly involved in the EPA negotiations, so individual ACP states were disadvantaged by limited trade negotiation capacity even while collectively there was a common interest in retaining fiscal autonomy over export taxes to support infant industry protection and industrial development. The EU negotiators have consistently taken the view that the issues to be negotiated were technical, and that WTO rules provided the baseline toward which EPAs must move. This perceptions–expectations gap between European and African sides widened as the EPA negotiations prompted renewed African concerns over provisions regarding rules of origin, national treatment, free circulation of goods and the 'non-execution' clause (the possibility of trade sanctions in the event of violations of democratic or human rights principles).

The various political actors excluded from the EPA negotiations, or relegated to the sidelines, included civil society and the national as well as regional interest groups (political, social and economic) with direct interest in the process and the eventual outcomes. The AU shared the concern of the individual African countries that the EPAs neglected the role of trade as an instrument of development, and reiterated the position when the issue of the EPA negotiations was raised at the second and third Africa–EU Summits, effectively placing the responsibility on the EU to be more flexible in the negotiations. Different options remain open to individual African countries as to the form of EPA to agree, or even to reject the agreement in favour of the Generalised System of Preferences offered to developing countries.

What remains unclear is how the EPA negotiations have contributed to the region-building processes across the Africa continent, and ultimately what kind of regional integration is likely to emerge from these inter-regional negotiations (Storey, 2008). The resort to bilateral negotiations between the EU and individual African countries, and the diverging interests of the latter, meant a missed opportunity to work together for the collective identification of interests in regional (that is, sub-continental) terms. Though regionalism has clearly emerged in Africa, both the form it is taking as well as the rationale and driving forces, varies across the continent (Asante, 1997; Engel and Gomes Porto, 2010). The continent-wide AU may be considered as a 'natural' region, the product of geography and of the political desire of African states for pan-African union (Murithi, 2005). This remains true even taking account of the divisions between those states that favour the rapid transformation to a United States of Africa and those that favour an incremental process of integration, with the RECs as key building blocks.

Important to the emergence of cooperative impulses among sovereign states is political will for joint action, and the political interest to sustain cooperation over the long term. Normative agendas do matter, particularly in terms of framing the rationale and in order to define the scope of cooperation. As the preceding discussion has indicated, African regionalism has its normative agenda – peace and security, development, poverty reduction and unity. But it is not at all clear that the normative agenda taking hold in the RECs owes much to the external influence of the EU. As this chapter has suggested, African countries have questioned, challenged and sometimes rejected the EU agenda in relation to regional integration – and done so at the level of the state, sub-regional and continental levels. To return to the discussion at the beginning of the chapter concerning why the EU promotes regional integration, it is clear that the policy adopted in the context of EU–Africa relations generally cannot be explained in terms of the export of the EU model, as happens in the case of enlargement. The EU model is very complex, incorporating a multilevel system of governance, a political community and a legal order. In neither the EPAs nor the JAES is there a proposal for the wholesale adoption of this system. Aspects of the European regulatory and governance system are of course embedded in the EPAs, and in the approach adopted within the JAES. The requirement to promote economic liberalisation under the conditions of the EPAs does imply some policy changes within the African signatory states, but this is far from the wholesale adoption of the *acquis communautaire*. The discussion above on the progress of EPAs would suggest that the failure of the European negotiators to engage politically with the RECs, and indeed with the individual states, means that the EU's influence on regional integration is significantly reduced.

In the JAES, the EU is probably more at home with this type of continent-to-continent initiative, and political dialogue at this level is what the EU political system does well, particularly in the hands of the European Commission with its long experience of community-building, large-scale programmatic planning and negotiations. The regular engagement between the AU and EU Commissions, the member states summitry and other inter-institutional communication processes allow multiple processes for the potential diffusion of ideas, for sociali-sation and persuasion. However, what matters is how (and whether) the ideas are diffused, and also identifying the channels that senders and receivers of ideas use (Hall, 1989). To the extent that channels and mechanisms for diffusion do not operate because they are weak, inef-fective or simply non-existent, then political dialogue becomes circular and limited in the capacity to influence. The institutional framework established to support the political dialogue in the JAES is top-heavy, mostly removed from the political interests of the state, and at a distance from other socio-economic interests, including the business community. Dialogue between the AU and EU Commissions may well fail to take account of political interests at the level of the state, or even the RECs. There is inadequate research on the politics of integration in Africa in terms of the interests of state (and non-state) actors, and of the relations between the states and the regional economic organisations, or the rela-tions between states and the AU institutions. Yet the evidence unearthed in the preparation of this chapter points to the need to understand more about these multilevel political processes in order to fully analyse the potential prospects and limitations of regional integration in Africa.

Conclusion

This chapter set out to examine how the EU conducts policy towards promotion of regional integration in Africa, taking as a focus of study the Joint Africa-EU Strategy and the EPAs. In particular, the Cotonou Agreement established a more contemporary approach infused with neo-liberal ideas wedded to new thinking about trade and development, and this is the context in which the EPAs are implemented. The JAES is ambitious in scope and scale, a grand union of the two continents for political dialogue, a strategic partnership of equals to address common interests and meet global challenges. The chapter outlined the operation of these two instruments of EU external policy, and concluded that while EU–Africa relations are strengthened and given new dynamism as a result, the potential for these two policies to shape regional integration in Africa is very much dependent upon the political dynamics inherent in the multilevel relations – from state to regional economic community, and to the continental level. There is evidence that the privileging of certain actors by the EU processes, and the marginalisation of other actors (state, sub-continental, private sector) can limit the integrative potential. State and non-state actors' interests matter and, though there is significant support for regional integration among African states, the top-down institutional framework established under the JAES can fail to diffuse ideas and to support integrative pressures further down at state level.

The chapter also identified the weaknesses in the political dialogue, where the differing perspectives of the EU and the AU can restrict the identification of common interests so that the two actors are effectively talking at cross purposes. In areas such as migration, human rights and security, issues where thematic partnerships were created under the JAES, the perceptions–expectations gap between the AU and EU can be large indeed. Security matters, for the EU, translate as the need to guarantee the internal security of Europe; for Africa, security threats can impact negatively on the prospects for economic and social development, and the African Peace and Security Architecture has emerged as a notable case of continental regional integration.

In other respects, the tendency towards constructing African regionalism through top-down institutional structures (loosely modelled on the EU) has produced a system where inter-state cooperation is fragmented, and the regional integration is limited (Draper, 2010). Distinguishing between internal relations among the states in a region, and the region's external relations, two observations are offered by way of understanding the implications of limited regional integration. Firstly, the top-down model of regionalism may have limited potential to deepen cooperation between sovereign states in the RECs, cooperation which over time creates trust and allows for the socialisation processes that lead these states to construct common (regional) interests, and facilitate a regional identity. In the absence of regional interests, the potential for informal regionalisation (increased levels of economic and social interaction) is thus reduced. The role that infrastructure can play in enhancing regionalisation has been recognised by international and African actors, and the chapter considered the prospects for efforts to promote infrastructure partnerships. Secondly, the degree to which the region can exhibit 'actorness' in external relations is an important measure of (internal) regional integration, as much as a reflection of the capacity to promote and defend the interests of the region. In the EPA negotiations, the evidence points to the quite disparate potential among the African regions to represent the interests of the member states, though it is also clear that the political dynamic shifted away from the EU as the deadline receded without reaching full agreement.

Notes

1 The growth of literature on comparative regionalism testifies to the influence of the EU as a regional political model, even though much of this literature also acknowledges the distinctiveness of the European approach and the differences of regionalism in Asia and Africa. See, for example, Choi and Caporaso (2002) and a special issue of the *Journal of European Integration*, 32 (6), 2010.

2 The institutional architecture to support the Joint Africa–EU Strategic Partnership embodies the kind of multilevel governance arrangements that have come to define the European governance model. At the apex of the structure is the triennial summit meeting between the AU and the EU, a gathering of the 80 member states that is expected to offer political guidance; the Ministerial Troika(s), comprising foreign ministers and *ad hoc* sectoral (state representatives) meet twice yearly, engaging in political dialogue, review and monitoring. Below this, senior officials from the member states undertake annual progress reviews, while direct implementation rests with the joint experts groups (one for each of the eight thematic partnerships) and the eight implementation teams. Supranational institutional linkages are conducted through the European Parliament and pan-African Parliament, and annual meetings of the Commissions, with the deployment of an EU delegation to the African Union in Addis Ababa. In a departure from previous institutional arrangements for EU–Africa relations, the Joint Africa–EU Strategy sought

to create closer engagement with non-state actors, particularly civil society organisations, business interest groups and with international partners.

3 The United Nations Economic Commission for Africa conducts regular reviews on the state of regional integration in Africa, and produces reports on the progress of regional community building. The UNECA perspective remains firmly lodged in the traditional model of regional economic cooperation, and it gives very little attention to the political aspects of region-building.

References

African Development Bank/OECD (2010) *African Economic Outlook, 2010.*

Aning, K. and Danso, K.F. (2010) 'EU and AU Operations in Africa: An African Perspective', in N. Pirozzi (ed.), *Ensuring Peace and Security in Africa: Implementing the New Africa–EU Partnership*, Rome: Istituto Affari Internazionali.

Asante, S.K.B. (1997) *Regionalism and Africa's Development: Expectations, Reality, and Challenges*, Basingstoke: Macmillan.

Bafoil, F. and Ruiwen, L. (2010) 'Re-examining the Role of Transport Infrastructure in Trade, Regional Growth and Governance: Comparing the Greater Mekong Subregion (GMS) and Central Eastern Europe', *Journal of Current Southeast Asian Affairs*, 2: 73–119.

Bickerton, C. (2009) 'Legitimacy Through Norms: The Political Limits to the European Union's Normative Power', in R. Whitman (ed.), *Normative Power Europe: Empirical and Theoretical Perspectives*, Basingstoke: Palgrave.

Börzel, T.A. and Risse, T. (2009) 'Diffusing (Inter-) Regionalism: The EU as a Model of Regional Integration', KFG Working Paper Series no. 7, September, Kolleg-Forschergruppe, Free University Berlin.

Choi, Y.J. and Caporaso, J. (2002) 'Comparative Regional Integration', in W. Carlsnaes, T. Risse and B.A. Simmons (eds), *Handbook of International Relations*, London: Sage, pp. 480–99.

Council of the European Union (2007) *The Africa–EU Strategic Partnership: A Joint Africa–EU Strategy*, 16344/07, Lisbon, 9 December.

Council of the European Union (2010a) *Joint Africa–EU Strategy Action Plan 2011–2013*, available at: *http://europafrica.files.wordpress.com/2008/07/second-final.pdf* (accessed 5 March 2012).

Council of the European Union (2010b) *Tripoli Declaration: 3rd Africa EU Summit*, Tripoli, 29/30 November, available at: www.consilium.europa.eu/uedocs/cms_data/docs/pressdata/EN/foraff/118118.pdf (accessed 5 March 2012).

Draper, P. (2010) *Rethinking the (European) Foundations of Sub-Saharan African Regional Economic Integration: A Political Economy Essay*, Working Paper 293, OECD: Paris.

Engel, U. and Gomes Porto, J. (2010) *Africa's New Peace and Security Architecture*, Aldershot: Ashgate.

European Commission (2008) *Regional Integration for Development in ACP Countries*, COM (2008) 604, 6 October.

European Commission (2009) *Implementation of the Joint Africa–EU Strategy and Its First Action Plan (2008–10): Input into the Mid-Term Progress Report*, SEC (2009) 1064.

European Commission (2010) *EU Development Policy in Support of Inclusive Growth and Sustainable Development Increasing the Impact of EU Development Policy*, Green Paper, COM (2010) 629, 10 November.

European Union (2000) 'Partnership Agreement between the Members of the African, Caribbean and Pacific Group of States of the One Part, and the European Community and its Member States, of the Other Part', signed in Cotonou on 23 June 2000, *Official Journal of the European Communities*, L 317/3, 15 December.

Farrell, M. (2009) 'EU Policy Towards Other Regions: Policy Learning in the External Promotion of Regional Integration', *Journal of European Public Policy*, 16 (8): 1165–84.

Fawcett, L. and Gandois, H. (2010) 'Regionalism in Africa and the Middle East: Implications for EU Studies', *Journal of European Integration*, 32 (6): 617–36.

Fawcett, L. and Hurrell, A. (1995) *Regionalism in World Politics: Regional Organization and International Order*, Oxford: Oxford University Press.

Finnemore, M. and Sikkink, K. (1998) 'International Norm Dynamics and Political Change', *International Organisation*, 52 (4): 887–917.

Gamble, A. and Payne, A. (eds) (1996) *Regionalism and World Order*, London: Macmillan.

Hall, P.A. (1989) *The Political Power of Economic Ideas: Keynesianism Across Nations*, Princeton, NJ: Princeton University Press.

Hill, C. and Smith, M. (eds) (2005) *International Relations and the European Union*, Oxford: Oxford University Press.

Hurrell, A. (2007) 'One world? Many worlds? The Place of Regions in the Study of International Society', *International Affairs*, 83 (1): 127–46.

Kotsopoulos, J. and Sidiropoulos, E. (2007) 'Continental shift? Redefining EU–Africa relations', Policy Brief, November, Brussels: European Policy Centre.

Lucarelli, S. and Manners, I. (eds) (2006) *Values and Principles in EU Foreign Policy*, London: Routledge.

Murithi, T. (2005) *The African Union: Pan-Africanism, Peace-Building, and Development*, Burlington, VT: Ashgate.

Nicolaïdis, K. (2010) 'The JCMS Annual Review Lecture: Sustainable Integration: Toward EU 2.0?', *Journal of Common Market Studies*, 48, Annual Review, 21–54.

ODI (2011) 'ODI Submission to the European Commission's Green paper on EU Development Policy in Support of Inclusive and Sustainable Growth', ODI, London.

Scheipers, S. and Sicurelli, D. (2008) 'Empowering Africa: Normative Power in EU–Africa Relations', *Journal of European Public Policy*, 15 (4): 607–23.

Smith, K.E. (2010) 'The EU in the World', in M. Egan, N. Nugent and W. Paterson (eds), *Research Agendas in EU Studies: Stalking the Elephant*, Basingstoke: Palgrave Macmillan.

Storey, A. (2008) 'Normative Power Europe? Economic Partnership Agreements and Africa', *Journal of Contemporary African Studies*, 24 (3): 331–46.

UNECA (2010) *Assessing Regional Integration in Africa IV: Intra-African Trade*, Addis Ababa: UNECA.

World Bank (1994) *Infrastructure for Development*, Washington, DC: World Bank.

World Bank (2009a) *Reshaping Economic Geography*, Washington, DC: World Bank.

World Bank (2009b) *Africa's Infrastructure: A Time for Transformation*, Washington, DC: World Bank/IBRD Publications.

Part III

Policies and partnerships

7

Foreign aid, donor coordination and recipient ownership in EU–Africa relations

Maurizio Carbone

The first decade of the 2000s was characterised by a number of important changes in the foreign aid policy of the European Union (EU). The new century started with the adoption of the Cotonou Agreement in June 2000 (European Union, 2000), which introduced a radical overhaul of the aid pillar in the long-standing partnership between the EU and the African, Caribbean and Pacific (ACP) group of countries. The new system combined an emphasis on both outcome conditionality, which allows the adjustment of resources on the basis of performance, and recipient ownership, which was meant to place developing countries 'in the driving seat' of the development process. The novelty of the new century, however, has been the EU's attempt to project a common vision on international development, with joint commitments made collectively by all EU development actors. In particular, in March 2002 and in May 2005 the member states jointly pledged to boost their volume of aid. In March 2006 and in May 2007, they committed to a number of actions to improve the effectiveness of aid, placing emphasis on the issue of donor coordination and complementarity. All these changes applied to the relations between the EU and Africa. In fact, the 2007 Joint Africa–EU Strategy (JAES) (Council of the European Union, 2007a) was celebrated by the two parties as starting a new era, based on a 'Euro-African consensus on values, common interests and common strategic objectives'. But the rhetoric of grand policy statements does not always match the practise on the ground. Not only did some of these commitments fail to be implemented, but various African 'partners' started to look for alternatives. Unsurprisingly, just after the 2007 Africa–EU Summit, the President of Senegal, Abdoulaye Wade, stated: 'China's approach to our needs is simply better adapted than the slow and sometimes patronising post-colonial approach of European investors, donor organisations and non-governmental organisations' (Wade, 2008).[1]

Against this background, this chapter explores the potential trade-offs between donor coordination and recipient ownership in the EU's aid relations with sub-Saharan Africa. This region is by far where the EU and its member states make their biggest efforts in terms of development assistance and also where aid is the most fragmented. This chapter, therefore, neither intends to assess the criteria that the EU uses to allocate its foreign aid, nor does it try to assess the impact of EU aid on African development.[2] Instead, it seeks to unravel the dynamics at play when aid strategies are negotiated between two (un)equal partners. To do so, the paper is divided into three sections. Firstly, it provides a concise discussion of the global agenda on aid effectiveness, focusing on the tensions between coordination and ownership. Secondly, it analyses the supranational programme managed by the European Commission within the context of the Cotonou Agreement, paying attention to the degree of involvement of African (both state and non-state) actors in the negotiations of two series of multi-annual development strategies (for 2002–07 and for 2008–13). Thirdly, it explores the EU as a collective donor, focusing on the efforts towards aid coordination and division of labour that have been made throughout the 2000s. The overall argument of this chapter is that the preoccupation of the EU (particularly the European Commission) with improving the quality of EU aid by emphasising donor coordination has fatally resulted in reduced ownership by African countries.

Promoting coordination and ownership: tensions and contradictions

The aid architecture in the 2000s has become significantly more complex. The substantial increases in the volume of aid made by the donors belonging to the DAC – from US$52 billion in 2000 to $128.5 billion in 2010 (DAC, 2012) – have been accompanied by the emergence of new donors (particularly China and India), which have made available considerable amounts of resources to a selected number of countries. The rise of these new donors has also challenged the consensus built around the 2005 Paris Declaration on Aid Effectiveness (OECD, 2008) which laid down a practical, action-oriented roadmap to improve the impact of aid on development. The two key principles underlying the Paris Declaration, which soon became the main reference document for aid agencies in donor countries and for treasury ministries in developing countries (Rogerson, 2005), were (donor) coordination and (recipient) ownership – and these principles were confirmed by the 2008 Accra Agenda for Action (OECD, 2008). But the (initial) implementation of this new consensus did not produce the expected results, which, together with the global economic crisis, led both donors and scholars to question foreign aid as a whole. Traditional donors, thus, started to pay more attention to the effects of aid, but at times they engaged in an ungrounded rush for tangible results in the short term. Unsurprisingly, some aid pessimists called for a radical reform of the aid industry (Easterly, 2006), if not even its 'death' (Moyo, 2009). The 2011 Busan Partnership for Effective Development Cooperation (OECD, 2011) acknowledged the role of new donors, seeking to integrate their experiences for a more inclusive aid system.[3]

Historically, the aid relationship has been characterised as an asymmetrical relationship between two unequal parties. Donors tend to make autonomous decisions based on their materialist interests or idealist motives – or a combination of the two. These decisions are generally taken independently of what other countries do, which has created a situation of 'aid darlings' and 'aid orphans'.[4] Recipient countries, often in desperate need of financial resources, have been left with no other choice than to accept the presence of multiple donors, which generally form 'cartels of good intentions' (Easterly, 2003) and impose various types of conditions. The World Bank, for most of the 1980s and 1990s, set the agenda on aid effectiveness. Initially, this meant that developing countries were induced to adopt reforms with the aim of improving their macroeconomic framework. Subsequently, when it was acknowledged that the involvement of recipients was a necessary condition to increase the impact of aid, ownership – which often implied a commitment by recipients to implementing 'jointly' agreed strategies – and participation – which translated into some forms of consultation of civil society – became the new mantra (Riddell, 2007). However, it was the DAC Paris Declaration that, for many, represented 'a significant juncture in the history of development assistance and co-operation' (Hyden, 2008: 259), not only because it included developing countries in the negotiation process, but also because it sought to make the aid relationship less asymmetrical. On the one hand, donors committed to making progress in donor coordination, by reducing the administrative burden they placed on recipient countries (that is, harmonisation) and by aligning their support with the partner country's national development strategies (that is, alignment). On the other hand, recipients would strive to take control and exercise leadership over their development policies and strategies (that is, ownership).

The costs of a lack of, and the obstacles to, donor coordination are evident, yet empirical evidence suggests that there has been even less donor coordination in the 2000s than in previous decades (Aldasoro *et al.*, 2010). In terms of costs, aid fragmentation wastes scarce resources because of overlapping of donor activities, leads to underfunding of less attractive countries and sectors, results in high administrative burden for developing countries, encourages corruption and ultimately retards economic growth (Easterly, 2003; Acharya *et al.*, 2006; Knack and Rahnam, 2007). As for the obstacles, they are of different types. In general, the 'blame' has been on donors, who may value their interests – be them of a commercial, political or even idealist nature – more than aid effectiveness and therefore seek to maintain a presence in their preferred countries and sectors. Moreover, aid bureaucracies generally prefer independent, uncoordinated activities because responsibility for failure is diffused; furthermore, the incentive structure within aid agencies for promoting donor coordination is very low (Bigsten, 2006). Increasingly, it has been shown that recipient countries have also become sceptical, fearing not only cuts in aid volumes, but a reduction of their policy space. Finally, the new aid architecture has been seen as jeopardising existing efforts: to the traditionally recalcitrant donors, such as the US and Japan, we need to add the emerging donors, who question a consensus they have not contributed to forging.

The concept of ownership has raised some controversy, not least because it has been used by donors in two different ways. The first sees it as 'commitment to policies', regardless of how those policies are chosen. The second sees it as 'control over policies', which implies that countries exercise leadership over the process and the outcome of choosing policies. Generally, donors start with the latter, but as soon as some disagreements emerge over policy choices they recede into the former (Whitfield, 2009). This is not surprising, because ultimately ownership implies a significant transfer of power over resources from donors to recipients. Donors have traditionally been reluctant to release control, but this phenomenon has become more prominent in recent times, when more emphasis has been placed on showing results: paradoxically, 'donor accountability requirements, set against shifting domestic political contexts, represent a major obstacle to ownership' (Hayman, 2009: 593). Moreover, tensions exist between ownership and principles of good governance: 'should partner governments be allowed to exercise power in their own way even if this contradicts principles of good governance?' (Hyden, 2008: 259). Finally, 'democratic ownership' – which implies the involvement of civil society, as well as national parliaments – has been proposed as the solution to government accountability problems. But this has not produced major improvements: in some cases, particularly in Africa, civil society is weak; in other cases, donors have faced the opposition of recipient governments, which do not wish to extend consultation to non-state actors, claiming that they have not much expertise to offer (Hyden, 2008).

The aid programme of the EU has not been extraneous to these dynamics. In fact, the EU has significantly contributed to the consensus that has emerged in the various forums on aid effectiveness (Holland and Doidge, 2012), but its implementation record has not been of the same level. These issues will be further discussed in the next two sections, but before we proceed we need to clarify the dual nature of the EU in development policy. On the one hand, the EU can be treated as a bilateral donor in that it provides financial assistance to developing countries through a substantial programme managed by the European Commission. On the other hand, it has increasingly tried to play the role of federator of the development policies of the 27 member states, which has not always been successful. In fact, the EU's ability to act cohesively and to shape the direction of international development has been weakened by the presence of '27 plus 1' development policies. These structural characteristics of EU development policy should be kept in mind when official discourse portrays the EU as the most generous donor in international development – providing about 55 per cent of the aid given by DAC members (DAC, 2012).

The EU as a bilateral donor

At the end of the 1990s, the supranational aid programme of the EU was considered one of the worst in the world (Short, 2000). In particular, criticism was directed at the criteria used to allocate aid, which did not target the poorest countries in the world, and at the excessive bureaucratic procedures to the delivery of aid, which led to a major gap between commitments and disbursements (DAC, 2002; Santiso, 2002; Arts and Dickson, 2004). To address such criticism, as well as the threats of aid repatriation by several EU member states, the European Commission launched an extensive reform of its overall external assistance programme. At the management level, it included, *inter alia*, the setting up of a single structure in charge of the implementation, monitoring and evaluation of all external aid instruments (EuropeAid); the increasing devolution of authority to the external delegations; and a reduction in the number of instruments to finance development activities. At the policy level, it established that the main objective of EU development policy was poverty reduction, the attainment of which was set out in the context of the Millennium Development Goals (MDGs), and that the EU should concentrate its efforts on a reduced number of sectors where it had an 'added value' to the activities of the member states (Dearden, 2008). This extensive period of reforms led to significant improvements in the delivery of aid, which were acknowledged by both academic indexes of aid agency performance (Knack *et al.*, 2011; Easterly and Williamson, 2011) as well as reviews conducted by donors (DAC, 2007, 2012; DFID, 2011).

The changes brought by the Cotonou Agreement into the aid pillar of the EU–ACP Partnership Agreement, more than the result of consensual negotiations with the partners, were a reflection of the general reform of EU external assistance. While it was established in Article 2 that 'the partnership shall encourage ownership of the development strategies by the countries and populations concerned' (European Union, 2000), the introduction of a rolling programming was seen as undermining 'any notion of equal partnership between the two parties' (Hurt, 2003: 172; see also Farrell, 2005). In fact, the new provisions on aid established that allocation would be based not only on needs, but also on performance in the implementation of projects and programmes, with the possibility to make adjustments following mid-term and end-of-term reviews. Another important innovation concerned the provisions for increased participation of non-state actors (for example civil society, social groups, business associations) in all phases of the development cooperation process (Carbone, 2008a). The two revisions of the Cotonou Agreement signed in June 2005 and in June 2010 did not alter the aid pillar significantly, but once again largely reflected the EU's inputs (Bretherton and Vogler, 2006). The first revision, in addition to making local authorities part of the programming exercise, introduced further flexibility into the aid disbursement process, which ultimately meant that the EU could unilaterally raise country allocations in case of exceptional circumstances (Mackie, 2008). For Kingah (2006: 68), these changes cast doubts on 'the appropriateness of using the term "partnership" rather than, for instance, a principal-agent relationship'. The second revision made the prevailing norms of aid effectiveness – enshrined in the Paris and Accra aid declarations – fundamental principles of the EU–ACP Partnership Agreement and central components of the programming exercise. In particular, to strengthen ownership, it was established that ACP parliaments should also be involved in the development cooperation process, with the view to having a more inclusive ownership (Carbone, 2013b).[5]

Multi-annual programming

Aid to sub-Saharan Africa is financed by the European Development Fund (EDF). Established in 1958, the EDF is replenished every five years by the member states and is not part of the EU budget. Proposals for its 'budgetisation' tabled by the European Commission and the European Parliament over the years have been resisted not only by most EU member states, but also by the ACP group, fearing that their privileged position in the EU's external relations could be jeopardised (Grimm, 2010). Aid programming consists of two main phases. Firstly, a draft Country Strategy Paper (CSP) and National Indicative Programme (NIP), which respectively provide an outline of the recipient country development strategy and identify the sectors of EU's intervention and a timetable for implementation, are prepared by the EU delegation in collaboration with recipient governments and other local actors. Secondly, the draft CSP and NIP are scrutinised by a inter-service Quality Support Group (iQSG) operating within the European Commission, then discussed by the member states in the EDF Committee, and finally adopted by the EU College of Commissioners (Carbone, 2008b). In the 2000s, two different generations of CSPs and NIPs were adopted.

In the case of the 9th EDF (2002–07), the European Commission seemed very eager to show that it was able to disburse money quickly and that coordination with the EU member states led to improvements in the quality of the CSPs and NIPs, whereas the 'goal of increasing ownership by partner countries of these processes ... [remained] elusive' (Mackie, 2007: 92). The improvements in the disbursement process, however, were in great part a consequence of the fact that budget support had become the EU's preferred mechanism to allocate aid – though some African representatives still saw in it a potential risk of shifted accountability, from people to donors (Corre, 2009). In an official analysis of the CSPs, the European Commission itself admitted that 'for the CSP to be a useful management tool for the Commission, through which specific policy objectives can be translated into practice, it was necessary to share the ownership of the drafting process and an appropriate balance had to be struck between country ownership and Commission control of the analysis and final programming document' (European Commission, 2002: 8). One of the key problems was that the National Authorising Officer (NAO), a senior government official appointed by each ACP state to represent it in all the operations financed through the EDF, was not able to cope with the new aid requirements, and, as a result, programming was mainly done by the EU delegations and a number of European technical assistants (Mackie, 2007; Interviews, May–June 2009).[6] An official evaluation of the CSPs for Malawi, for example, stated that, despite the choice of sectors being in line with country needs, 'there are major doubts about the government ownership of underlying policies, with the general perception that policies are externally driven' (cited in Carbone, 2008b: 223).

More critical remarks (and, of course, this is not a surprise) came from civil society. Analyses of European non-governmental organisations

(NGOs) concentrated mostly on the limited attention that the CSPs paid to the social sectors. In particular, it was pointed out how, despite the fact that sub-Saharan Africa was the furthest region from achieving the MDGs, scarce attention was devoted to poverty eradication and human development (e.g. health and education in particular, but also gender equality and child mortality), whereas transport and good governance were identified as priority sectors in a large number of the CSPs (Van Reisen, 2007; Eberlei and Auclair, 2007).[7] Reports of African NGOs focused more on ownership and participation. In particular, they highlighted the fact that in some cases European consultants were hired to prepare the draft CSPs, on the basis of existing documents and strategies of other donors. Moreover, the participation of non-state actors – which in a large number of cases was equated with consultation – was hampered by the short time set aside for it, the limited range of civil society involved (often supporting the EU's views), the lack of institutional mechanisms within the EU delegation to facilitate the process, as well as the divisions within the sector itself (Carbone, 2008b). Critical accounts were also offered by the academic literature, which pointed to the fact that although the EDF is more directly attuned to development than other aid instruments for other regions, it 'has always served the EU's interests, regardless of the rhetoric of partnership. It has been refined in recent years and while it still serves as a diplomatic sweetener, its main function is a structural intervention to promote economic development and liberalization' (Holden, 2009: 147; see also Hurt, 2003).[8]

Two important aspects of the 9th EDF programming process (and this practise continued also with the 10th EDF) were the increased use of intra-ACP funds and the significant amount of resources devoted to regional integration. Firstly, numerous ACP–EU facilities were created within a few years in various areas, for instance in water, energy, migration, natural disasters, peace and infrastructure. But while these were an efficient way for EU member states to deliver increased volumes of aid faster, they were criticised by the ACP group because in most cases they were initiated by the EU. Of all the facilities, in fact, only one (Natural Disaster Facility) was the result of an ACP initiative, while another (African Peace Facility) came as a product of a dialogue between the EU and the AU (Mackie, 2007: 94; Corre, 2009). Secondly, the EU significantly strengthened the amount of aid allocated to regional entities, which was in line with its attempt to promote regional integration. However, not only was a lot of that assistance given to facilitate the conclusion of the highly controversial Economic Partnership Agreements (EPAs), but considering that the degree of accountability between regional authorities and the national level in Africa is weak (not to mention the asymmetrical relations *vis-à-vis* the EU), aid for regional integration resulted in the by-passing of recipient ownership (Farrell, 2009).

This sort of trade-off between aid effectiveness and recipient ownership continued with the second generation CSPs (2008–13). Once again, the assumption within the European Commission was that increased emphasis on budget support, which amounted to about 50 per cent of all operations, would result in greater ownership (European Commission, 2009a). Yet, European NGOs even argued that there was 'a very distinct deprioritisation of the MDGs' (Van Reisen, 2007). More significantly, African governments lamented the fact that the EU still set priorities for them and sought to impose its agenda (in trade, migration, and anti-terrorism) under the guise of governance (REPAOC Coordination SUD, 2007; Van Reisen, 2007). For instance, in Cameroon the European Commission presented as 'non-negotiable' the need to have good governance as one of the focal sectors. When dialogue existed, it was conducted with the NAO, while other members of the government and ministries were rarely given the opportunity to offer their perspective. Moreover, African parliaments, meant to increase democratic ownership, were totally absent from the discussions. Similarly, several African NGOs claimed that no lessons had been learnt from the previous programming exercise: in most cases, there was lack of transparency in the selection of participants and inadequate preparatory information, and limited feedback on results was given (Carbone, 2008a). Of course, some exceptions existed: in Zambia, the EU took Zambia's National Development Plan as the guiding document for the 10th EDF; in Burundi, the government was very active in proposing strategies in line with the PRSP process, though NSAs were not involved; in Benin, the strategy was widely shared by Benin's government and NSAs, and the identification of the sectors was supported by everybody (Sebban, 2006).[9]

The CSPs for 2008–13 included two important novelties: the MDG contracts and the governance incentive tranche, which were mainly designed to promote coordination and strengthen recipient ownership. The MDG contracts represent a longer and more predictable form of budget support – the time horizon is six years, twice as long as the classic budget support arrangements – meant to reward performance against MDG-related indicators (particularly in health, education, and access to water). Considering the high risks, it was initially agreed for eight 'strong performers' – Burkina Faso, Ghana, Mali, Mozambique, Rwanda, Tanzania, Uganda and Zambia – that showed a commitment to monitoring and achieving the MDGs and had active donor coordination mechanisms to support performance review and dialogue. The logic of the contract was in that it sought 'to improve the "predictability of partnership" between the EC, its Member States and partner countries' and at the same time, 'be sufficiently distinctive and attractive to partner countries as to create incentives for non-MDG Contract countries to qualify for it' (European Commission, 2011: 6–7). The MDG contract was welcomed positively by civil society organisations for its attempt to implement outcome-based conditionality and its focus on results, and for this the EU was invited to extend it to other countries (Eurodad, 2008) – though it should be noted that several EU member states had initially opposed the proposal, and hence it was decided to limit it to strong performers.

The other component of the new approach to aid introduced by the 10th EDF programming exercise was the governance incentive tranche – and 2.7 billion euro (about one tenth of the total) were reserved for this initiative. With this incentive-based approach to democratic governance, additional funding would be allocated to ACP countries on the condition that they put some governance measures in place. While the EU emphasised that the GIT would ultimately enhance ownership, the general criteria to allocate funds were not discussed with the stakeholders. Moreover, some of these criteria proved controversial in that they seemed to clearly project the EU's interests: re-admission of illegal migrants, the fight again terrorism, private sector-friendly policies. Unsurprisingly, the results did not match expectations. The EU assumed that all countries would engage in reforms in order to get the tranche, whereas most African countries perceived it as simply making funds conditional on good governance. Moreover, the actual disbursement of the tranches shows that the large majority of countries (around 70 per cent) received an equal amount of incentives (Molenaers and Nijs, 2009; Carbone, 2010).

The EU as a collective donor

One of the most important novelties in the 2000s was the EU's attempt (particularly the European Commission) to promote a common approach to international development. The first achievement, in this respect, was the decision that the EU's member states took in March 2002 to increase their collective volume of aid from 0.33 to 0.39, as a percentage of their collective national income. This commitment may seem modest, but its consequences were remarkable in that a number of other international donors followed suit. Moreover, considering that the 2006 target was met before the deadline, in May 2005, the EU member states committed to a more ambitious target, that of reaching 0.56 per cent by 2010 and 0.7 per cent by 2015. In that context, they pledged to devote half of those increases to sub-Saharan Africa. They also launched a new ambitious agenda on policy coherence for development, as well as aid effectiveness (which will be examined in detail in the next section). All these initiatives attempted to combine two (not necessarily) complementary goals. On the one hand, there was the 'donor-driven' attempt for the EU to act as a single development actor and promote its values in international development, with the aim to differentiate its approach from those of the Bretton Woods Institutions, the US and increasingly China. On the other hand, there was the 'recipient-driven' attempt to make foreign aid more effective by eliminating waste due to overlapping of efforts, especially in the aid darlings, and redistributing efforts towards the aid orphans (Carbone, 2007).

The attempt to project a common vision in international development saw its most important manifestation in the adoption of the European Consensus on Development in December 2005 (European Union, 2006c). The novelty of the Consensus was that it committed the EU member states not only as participants in the supranational development policy but also as bilateral donors. In particular, the Consensus reiterated that the primary objective of EU development policy is poverty eradication and reaffirmed the commitments made by all member states to delivering more and better aid. The adoption of the EU's Strategy for Africa (Council of the European Union, 2005) offered an opportunity to test the potentials of the EU as a single development actor. The strategy, however, was widely criticised because the EU failed to engage with African actors and because no commitments for additional financial resources were made. The subsequent Joint Africa–EU Strategy (December 2007), which was the result of a more participatory process, generated a new consensus on common values and interests between two allegedly equal partners, including a very detailed action plan with eight 'Africa–EU Partnerships'(Council of the European Union, 2007a).[10] Once again no additional financial resources were allocated – thus, the implementation of the JAES mainly depended on the EDF – with the EU member states simply reiterating their 2005 aid pledges. The 0.56 aid target for 2010, however, was not met: in fact, the EU average was 0.44 per cent (DAC, 2012). This was hardly a surprise, particularly in light of the global crisis that hit various countries – though it should be noted that some EU member states, most notably the UK, decided to ring-fence foreign aid.

Joint programming and division of labour

The commitment to boost volume of aid of March 2002 and May 2005 were complemented by an ambitious agenda on aid effectiveness. This new agenda was built around three axes: firstly, providing a transparent mapping of activities by all European donors; secondly, making progress on the issues of harmonisation and alignment, including joint programming; third, promoting more co-financing and division of labour between all EU donors (European Commission, 2006a). While the first commitment was easily met – in fact donor atlases which highlighted the unbalanced distribution between aid darlings and aid orphans were published in various years since 2006 (DAC, 2007, 2012) – the other two commitments presented numerous challenges. In particular, the Council adopted a common format for a Joint Programming Framework (JPF), which was meant to allow the Commission and the member states to share analyses of the development situation of the partner country and adopt a joint response strategy, selecting objectives and focal areas and dividing tasks between EU donors, in concert with the partner country. The ultimate aim of joint multi-annual programming was not only to increase the synchronisation of the programming processes of the EU member states and the European Commission, but also to facilitate the gradual alignment of EU policies with the development plans of recipient countries (European Commission, 2006b).

The initial implementation of the joint multi-annual programming exercise turned out to be a missed opportunity. Despite the fact that it was launched in a selected number of ACP countries – most of which were in Africa – with good prospects of success, the outcome was less positive than anticipated. This was mainly for two reasons. Firstly, the European Commission showed little flexibility: pressure was high to finalise the process in accordance with the timeline identified in the 10th EDF programming process. Secondly, a limited number of EU member states took part in this process. Ultimately, the EU common framework for CSPs was used only in three cases (i.e. Sierra Leone, Somalia and South Africa). Experience on the ground showed that the process was complex, because of the heterogeneous nature of donor programming and mechanisms, insufficient communication between headquarters and in-country staff, capacity constraints that prevented all EU donors from participating in discussions on a regular basis and tensions with other donor-wide harmonisation processes (European Commission, 2009b, 2010; Interviews, May–June 2009 and May–June 2010). In fact, some recipient countries stated that synchronisation should take place on a partner country cycle as opposed to establishing a single, global EU cycle. This could go against increasing EU visibility, but would be more in line with principles like alignment and ownership (Interviews, May–June 2009; HTSPE, 2011).

Probably the most important commitment on aid effectiveness was enshrined in the *Code of Conduct of Complementarity and Division of Labour*, adopted by the Council in May 2007 (Council of the European Union, 2007b). The Code of Conduct was presented as a voluntary document, embedded in the principles of ownership – which actually meant that developing countries were responsible for coordinating donors, and if they did not have the capacity they would receive adequate support – and inclusiveness – it was open also to non-EU donors. The idea behind it was that member states would promote: 'cross-country complementarity', by concentrating their activities on a reduced number of priority countries, making sure that adequate funding is given to aid orphans; 'in-country complementarity', by concentrating their activities on no more than three sectors per recipient, and either delegating resources to other donors or leaving the sector. EU member states committed to making choices on the basis of a self-assessment of their comparative advantage, in consultation with other member states, international donors and in concert with recipient countries. The approval of the Code was greeted by some as 'a potential revolution in the field of EU development policy', and by others as 'another rhetorical commitment made by the EU, without consulting the developing countries, and bound to fail' (Interviews, March 2008). To speed up the process, the EU decided to launch a Fast Track Initiative (FTI) on Division of Labour in about 30 countries worldwide. The aim of the FTI was to support selected developing countries in the implementation of in-country division of labour.

The initial implementation of division of labour produced mixed results at best, and some practitioners on the ground even talked of 'aid effectiveness fatigue' (Interviews, June 2010). The European Commission itself acknowledged that even where some progress in donor coordination was recorded, for instance in a few countries in Eastern and Southern Africa, recipient ownership was still a problem, since in less than 30 per cent of the cases recipient governments took an active role (European Commission, 2011). The EU member states resisted donor coordination and division of labour for various reasons, most notably: risk associated with loss of visibility in partner countries, an excessive number of meetings, limited engagement of non-EU donors and perceived hesitation of recipients. More specifically, national aid agencies on the ground were sceptical, being under significant pressure to show tangible results back at home and the lack of incentives to promote donor coordination given by their headquarters (Carbone, 2013a). The European Commission, rather than a driving force for EU coordination, was in most cases (e.g. Mozambique, Tanzania, Zambia) seen as the 28th EU donor (Delputte and Söderbaum, 2012). Donor coordination was also resisted by recipient countries. In some cases, this was a choice: recipients feared a potential loss of resources and/ or the imposition of stricter conditionalities (for example, Ethiopia). In other cases, it was a consequence of the limited capacities to lead the coordination process. A lot of pressure was placed on central ministries (for example, treasury, development planning), while sector ministries (for example, health, education, agriculture), which until then had dealt directly with donors, were often excluded from aid negotiations – and this redistribution of power, of course, created tensions within recipient governments. It should be also noted that in those countries where there was a limited presence of European donors (that is, Burundi, Central African Africa), coordination was not considered an important concern (Interviews, May–June 2009 and May–June 2010).

In general, coordination and division of labour were often initiated and promoted by donors. Surprisingly, a senior official within the European Commission has argued that 'Donors continue to develop their own policy instruments ... in the belief that a country's own policy will run the risk of not producing the desired results. This is then no longer a matter of ownership, but of "informed" ownership (informed by donors, naturally)' (Petit, 2011: 3). Of course some exceptions exist, as in the case of Rwanda, which has developed a very assertive aid coordination policy. In those countries that have been actively involved in the division of labour (for example, Tanzania, Uganda, Mozambique), governments feared that increased dialogue with donors had become too political, and thus the risk was that of reduced ownership (Odén and Wohlgemuth, 2011). Interestingly, some observers have noted that, while the EU's emphasis on ownership is laudable, 'the burden of implementing this rhetoric might be too much for the capacity of many developing countries and regional bodies if the responsibility falls squarely on their shoulders' (Grimm, 2010: 59; see also Corre, 2009). The development of national capacities, in fact, requires not only time, but also major investments (also by the donor community). But somehow ironically the increased pressure to show results (in the short term) means that some donors have become less willing to accept delays in their disbursement of aid for increased ownership (Odén and Wohlgemuth, 2011).

Conclusion

This paper has explored the changing aid relationship between the EU and Africa in the first decade of the twenty-first century, emphasising the existence of a paradox: on the one hand, the EU has, on various occasions, manifested a firm intention to improve the quality of its foreign aid, trying to make concrete progress on the issue of donor coordination and division of labour; on the other hand, it has failed to fully listen to the voices of the recipient countries in aid negotiations. This failure to promote recipient ownership may be due to a capacity gap in some African countries, which have found successfully leading in their interactions with donors problematic. However, this chapter has found two additional, interlinked explanations. The first is the rush by the European Commission to show that it was able to disburse and spend aid quickly. The second is that in the new emphasis on coordination and division of labour, the EU has been more preoccupied in showing that it could shape the direction of international development, rather than with what happened on the ground. Policy evaporation – that is the gap between policy statements and reality on the ground – has thus become a characteristic if not a feature of EU development policy.

More specifically, in the management of the EU supranational aid programme (within the context of the EU–ACP Partnership Agreement), the European Commission had its hands tied not so much by a mandate of the EU member states, but more by its past management failures. The European Commission, nevertheless, was a hostage to the EU member states, not only because it had to face their continuous pressures to deliver aid faster and even their threats of aid repatriation, but also because it was not given adequate resources to deal with the increased administrative burden coming from the new programming process. The empirical analysis of the programming process for the 9th and 10th EDFs shows that the European Commission concentrated mostly on coordination with the member states, rather than on ownership. It also shows that in the case of recipient countries the focus was mainly on the treasury department within governments, while other ministries, parliaments and civil society were not adequately involved. Of course, most African countries suffered from lack of adequate bureaucratic resources, but the EU did not do much to enhance capacity building.

Similar conclusions can be drawn when we consider the EU as a collective donor. The commitments to, and the initial experience with, joint programming and division of labour suggest that the EU took the issue of aid effectiveness very seriously. But in the limited cases where progress was registered, there was a sort of trade-off between coordination and recipient ownership: African countries, with the exception of a few cases, found their policy space for negotiations with the EU progressively reduced. In other words, more Europe meant less ownership. This in a way broadly confirms more global trends about aid harmonisation. In fact, it has been argued that the limited development gains resulting from the reduction of transaction costs in the aid allocation process are outweighed by the asymmetrical conditions under which discussions about aid effectiveness take place, which exacerbate the imbalances of power between donors and recipients. Contrary to the stated objectives, the effect of aid co-ordination and harmonisation often resulted in a further circumscribing of national autonomy over development processes.

Notes

1 For an analysis of the EU's reaction to China's rise in Africa, see Carbone (2011).

2 One of the arguments that has been put forward is that EU aid has become increasingly more securitised, with the fight against international terrorism taking a significant portion of resources. Some scholars have been more assertive (Woods, 2005; Hadfield, 2007; Orbie and Versluys, 2008), others less so (Olsen, 2008; Carbone, 2011).

3 Paris, Accra, and Busan are the cities where, respectively the second, third, and fourth forum on aid effectiveness were held. The first forum was held in Rome in 2003.

4 Examples of 'aid darlings' (countries that receive large quantities of aid) are Mozambique, Tanzania, Rwanda, Ghana or Burkina Faso). Examples of 'aid orphans' (countries that are generally overlooked by donors) are Chad, Burundi, Guinea and the Central African Republic.

5 The adoption of the Cotonou Agreement and the improvements in aid management did not lead to significant change in the amounts of resources that EU member states delegated to the European Commission for ACP countries. The initial figure (14.33 billion euro) proposed by the European Commission for the 2002–07 period was decreased by the member states to 13.8 billion euro, to the consternation of African countries. Similarly, the European Commission's initial proposal for the 2008–13 period was also resisted by some EU member states. The final amount, which was about 95.92 per cent of the original proposal, failed to take into account the 2002 and 2005 commitments of the member states to increase their volume of aid. Needless to say, the African countries expressed dissatisfaction with this failed increase (Carbone, 2008b).

6 A number of formal and informal interviews were conducted in Brussels in January–March 2008 and in March–April 2010, in Senegal and Gambia in May–June 2009, in Ethiopia and Djibouti in May–June 2010. Interviewees have requested full anonymity.

7 Support to the transport sector, which appeared in a larger number of CSPs, raised some controversies. On the one hand, officials within the European Commission pointed to the fact that transport affects economic growth, which ultimately leads to poverty eradication. On the other hand, European NGOs claimed that in a large number of cases the EU supported the

building of international roads, which, contrary to rural roads, are generally not driven by pro-poor interests (Van Reisen, 2007; Eberlei and Auclair, 2007).

8 The MTRs for the 9th EDF, conduced in 2004, turned out to be a missed opportunity for a more participatory process, mostly because not adequate time was given to involve recipient governments and non-state actors (Laryea, 2004). Even the European Commission admitted that: 'There is obviously a tension between new policy commitments defined unilaterally by the EU and the principle of country ownership of national development strategies and donor support to them' (European Commission, 2005: 15).

9 The MTRs for the 10th EDF, conducted in 2009–10, was an opportunity to adjust CSPs and NIPs also in light of the global economic crisis. In general, many African NGOs expressed the view that also in this case the MTR process represented another missed opportunity to increase ownership of recipient countries (CONCORD, 2010).

10 The eight partnership areas are: peace and security; democratic governance and human rights; trade and regional integration (including the implementation of the EU–Africa Partnership for Infrastructure, launched in 2006); Millennium Development Goals; energy; climate change; migration, mobility and employment; and science, information society and space.

References

Aldasoro, I., Nunnenkamp, P. and Thiele, R. (2010) 'Less Aid Proliferation and More Donor Coordination? The Wide Gap between Words and Deeds', *Journal of International Development*, 22: 920–40.

Acharya, A. Fuzzo de Lima, A. T. and Moore, M. (2006) 'Proliferation and Fragmentation: Transactions Costs and the Value of Aid', *Journal of Development Studies*, 42: 1–21.

Arts, K. and Dickson, A.K. (eds) (2004) *EU Development Cooperation: From Model to Symbol*, Manchester: Manchester University Press.

Bigsten, A. (2006) 'Coordination et utilisations des aides', *Revue d'économie du développpement*, 20: 77–103.

Bretherton, C. and Vogler, J. (2006) *The European Union as a Global Actor*, 2nd edition, London: Routledge.

Carbone, M. (2007) *The European Union and International Development: The Politics of Foreign Aid*, London: Routledge.

Carbone, M. (2008a) 'Theory and Practise of Participation: Civil Society and EU Development Policy', *Perspectives on European Politics and Society*, 9 (2): 241–55.

Carbone, M. (2008b) 'Better Aid, Less Ownership: Multi-annual Programming and the EU's Development Strategies in Africa', *Journal of International Development*, 20 (2): 118–229.

Carbone, M. (2010) 'The European Union, Good Governance and Aid Coordination', *Third World Quarterly*, 31 (1): 13–29.

Carbone, M. (2011) 'The European Union and China's Rise in Africa: Competing Visions, External Coherence, and Trilateral Cooperation', *Journal of Contemporary African Studies*, 29 (2): 203–21.

Carbone, M. (2013a) 'Between EU Actorness and Aid Effectiveness: The Logics of EU Aid to Sub-Saharan Africa', *International Relations*, 27.

Carbone, M. (2013b) 'The EU and the ACP', unpublished paper.

CONCORD (2010) 'Civil Society Involvement in the Review of the 10th European Development Fund', Cotonou Working Group, Briefing paper, February.

Corre, G. (ed.) (2009) *Whither EC Aid?*, Maastricht: ECDPM.

Council of European Union (2005) 'The EU and Africa: Towards a Strategic Partnership', 15961/05 (Presse 367), 19 December.

Council of the European Union (2007a), *The Africa–EU Strategic Partnership: A Joint Africa–EU Strategy*, Lisbon, 16344/07 (Presse 291), 9 December.

Council of the European Union (2007b) *EU Code of Conduct on Complementarity and Division of Labour in Development Policy*, available at: *http://register.consilium.europa.eu/pdf/en/07/st09/st09558.en07.pdf* (accessed 5 March 2012).

Dearden, S. (2008) 'Delivering the EU's Development Policy: Policy Evolution and Administrative Reform', *Perspectives on European Politics and Society*, 9 (2), 114–27.

Delputte, S. and Söderbaum, F. (2012) 'European Aid Coordination in Africa: Is the Commission Calling the Tune?', in S. Gänzle, S. Grimm and D. Makhan (eds), *The European Union and Global Development: An 'Enlightened Superpower' in the Making*, Houndmills: Palgrave Macmillan, pp. 37–56.

Development Assistance Committee (DAC) (2002) *European Community: Peer Review 2002*, Paris: OECD.

Development Assistance Committee (DAC) (2007) *European Community: Peer Review 2007*, Paris: OECD.

Development Assistance Committee (DAC) (2012) *European Union: Peer Review 2012*, Paris: OECD.

DFID (2011) *Multilateral Aid Review*, London: Department for International Development.

Easterly, W. (2003) 'The Cartel of Good Intentions: The Problem of Bureaucracy in Foreign Aid', *Journal of Policy Reform*, 5: 223–50.

Easterly, W. (2006) *The White Man's Burden: Why the West's Efforts to Aid the Rest Have Done So Much Ill and So Little Good*, Oxford: Oxford University Press.

Easterly, W. and Williamson, C.R. (2011) 'Rhetoric *versus* Reality: The Best and Worst of Aid Agency Practises', *World Development*, 39: 1930–49.

Eberlei, W. and Auclair, D. (2007) *The EU's Footprint in the South: Does European Community Development Cooperation Make a Difference for the Poor?*, Brussels: CIDSE.

Eurodad (2008) *Outcome-Based Conditionality: Too Good to be True?*, February 2008.

European Commission (2002) *Progress Report on the Implementation of the Common Framework for Country Strategy Papers*, SEC (2002) 1279, 26 November.

European Commission (2005) *Progress Report on the Mid-Term Review of the First Generation of the Country Strategy Papers*, SEC(2005) 1002, 14 July.

European Commission (2006a) *Annex to the Communication on 'Financing for Development and Aid Effectiveness'*, SEC (2006) 294, 2 March.

European Commission (2006b) *Increasing the Impact of EU Aid: A Common Framework for Drafting Country Strategy Papers and Joint Multiannual Programming*, COM (2006) 88, 2 March.

European Commission (2006c) *The European Consensus on Development*, Luxembourg: Office for Official Publications of the European Communities.

European Commission (2009a) *iQSG Progress Report on Second-Generation Country Strategy Papers 2007/8–2013*, SEC (2009) 431, 30 March.

European Commission (2009b) *Aid Effectiveness after Accra: Where Does the EU Stand and What More Do We Need to Do?*, SEC (2009) 443, 8 April.

European Commission (2010) *Aid Effectiveness – Annual Progress Report 2010*, SEC (2010) 422, 21 April.

European Commission (2011) *EU Accountability Report 2011 on Financing for Development: Review of Progress of the EU and Its Member States*, SEC (2011) 502, 19 April.

European Union (2000), 'Partnership Agreement between the Members of the African, Caribbean and Pacific Group of States of the One Part, and the European Community and its Member States, of the Other Part', Signed in Cotonou on 23 June 2000, *Official Journal of the European Communities*, L 317/3, 15 December.

European Union (2006) 'Joint Statement by the Council and the Representatives of the Governments of the Member States Meeting within the Council, the European Parliament and the Commission on European Union Development Policy: "The European Consensus"' *Official Journal of the European Union*, C 46/1, 24 February.

Farrell, M. (2005) 'A Triumph of Realism over Idealism? Cooperation between the European Union and Africa', *Journal of European Integration*, 27 (3): 263–83.

Farrell, M. (2009) 'EU Policy Towards other Regions: Policy Learning in the External Promotion of Regional Integration', *Journal of European Public Policy*, 16 (8): 1165–84.

Grimm, S. (2010) 'EU Policies toward the Global South', in F. Söderbaum and P. Stålgren (eds), *The European Union and the Global South*, Boulder, Co London: Lynne Rienner Publishers, 43–61.

Hadfield, A. (2007) 'Janus Advances? An Analysis of EC Development Policy and the 2005 Amended Cotonou Partnership Agreement', *European Foreign Affairs Review*, 12 (1): 39–66.

Hayman, R. (2009) 'From Rome to Accra via Kigali: "Aid Effectiveness" in Rwanda', *Development Policy Review*, 27: 581–99.

Holden, P. (2009) *In Search of Structural Power: EU Aid Policy as a Global Political Instrument*, Aldershot: Ashgate.

Holland, M. and Doidge, M. (2012) *Development Policy of the European Union*, Houndmills: Palgrave Macmillan.

Hurt, S. (2003) 'Co-operation and Coercion? The Cotonou Agreement between the European Union and ACP States and the End of the Lomé Convention', *Third World Quarterly*, 24 (1): 161–76.

Hyden, G. (2008) 'After the Paris Declaration: Taking on the Issue of Power', *Development Policy Review*, 26 (3): 259–74.

HTSPE (2011) 'Joint Multi-annual Programming: Final Report', study prepared for the European Commission.

Kingah, S. (2006) 'The Revised Cotonou Agreement between the European Community and the African, Caribbean and Pacific States: Innovations on Security, Political Dialogue, Transparency, Money and social Responsibility', *Journal of African Law*, 50 (1): 59–71.

Knack, S. and Rahman, A. (2007) 'Donor Fragmentation and Bureaucratic Quality in Aid Recipients', *Journal of Development Economics*, 83: 176–97.

Knack, S., Rogers, F.H. and Eubank, N. (2011) 'Aid Quality and Donor Rankings', *World Development*, 39: 1907–17.

Laryea, G. (2004) 'The 2004 mid-term reviews: faultlines and opportunities. ACP–EU', *Civil Society Network Newsletter*, 3: 1–6.

Mackie, J. (2007) 'EDF Management and Performance', in G. Laporte (ed.), *The Cotonou Partnership Agreement: What Role in a Changing World?* Maastricht: ECDPM, pp. 89–96.

Mackie, J. (2008) 'Continuity and Change in International Co-operation: The ACP–EU Cotonou Partnership Agreement and its Firs Revision', *Perspectives on European Politics and Society*, 9 (2): 143–56.

Molenaers, N. and Nijs, L. (2009) 'From the Theory of Aid Effectiveness to the Practice: The European Commission's Governance Incentive Tranche', *Development Policy Review*, 27 (5): 561–80.

Moyo, M. (2009) *Dead Aid: Why Aid Is Not Working and How There Is a Better Way for Africa*, New York: Farrar, Straus & Giroux.

Odén, B., and Wohlgemuth, L. (2011) 'Where is the Paris Agenda Heading? Changing Relations in Tanzania, Zambia and Mozambique', European Centre for Development Policy Management, Briefing Note, 21.

OECD (2008), *The Paris Declaration on Aid Effectiveness and the Accra Agenda for Action*, Paris: OECD, available at: www.oecd.org/development/ aideffectiveness/34428351.pdf (accessed 5 March 2012).

OECD (2011), *Busan Partnership for Effective Development Co-Operation*, Paris: OECD, available at: http://www.oecd.org/dac/aideffectiveness/49650173. pdf (accessed 5 March 2012).

Olsen, G.R. (2008) 'The Post September 11 Global Security Agenda: A

Comparative Analysis of United States and European Union Policies Towards Africa', *International Politics*, 45: 457–74.

Orbie, O. and Versluys, H. (2008) 'The European Union's International Development Policy: Leading and Benevolent?', in J. Orbie (ed.), *Europe's Global Role: External Policies of the European Union*, Aldershot: Ashgate, 67–90.

Petit, B. (2011) 'The technocratic trivialisation of aid', Foundation for International Development Study and Research, Working paper 'Development policies' series/P16-Fra, March.

REPAOC- Coordination SUD (2007), *10th European Development Fund Programming for West Africa: Towards a Democratisation of ACP–EU Relations?*, available at: www.ong-ngo.org/old/IMG/pdf/Rapport_FED_2009_-_synthese_en_EN.pdf (accessed 30 June 2012).

Riddell, R.C. (2007) *Does Foreign Aid Really Work?*, Oxford: Oxford University Press.

Rogerson, A. (2005) 'Aid Harmonisation and Alignment: Bridging the Gaps between Reality and the Paris Reform Agenda', *Development Policy Review*, 23: 531–52.

Santiso, C. (2002) 'Reforming European Foreign Aid: Development Cooperation as an Element of Foreign Policy', *European Foreign Affairs Review*, 7: 401–22.

Sebban, F. (2006) *We Decide, You "Own": An Assessment of the Programming of European Community Aid to the ACP Countries Under the 10th EDF European Development Fund (EDF)*, Brussels: Eurostep.

Short, C. (2000) 'Aid that doesn't help', *Financial Times*, 23 June.

Van Reisen, M. (2007) *2015-Watch: The EU's Contribution to the Millennium Development Goals. Halfway to 2015, Mid-Term Review*, Copenhagen: Alliance2015.

Wade, A. (2008) 'Time for the west to practice what it preaches', *Financial Times*, 23 January.

Whitfield, L. (ed.) (2009) *The Politics of Aid: African Strategies for Dealing with Donors*, Oxford: Oxford University Press.

Woods, N. (2005) 'The Shifting Politics of Foreign Aid', *International Affairs*, 81 (2): 393–409.

8

EU human rights and democracy promotion in Africa: normative power or realist interests?

Gordon Crawford

Under the slogan of 'one Europe, one Africa', the Joint Africa–EU Strategy (JAES), adopted at the second EU–Africa Summit in Lisbon (December 2007), introduced two significant changes (Council of the European Union, 2007). Firstly, Africa is treated by the EU as a single continent for the first time. This addresses the long-standing fragmentation of EU–Africa relations, split between the Cotonou Partnership Agreement (CPA) for sub-Saharan Africa and the Euro-Mediterranean Partnership (EMP) and later the European Neighbourhood Policy (ENP) Agreements and Union for the Mediterranean for North Africa, although the JAES is intended to complement rather than replace these Agreements. Secondly, the JAES has the stated aim of establishing a 'strategic relationship' between the EU and Africa, intended to enhance intercontinental dialogue and co-operation and ostensibly based on the principles of equality, partnership and African ownership. Such language is very much in line with the notion of the EU as a 'normative power', that is a distinctive international actor that is driven by principles and values, a characterisation found in academic literature (Manners, 2002; Lightfoot and Burchell, 2005) and in the EU's self-presentation of its international role. Yet such claims need to be empirically investigated (Sjursen, 2006a: 170), and this chapter makes such a contribution. It poses the question: to what extent does the EU act as a normative power in its relations with Africa, or, alternatively, do realist self-interests predominate?

This question is addressed through examining the EU's human rights and democracy promotion policy in Africa. This is an appropriate 'test' because, firstly, human rights, democracy and the rule of law are commonly regarded as amongst the core principles of 'normative power in Europe' (Manners, 2002; Sjursen, 2006a: 171). Secondly, the promotion of human rights and democracy are 'essential elements' in the preexisting agreements (CPA, EMP/ENP) with African countries, while 'democratic governance and human rights' is a focal area within the JAES. Thirdly, the values associated with human rights and democracy are universally accepted (see below), and therefore less controversial. Hence investigation here examines EU policy performance in the area of human rights and democracy promotion within both the existing agreements and the new structures of the JAES.

The chapter unfolds as follows. The second section outlines the characterisation of the EU as a normative power, while the third examines the rise of human rights and democracy promotion within EU foreign and development policies as a key element of its purported normative core, including within the JAES. The fourth part examines human rights and democracy policy implementation within the Cotonou and EMP/ENP Agreements and during the initial years of the JAES, with findings of a chasm between the policy rhetoric and the reality of policy implementation. Finally, the concluding part explores theoretically based explanations that may account for such findings, and answers the question posed.

Normative power Europe?

The characterisation of the EU as a normative power in international affairs has become influential in recent years, following Manners (2002). This suggests that the EU is a distinctive international actor which aims to influence other states and international affairs through norm diffusion and socialisation. This conceptualisation implies that the norms promoted are positive ones and that 'normative power Europe' is a 'force for good' (Sjursen, 2006b: 235), with the key principles that make up the normative core inclusive of human rights, democracy and the rule of law (Manners, 2002; Sjursen, 2006a: 171). Indeed, this normative core is perceived as stemming from the very values of human rights and democracy that are constitutive of the EU's fundamental essence (Manners, 2002: 253). This characterisation of 'normative power Europe' is also one that corresponds with the EU's self-presentation of its foreign and development policies.

The notion of 'normative power Europe' has led to considerable debate. Authors such as Sjursen (2006b) and Bicchi (2006) have contested the vagueness of the concept and noted that it is not sufficient to demonstrate that the EU promotes 'norms and values' in its international actions, but that the nature of the norms themselves need to be assessed and justified. Sjursen (2006b: 242–3) invokes Habermas's (1990) 'principle of universalization' to accord legitimacy to only those norms that would gain universal approval, otherwise normative power Europe could merely be an expression of 'Eurocentric cultural imperialism'. Of the norms and principles that the EU purports to promote, human rights and democracy are probably the most universally accepted, confirming human rights and democracy as an appropriate proxy for investigating the EU's claim to be a normative actor.

A further challenge to the EU as a normative power has come from sceptical and/or realist perspectives. Initially a sceptical position was expressed by Youngs (2004). Rather than simply doubting the EU as a normative, value-driven foreign policy actor, he emphasised that rationalist security concerns and normative values can be combined, and demonstrated this with reference to EU human rights policies. Hyde-Price (2006) articulated a stronger critique of the whole 'liberal-idealist' notion of the EU as a normative power and asserted an alternative theoretical account based on neo-realism. This has little time for explanations based on constructivist notions of ideational diffusion, but explains EU foreign and security policy as a response to systemic changes in the structural distribution of global power, combined with the utilisation of EU institutions by its member states as a collective instrument to shape the external environment in ways favourable to their own national interests. Subsequently, authors such as Cavatorta *et al.* (2008) and Seeberg (2009) have applied a realist perspective to explain the EU's external policies in the Middle East and North Africa (MENA), arguing that this 'contributes to a better understanding of what the EU does abroad' (Cavatorta *et al.*, 2008: 357).

The claims that the EU acts as a normative power, as well as the theoretical challenges to such claims, all point to the need for in-depth investigation. The contribution by Scheipers and Sicurelli (2008: 620) concluded positively that the 'fields of human rights and environmental protection provide evidence that the EU is constructing ... its identity as a normative power', with EU advocacy for the establishment of the International Criminal Court (ICC) to African nations taken as a proxy for the promotion of human rights norms. This contribution intends to extend such an investigation, though not by focusing on a single human rights case study such as, the ICC, but by examining the scale of EU activities to promote democracy and human rights in its agreements with African nations, including through the JAES.

Human rights and democracy in EU–Africa relations

The EU's self-presentation as a normatively driven and benevolent actor in international affairs is especially evident from policy statements concerning its relations with developing countries (Sicurelli, 2010: 21). In particular, the language of the JAES is strongly suggestive of the predominance of a normative agenda, inclusive of the stated principles of equality, partnership and African ownership, as well as the focus on human rights and democratic governance. It should also be recalled, however, that these are not new discourses, but ones that have permeated Europe's Africa policies for some decades.

The EU–ACP Partnership Agreement

For 25 years, EU–Africa relations were expressed primarily through the Lomé Conventions (1975–2000) between the then European Community (EC) and the African, Caribbean and Pacific (ACP) group of countries, although Africa in ACP terms has always been restricted to sub-Saharan Africa. During the 1970s and 1980s, this was the flagship of the EC's development programme, portrayed as a progressive aid and trade agreement characterised by equality and partnership.

Yet, the introduction in 1989 of a 'human rights clause' (Article 5) into Lomé IV, enabling the EC to suspend co-operation in circumstances of alleged human rights violations, led to specific questioning of the partnership principle. Although the ACP states did not object to the clause in principle, they became increasingly dissatisfied with its unilateral implementation by the EC in practice. Nonetheless, this political dimension was strengthened in the mid-term review of Lomé IV in 1995, with respect for human rights, democratic principles and the rule of law all becoming 'essential elements' (Crawford, 1996). Such trends continued in the Cotonou Agreement, the successor to Lomé from 2000. However, to address ACP concerns, a new consultation procedure (Article 96) was introduced to deal with alleged violations of essential elements.[1] Additionally, Cotonou introduced regular political dialogue between the EU and the ACP, described as a 'key element in the new partnership' (David, 2000: 14), and inclusive of 'regular assessment of the developments concerning the respect for human rights [and] democratic principles' [Article 8(6)].

Euro-Mediterranean Partnership and European Neighbourhood Policy

Issues of human rights and democracy have also featured strongly in the EU's relations with North Africa, at least at the level of rhetoric. From 1995, EU relations have been governed by the Euro-Mediterranean Partnership (EMP, also known as the Barcelona process) and, since 2004, by the European Neighbourhood Policy (ENP).[2] Political and security dialogue was one of the three pillars of the EMP from its inception, aimed at 'deepening political dialogue' and striving towards a 'Euromed region, underpinned by sustainable development, rule of law, democracy and human rights' (European Commission website).[3] Additionally, human rights and democracy clauses were incorporated into the bilateral Association Agreements between the EU and individual countries, enabling the EU to invoke this clause and suspend aid or trade benefits in the event of alleged violations of human rights or democratic principles.

The ENP similarly states that it builds on 'mutual commitment to common values principally within the fields of the rule of law, good governance, (and) the respect for human rights', with political reform stated as a key area for specific action in the individual country Action Plans (European Commission, 2004: 3).[4]

Joint Africa–EU Strategy

The EU's self-presentation as a normative actor assumed a continental dimension with the declaration of the JAES at the Lisbon Summit in December 2007. Normative characteristics are prominent with 'respect for human rights, democratic principles and the rule of law' (Africa–EU Strategic Partnership, 2007a: paragraph 6), described as integral to the 'shared vision and principles'. Of the four stated objectives, one is 'to strengthen and promote peace, security, democratic governance and human rights'. The JAES is implemented through successive Action Plans, the first adopted in Lisbon for 2008–10 (Africa–EU Strategic Partnership, 2007b). This outlined eight thematic 'partnership' areas, including 'Democratic Governance and Human Rights' (DGHR), with 'priority actions' identified for each.[5]

To implement the new partnership, a complex institutional architecture has been created at 'continental, regional, national and local levels' (Africa–EU Strategic Partnership, 2007a: paragraph 93), bringing the EU and AU into regular and close contact and potentially facilitating processes of norm diffusion. As well as the triennial Africa–EU Summit meetings of heads of state and government, dialogue between summits is maintained through regular meetings of senior ministers and officials on a troika basis, most significantly the biannual Africa–EU Ministerial Troika.[6]

Regarding implementation of the Action Plan, the two most significant pieces of the new institutional architecture are the 'informal Joint Expert Groups' (iJEGs) and the Joint AUC–EC Task Force (JTF), both reporting to the biannual Ministerial Troika. The JTF is responsible for the overall coordination and monitoring of the JAES and Action Plan, with six-monthly meetings. The iJEGs have responsibility for the implementation of each thematic partnership, and 'bring together African, European and international key-actors (including civil society organisations) with the necessary competence and commitment to work on the priority action concerned' (Africa–EU Strategic Partnership, 2007b: 3), with support provided by Implementation Teams (ITs) of government and Commission officials from both sides. Finally, reporting requirements are annual, with a joint report prepared by the two Commissions and the EU Council Secretariat.

Overall, the impression is created of the establishment of robust institutional structures. But are they functioning effectively and have facilitated constructive dialogue on issues of human rights and democratic governance? This is assessed below.

Assessing EU human rights and democracy policy implementation in Africa

Has the EU matched its words with actions and demonstrated a commitment to the norms of human rights and democracy in practice? EU policy implementation has always entailed a two-pronged approach: 'negative' in terms of the implicit aid suspension contained within human rights and democracy clauses, and 'positive' in terms of the provision of human rights and democracy assistance, with preference expressed for a positive approach (European Commission, 2001: 9). This section considers the extent of EU human rights and democracy promotion in Africa. It examines positive and then negative measures within the existing arrangements, and then turns to the initial implementation of the JAES.

Human rights and democracy assistance

Cotonou Partnership Agreement
The financial instrument of the CPA is the European Development Fund (EDF), with the 10th EDF running from 2008 to 2013. What proportion of funds has been provided to the human rights and democracy sector in sub-Saharan African countries? Calculations are made here from information available in the public domain.

Table 8.1 provides a breakdown of EDF expenditure for the category of 'government and civil society' from 2007 to 2009 in all ACP countries.[7] This expenditure category captures all human rights and democracy assistance, but also includes expenditure on other significant areas such as government capacity building and development NGOs. Therefore, despite being relatively low, these percentages are

Table 8.1. Government and civil society expenditure in ACP countries

Year	% of total expenditure
2007	15.65%
2008	8.11%
2009	11.85%

Sources: European Commission (2008: 168); European Commission (2009a: 162); European Commission (2010: 187).

considerably higher than the amounts expended solely on human rights and democracy. While the annual expenditure varies somewhat, support for government and civil society averages at almost 12 per cent over the three years. It is estimated that the human rights and democracy component is likely to be less than 5 per cent, given the large-scale nature of support to government capacity building.

Table 8.2 examines the sectoral breakdown under the 10th EDF (2008–13) for sub-Saharan African countries and highlights the allocations to the 'governance' sector as a percentage of total EDF. Here, human rights and democracy support is concealed within the wider 'governance' category, which includes substantial expenditure on public sector management and anti-corruption activities. Additionally, support to non-state actors (NSA) has been included to ensure that no financial allocations in the human rights and democracy area are omitted, although NSA is also a wider category that encompasses support to all civil society organisations, not solely human rights and democracy NGOs. Therefore, while the figures in Table 8.2 do capture all human rights and democracy support, the percentages for total governance and NSA support (final column) are again significantly higher than those solely covering human rights and democracy assistance. It should also be noted that the EU has encouraged allocations to the 'governance' sector through the introduction of the 'Governance Incentive Tranche' (ECGIT) in the 10th EDF. This is a separate fund, though taken from the overall EDF, disbursed to those ACP states which commit themselves to governance reforms (Carbone, 2010). ECGIT funds are not included in the figures below, however, partly because the additional allocations can be expended on any sector, 'possibly but not necessarily governance' (Molenaers and Nijs, 2009: 568).

Of the 43 sub-Saharan African countries here, eight have allocated more than 20 per cent of their EDF to the governance sector plus non-state actors combined. Additionally, half the countries (22) have allocated over 10 per cent, while the other half (21) have allocated under 10 per cent. The average allocation is 11.4 per cent. This could be seen as evidence of greater prioritisation of the governance sector, doubtless encouraged by the promise of additional ECGIT funds. However, to what extent does an increased focus on 'governance' correspond to a promotion of human rights and democracy? The EU's category of 'governance' is broad and does not disaggregate the specific support for human rights and democracy within it. Nonetheless, an examination of Country Strategy Papers (CSPs) does provide an indication of the degree of human rights and democracy support. This analysis is undertaken here for the three countries with the largest sectoral allocations to governance, Central African Republic (53.06 per cent), Mauritania (42.31 per cent) and Chad (34 per cent).

It is interesting that these three countries should have such high governance allocations, given that all three score very poorly on Freedom House's political rights and civil liberties indicators, not even meeting the minimal criteria for an 'electoral democracy' (Freedom House, 2010). They are some of the most undemocratic countries in the subcontinent, having held seriously flawed elections and having severe and wide-ranging democratic deficits (ibid.). In itself this may suggest that the governance dimension has little to do with democracy support, and this is confirmed by a brief examination of the CSPs. In the Central African Republic (CAR), the four elements of the governance sector are described as: '1) restoring basic social services; 2) re-establishing security; 3) restoring State authority; and 4) meeting specific economic and environmental needs' (CAR CSP).[8] This would appear to focus almost entirely on building government capacity and service provision. Despite EU aid suspension to the CAR on democracy and human rights grounds from 2003 to 2005 during the period of military rule, the restoration of aid following elections in 2005 does not appear to focus on building democratic institutions or ensuring respect for civil and political rights. Mauritania has also been characterised by authoritarian and military rule and EU aid suspension. Elections in 2007 entailed a move towards a democratic regime and led to a restoration of EU aid. Governance is stated as the first focal sector of the current country strategy and entails support for 'decentralisation, modernisation of the State and stronger citizen involvement' (Mauritania CSP). Although some democratic elements are visible here, the focus on state capacity building is again evident. In Chad, the country strategy is stated as focused on 'a more "political" type of cooperation based on good governance'. Governance support is in the areas of 'safety and legal certainty, public finance, the institutions and the democratic process, decentralisation, and more generally local governance' (Chad CSP). Democratic processes and institutions are more evident here, but the overall good governance programme is also inclusive of wider measures to strengthen public financial management and the legal framework.

Table 8.2. Governance allocations from 10th EDF (2008–13) to sub-Saharan Africa[1]

Sub-Saharan African countries	Funds allocated to Governance sector and non-state actors (2008–13) million euros		Total National Indicative Programme under 10th EDF million euros	Governance and NSA support as percentage of total 10th EDF %
	Governance	NSA		
Angola	27.0	3.0	172.9	17.35
Benin	60.0	12.0	267.0	26.97
Botswana		5.0	56.3	8.88
Burkina Faso	42.3		423.0	10.00
Burundi		4.0	150.5	2.66
Cameroon	20.0	3.0	190.8	12.05
Cape Verde			38.4	0.00
Central African Republic	58.0		109.3	53.06
Chad	84.8		249.4	34.00
Comoros	6.0		36.0	16.66
Congo Brazzaville	6.0	1.0	77.4	9.04
Djibouti	2.5	0.6	36.9	8.40
Ethiopia	25.0	10.0	536.6	6.52
Gabon	0.7		40.8	1.72
Gambia	8.0	2.0	63.2	15.82
Ghana	74.0	8.0	282.1	29.07
Guinea-Bissau	1.0		77.2	1.30
Ivory Coast	25.0		218.4	11.45
Kenya	4.6	4.6	306.0	3.01
Lesotho	8.2		108.8	7.54
Liberia	7.0	1.0	119.7	6.68
Madagascar		4.7	461.7	1.02
Malawi	15.0		349.0	4.29
Mali	46.9		426.6	10.99
Mauritania	47.5	5.3	124.8	42.31
Mauritius		4.0	39.6	10.01
Mozambique	8.0		482.3	1.66
Namibia	2.0	5.0	82.6	8.47
Niger	42.0		366.4	11.46
Nigeria	125.0	15.0	579.7	24.15
Rwanda	8.5		223.2	3.81
Sao Tome & Principe	2.3		13.3	17.29
Senegal			230.5	0.00
Seychelles	0.8		4.7	17.02
Sierra Leone	20.0		193.4	10.34
Somalia	60.0		212.3	28.26
Sudan	30.0	2.0	258.0	12.40
Swaziland	3.0	0.5	52.6	6.65
Tanzania	3.0	20.0	443.6	5.18
Togo	13.0	4.8	84.0	21.19
Uganda	10.0	2.0	350.9	3.42
Zambia	8.0	4.5	379.9	3.29
Zimbabwe		7.8	129.6	6.02
TOTAL	905.1	129.8	9049.4	11.44

Table 8.2. (continued)

Note: [1] The full title given to this sector in the EU's sectoral breakdown of the 10th EDF is 'governance, democracy, human rights and support for economic and institutional reforms'. Within this, financial allocations per country are provided in three columns: (1) governance; (2) economic and institutional reforms; (3) non-state actors. The figures from columns (1) and (3) are reproduced in Table 2, then added together and calculated as a percentage of total EDF allocation per country

Source: '10th EDF sectoral breakdown' spread sheet ('domaines de concentration'), available at: http://ec.europa.eu/development/geographical/maps/launch_en.cfm and go to 'Overview per country/sector'.

Therefore, despite 'governance' being a key focal area in the three country cases, the democracy and human rights element is relatively minor and subordinate to state capacity building. The greater proportion of governance expenditure pertains to strengthening government capacity whatever the nature of the political regime, democratic or otherwise. It is also significant that there is no allocation to non-state actors in either Chad or the CAR, and therefore no support to civil society groups who may play a key role in democratisation processes and highlighting human rights abuses. With reference to Table 8.2, it can be stated with confidence that democracy and human rights support constitutes a small proportion of the total 'governance and NSA support' in many cases.

Euro-Mediterranean Partnership and European Neighbourhood Policy

There is a degree of consensus amongst analysts that EU democracy promotion activities in North Africa have failed, due partly to a lack of political will on the EU side (Bicchi, 2010: 977; Pace, 2009; Youngs, 2002).[9] EU efforts to promote political reform through the EMP and ENP have remained insignificant and derisory, especially in comparison with the normative expectations generated by the rhetoric of both programmes.

Under the EMP, human rights and democracy assistance amounted to a tiny proportion of total EU funds for the region. Yacoubian (2004: 7) estimated the amount of funding for the political pillar to be no more than 1 per cent in the early years, while Amirah-Fernández and Menéndez (2009: 329) state that human rights and democracy projects received less funding than family planning or drug eradication programmes, amounting to only 2 per cent of total European aid to the Mediterranean region.[10]

Since 2007, the European Neighbourhood Partnership Instrument (ENPI) has become the financial instrument for EU assistance to ENP

Table 8.3. Government and civil society expenditure in ENPI South countries

Year	% of total expenditure
2007	9.73
2008	19.49
2009	5.30

Sources: European Commission (2008: 166); European Commission (2009a: 160); European Commission (2010: 187).

countries. Table 8.3 provides a sectoral breakdown of ENPI South (Middle East and North Africa) expenditure for the category of 'government and civil society', calculated as a percentage of overall expenditure.

While annual expenditure varies considerably, the average expenditure on government and civil society in the first three years of the five-year financial envelope is approximately 11.5 per cent, similar to that for ACP countries (see Table 8.1). However, it can be safely assumed that a large proportion of this goes to autocratic governments, aimed at strengthening the overall capacity of government ministries, while support to human rights organisations remains extremely limited, given that ENPI expenditure has to be government approved. As stated for the ACP countries, the human rights and democracy component is almost certainly less than 5 per cent. Cavatorta *et al.* (2008: 374) similarly note that the EU rhetoric about 'the role of an independent and active civil society to foster democracy' has been a 'low-level priority' in practice, with the EU shunning the most popular civil society movements, perceived as supportive of radical Islamism (2008: 374).

The findings of another assessment are highly pertinent. After two years of the ENP, Emerson *et al.* (2007) conducted an evaluation of democracy promotion in both ENP South and East, and noted the following. Firstly, the Commission was 'surprisingly silent' on the subject despite 'the ENP in principle giv[ing] first place to democracy promotion, human rights, the rule of law and the development of civil society' (ibid.: 15). Secondly, a new emphasis on the term governance, interpreted as 'an attempt to gloss over' the omission of democracy and human rights issues, while noting that 'governance' can 'cover many things' (ibid.). This is essentially the same finding as highlighted here in relation to sub-Saharan African countries under EDF. Thirdly, at the individual country level, their view was that 'the "joint ownership" of the Action Plans has emasculated the "democracy priority"', including in Morocco, Tunisia and Egypt (ibid.), an interesting point which is discussed further below in relation to the JAES. They concluded that the 'weak performance of the ENP in the democracy field is coming at a high cost to the political credibility and reputation of the EU' (ibid.).[11]

Political conditionality and sanctions

Regarding the application of the human rights and democracy clause and the implementation of aid sanctions, there is a stark contrast between the EU's responses in sub-Saharan Africa and North Africa respectively. In sub-Saharan Africa, the conditionality clause in the Lomé and Cotonou Agreements has been applied for varying periods, and often on more than one occasion, in ten countries (Del Biondo, 2011).[12] In contrast, no North African country has been subject to the human rights clause under the EMP or ENP (Reynaert and Del Biondo, 2009: 9), despite the autocratic nature of these regimes (prior to 2011). Under the EMP, Yacoubian (2004: 7) showed that the EU was 'deeply reluctant' to use conditionality, with the human rights clause 'rarely invoked'.[13] Further, Youngs (2008: 6) argued that 'most conspicuously, democracy-related conditionality has not been a part of the European policy mix in the Middle East', citing the example of Egypt, amongst others, where European criticism has been 'all but inaudible'. Indeed, rather than invoking the human rights and democracy clauses, Amirah-Fernández and Menéndez (2009: 332) noted the EU's 'indulgence towards authoritarian regimes' in the Middle East and North Africa region, with 'punitive actions increasingly waning in the face of democratic shortfalls'.

What explains this incoherence and inconsistency in policy implementation between sub-Saharan Africa and North Africa respectively? And what are the implications for the EU as a normative or realist actor?

The application of the human rights clause in sub-Saharan Africa under EU–ACP Agreements could be interpreted as confirming a normative approach, especially when based on the elaborate consultation process that the Cotonou Agreement entails (Reynaert and Del Biondo, 2009: 8). Looking more closely at the countries where sanctions have been applied, however, the sheen of 'normative power Europe' rapidly tarnishes with awareness that aid sanctions have generally been implemented in countries where there are few EU strategic and economic interests and not elsewhere. Del Biondo (2011) seeks to explain why the human rights and democracy clause has not been invoked in five 'non-cases' in sub-Saharan Africa (Chad, Ethiopia, Kenya, Nigeria and Rwanda), all characterised by seriously flawed elections. Security interests are the most significant explanatory factor, with Ethiopia and Kenya being strategically important allies of the EU in the 'fight against terror' and Islamist influence in the Horn of Africa, while political stability is prioritised over democratisation in Rwanda (Del Biondo, 2011). Regarding Chad, despite scoring almost the lowest possible Freedom

House ratings on political rights and civil liberties,[14] the strategic interests of France as the former colonial power are foremost, particularly concerned to maintain its large military base (ibid.: 15). Economic interests feature prominently in the case of Nigeria as an important oil producer (ibid.: 16). Therefore rationalist self-interests are foremost in explaining why conditionality was *not* imposed in such cases, whereas the lack of such interests partly explains why the human rights clause was readily invoked in others.

If the EU's pursuit of realist interests is somewhat opaque in sub-Saharan Africa, it is clearly evident in North Africa. The pursuit of security interests, inclusive of anti-migration measures and opposition to radical Islam, and of economic interests, especially through economic and trade integration, would appear to be the main drivers of regional engagement. Here we focus on the prioritisation of European security and energy interests respectively over political reform.

Security was a key objective of the original Barcelona Declaration, described as the creation of a 'zone of peace, stability and security in the Mediterranean' (Euro-Mediterranean Ministerial Conference 1995). The subsequent European Security Strategy (ESS) added a strong linkage between security and democracy, stating that 'the best protection for our security is a world of well-governed democratic states' (Council of the European Union, 2003: 10), reaffirming and enhancing a normative approach that prioritises the democratisation of the EU's near neighbours as a means to achievement of its own security objectives. Yet, in practice, the emphasis on democratisation has weakened. It has become increasingly apparent that *political stability* in North Africa and the Middle East, not democracy, is prized as the key to European security, and this led to the strengthening of existing authoritarian incumbents in Arab 'partner' countries. Cavatorta *et al.* (2008: 368) have argued convincingly, citing Morocco, that traditional security concerns have replaced the promotion of normative values, especially in the context of a perceived threat from radical Islam, with support provided to existing autocratic rulers rather than to political opposition or civil society groups.

In relation to energy security, Amirah-Fernández and Menéndez (2009: 332) have emphasised how EU energy-related interests have intensified in recent years, with the rise of oil and gas prices and the intent to reduce dependence on Russia, resulting in further marginalisation of a democratic reform agenda. In particular, they have noted the EU's establishment of a strategic energy partnership with Algeria in the context where Algeria had not yet signed an action plan that commits the government (in theory at least) to undertake political reforms. As they remark, this 'renders the idea of a rewards-based policy in return for reforms highly ineffective' (Amirah-Fernández and Menéndez, 2009: 332), and, one could add, renders impotent any idea of a serious human rights and democracy promotion policy in North Africa.

Implementation of Joint Africa–EU Strategy

Normatively, the JAES promises much in terms of both process – 'a real partnership characterised by equality' – and content, with democratic governance and human rights being a key thematic partnership. But to what extent have human rights norms and democratic principles been strengthened in practice? Through an analysis of key documents, this sub-section assesses the initial progress made in establishing the Democratic Governance and Human Rights (DGHR) thematic partnership, and outlines its shortcomings.

Analysis of the two assessment reports for 2008 and 2009 (Africa–EU Ministerial Troika, 2008, 2009b) indicates that policy implementation in the DGHR thematic area has been slow and limited at best. By late 2009, the key implementing body, the iJEG for DGHR, was barely active, and 'priority actions' mostly remained at the proposal and planning stage, with few concrete accomplishments to the partnership thus far.

The first Progress Report, submitted to the 11th Africa–EU Ministerial Troika in November 2008, indicates that very little happened in the first year. The strong impression is that meetings only occurred, especially on the AU side, when it became urgent and necessary to do so. For instance, the first iJEG meeting only took place immediately before the Ministerial Troika, on 18 November 2008, and put forward plans and proposals for possible activities in the three 'priority action' areas, largely a restatement of the contents of the Action Plan. Prior to that the EU and AU Implementation Teams (ITs) had met and worked on 'position papers',[15] though it is notable that the AU IT only convened in November 2008, immediately preceding the iJEG meeting and Ministerial Troika. What is most significant, however, is the composition of the ITs. The African IT was chaired by Egypt (under President Mubarak at that time), with the involvement of 13 African countries, including Algeria, Zimbabwe, Nigeria and Kenya.[16] This is hardly a list of governments that are distinguished by their respect for human rights and democratic principles. Cross-checking with Table 8.2 (above), it is also noteworthy that key participating governments had only committed very low amounts of EDF allocations to 'governance' support, somewhat undermining the credibility of their active participation in the DGHR 'partnership'.[17]

In order to progress their activities, all eight thematic partnerships were required to submit road maps for implementation of the First Action Plan (2008–10) to the 12th Africa–EU Ministerial Troika, meeting in April 2009. It is clear from the document submitted (Africa–EU Ministerial Troika, 2009a) that the road maps varied considerably in depth and detail, with the one on DGHR being particularly thin. It proposed a few activities in the three priority action areas, mostly for discussion at the next iJEG, but with little substance or detail.[18]

From monitoring and evaluation reports submitted for 2009 (European Commission, 2009b; Africa–EU Ministerial Troika, 2009b), it is again clear that progress in the DGHR partnership was very limited. While implementation of the activities outlined in the roadmap of April 2009 were dependent on further discussion at the 3rd iJEG meeting, this meeting had not yet occurred over half a year later (Joint AUC–EC Task Force, 2009: 15). Given that the iJEG is the main implementing body, this means that progress has been extremely slow. The reports did try to present the DGHR partnership in the best possible light, of course, and thus highlighted a number of activities in the three priority areas. Yet a closer look at the text reveals that almost all of these activities remained future-oriented, many simply repeating what was in the original Action Plan of late 2007. With the exception of electoral observation cooperation and a session of the AU–EU Human Rights Dialogue, held in Brussels in April 2009, activities had *not yet happened* (as of October 2009) and remain phrased in the future tense. The notions of 'priority' and 'action' seem to be severely compromised.

Therefore, while the concept of 'normative power Europe' is predicated on the assumption that the EU is genuinely interested in the promotion of human rights and democracy principles, with the JAES declared to be a strategic arena where such norm diffusion could occur, there is little evidence thus far of serious intent on either the European or African side to develop an active and meaningful thematic partnership in this area.

Conclusion

This chapter posed the question: To what extent does the EU act as a normative power in its relations with Africa, or, alternatively, do realist self-interests predominate? It has investigated this question through examining the implementation of human rights and democracy promotion policies both in existing agreements and in the new JAES institutional structures. Findings are of a chasm between policy rhetoric and the reality of implementation, with the EU's self-presentation of itself as a normative actor in international affairs not substantiated in practice. Evidence here, inclusive of both positive and negative measures, indicates that human rights and democracy policy implementation has been virtually absent in North Africa in relation to the EMP/ENP and has remained limited in sub-Saharan Africa under Cotonou. Regarding the JAES, the first two years have been characterised by extremely slow progress in implementation of the 'Action Plan' in the DGHR sphere. Such findings are different from those of Scheipers and Sicurelli who found 'evidence of EU normative commitment' (2008: 620) in its relations with Africa, albeit qualified, through their case-study of the International Criminal Court, focusing on the EU's attempts to persuade sub-Saharan African states to ratify the ICC statute. This difference in findings is perhaps not surprising given that the evidence base is different, with the ICC case limited to one particular measure of human rights advocacy where the EU can promote its own image with little or no financial cost. Indeed, Scheipers and Sicurelli themselves recognise the 'possible contradictions between how the EU represents itself and what the EU does' (2008: 620). Investigation here has sought to examine more generally what the EU *does* to promote democracy and human rights in Africa, and found its practice to be wanting. In this concluding section, various explanations are considered that may account for this chasm between rhetoric and reality. We commence with explanations that remain within the framework of the EU as a normative actor, before undertaking a paradigm shift to suggest that a realist interpretive lens may provide a more plausible explanation.

Three hypothetical reasons for the rhetoric–reality gap are possible that remain sympathetic to the EU, focusing on time-scale, the nature of partnership and resistance by African autocratic leaders. The first argument suggests that norm diffusion is a gradual process that requires longer-term engagement (see Cavatorta *et al.*, 2008: 372), and thus we should not rush to judgement. The weakness of this argument, however, is that the promotion of human rights and democracy is not a new policy, with two decades of implementation in the case of the ACP countries and over 15 years of experience with the EMP and ENP since the Barcelona Declaration in 1995. Available information points to a consistently low prioritisation in country programmes in both sub-Saharan Africa and North Africa, not one that is gradually increasing.

A second argument focuses on the significance of the nature and quality of the partnership. This suggests that an approach that aims at 'partnership, equality and African ownership' involves a mutual commitment by EU and African state actors to constructive engagement and thus offers opportunities for norm diffusion through dialogue, argument and persuasion. The cultivation of such a high quality partnership may take some time, but ultimately will result in more effective norm diffusion and socialisation. Further, it is argued that an overzealous approach by the EU to issues of human rights and democracy, inclusive of negative measures and sanctions, could result in a backlash against democracy promotion, similar to that experienced by the US (see Carothers, 2006). Yet, two weaknesses can be highlighted in this argument. One weakness is that the promotion of human rights and democracy can actually be further compromised by the principles of partnership, equality and ownership that are stated to be at the centre of the JAES. Clearly if the prioritisation of human rights and democracy issues is not shared by some parties to the agreement, then the prospects for a serious commitment to addressing such issues actually becomes less likely in a strategy based on 'partnership, equality and joint ownership' with relatively autocratic state actors, with such a partnership becoming 'emasculated' (Emerson *et al.*, 2007: 15). The second weakness in the argument suggests that the EU's 'indulgence' (Amirah-Fernández and Menéndez, 2009: 332) and indeed 'strengthening' (Cavatorta *et al.*, 2008: 368) of authoritarian regimes is quite distinct from a 'softly, softly' approach associated with partnership, and indicates that hidden agendas are in operation.

A third (and related) argument could suggest that the EU has attempted to remain true to its principles but has been outmanoeuvred by certain African autocratic leaders, who have captured the African leadership of the DGHR thematic area in order to undermine it. The fact that civil and political rights violating governments have placed themselves at the centre of a policy that purports to promote human rights and democracy, inclusive of Egypt (under President Mubarak) as the chair of the African side along with other civil and political rights-violating governments such as Zimbabwe and Algeria as influential participants, would seem to support such a view.[19] But a response would question why the EU has apparently allowed this to happen if it is genuinely committed to the promotion of democracy and human rights. It is difficult to comment on what has occurred behind diplomatic closed doors, but the lack of demonstrable opposition to a hijacking of the democracy and human rights agenda begs broader questions concerning the seriousness of intent on the EU side.

Therefore a paradigm shift is undertaken here and we ask whether a realist and materialist interpretive framework provides a more convincing explanation. The evidence presented has suggested that this is indeed the case. Despite the high profile rhetoric surrounding human rights and democracy as priority objectives, this has not led to significant policy implementation. The new institutional architecture of the JAES may provide the potential for greater political dialogue and norm diffusion, but the lack of commitment to the DGHR thematic partnership does not suggest that democracy and human rights issues will feature significantly in the 'strategic relationship'. These different sources of evidence all suggest that a realist agenda operates at the heart of EU policy. The predominance of EU economic and security interests is most evident in North Africa. Here the purported commitment by both sides in the EMP and ENP to 'respect for human rights and fundamental freedoms' and the existence of a human rights clause in all bilateral agreements are examples of the use and abuse of political rhetoric to conceal unprincipled and disreputable practices, as evidenced by the EU's ongoing support for autocratic and human rights-violating regimes. While there may appear to be elements of a more normatively driven aid sanctions policy in sub-Saharan Africa through the Cotonou Agreement, EU strategic interests are also discernible since aid sanctions have generally been implemented in countries where there are few EU strategic and economic interests and not elsewhere. And perhaps the dominance of such realist concerns to the detriment of normative values should not surprise us. Historically, access to Africa's human and mineral resources has been a key driver of European relations with the continent since the transatlantic slave trade and colonialism onwards. In the contemporary situation, there is an intensification of competition for Africa's resources from rising new powers like China and Brazil. Therefore, the establishment of the JAES, in addition to existing regional programmes, can be seen as an attempt to secure strong intercontinental linkages, inclusive of a multiplicity of EU–AU institutional connections which will be beneficial to EU material interests.

Yet, if this is the case, it still leaves one unanswered question. If the EU does *not* have a serious intent to promote human rights and democracy, in common with some African leaders, then why are such principles consistently ranked so highly in the list of common objectives, including in the JAES? Here the paradox at the heart of the EU's human rights and democracy policy in Africa becomes apparent: the rhetoric in itself can serve the purposes of both sides. For the EU, the ideational aspect plays an important role, one which serves to strengthen its self-presentation as an ethical and normatively driven actor in international affairs, while the pursuit of material self-interests continues in an underhand way. For many African governments, especially autocratic ones, it enables them to make a rhetorical commitment to values that are increasingly accepted as universal and thus to gain greater acceptance in global fora, while reducing any pressure to undertake political reforms in the direction of democratisation. Both sides know that as long as the policy rhetoric is maintained, and a few cursory actions taken, then serious policy implementation can safely be put to one side. Ironically, perhaps this unspoken compact indicates that an element of partnership does exist in EU–Africa relations, though not one that is likely to promote greater respect for human rights and democratic principles.

Acknowledgements

I would like to thank Egle Cesnulyte and Anne Flaspoeler for research assistance and the editor, Maurizio Carbone, for very useful comments.

Notes

1 A similar human rights and democracy clause has been inserted into other regional and bilateral agreements, standardised since May 1995 (Crawford, 1998).
2 The signatories of the Barcelona Declaration in 1995, establishing the EMP, included all North African countries, except Libya. The ENP involves 16 of the EU's neighbours to the south in North Africa and the Middle East and to the east in Eastern Europe.
3 See http://ec.europa.eu/external_relations/euromed/political_en.htm (accessed 30 June 2012).
4 From 2004, the management of Association Agreements was effectively transferred to ENP, while EMP remained as a multilateral forum for dialogue. In 2008, the EMP was re-launched as the Union of the Mediterranean.
5 Those for DGHR are to: enhance dialogue at global level and in international fora; promote the African Peer Review Mechanism and support the African Charter on Democracy, Elections and Governance; strengthen cooperation in the area of cultural goods (Africa–EU Strategic Partnership, 2007b: 2).
6 On the EU side, the Troika consists of the current and incoming EU Presidencies, the European Commission and the EU Council Secretariat, while on the African side the Troika consists of the current and outgoing AU Presidencies and the AU Commission (Africa–EU Strategic Partnership, 2007a: paras. 100–2).
7 These data were only available in the Annual Reports from 2008 onwards.
8 Country Strategy Papers are available at the European Commission Directorate General for Development website: http://ec.europa.eu/development/geographical/methodologies/strategypapers10_en.cfm (accessed 30 June 2012).
9 These views were confirmed by the so-called 'Arab Spring' of 2011.
10 A separate MEDA Democracy Fund was established in 1996 to fund small democracy-related projects, mainly supporting civil society groups, but discontinued in 2001.
11 In addition to sectoral assistance from the mainstream budgets for

sub-Saharan Africa (EDF) and North Africa (ENPI), the European Instrument for Democracy and Human Rights (EIDHR) is another potential source of human rights and democracy support for African countries. This is a global fund disbursed by the European Commission to NGOs and inter-governmental organisations, without the need for consent from 'third-country' governments (European Commission, 2009a: 94). Regional breakdowns of expenditure undertaken by the author indicated that African countries received almost approximately one-third (31 per cent) of micro-project funding from 2002 to 2006 (an average of 9 per cent to North African countries and 23 per cent to sub-Saharan Africa countries), considered to be a reasonable share of available funds. Analysis of the other two main categories of EIDHR funding also indicated that a relatively fair share of funds has been disbursed to Africa.

12 Togo (1998 and 2004), Niger (1999 and 2009), Guinea Bissau (1999 and 2004), Côte d'Ivoire (2000 and 2001), Liberia (2001), Zimbabwe (2002), the Central African Republic (2003), Republic of Guinea (2004 and 2009), Mauritania (2005 and 2008) and Madagascar (2009) (Del Biondo, 2011).

13 Only Tunisia had its aid reduced under MEDA on human rights grounds, due to a particular instance where the Tunisian government was perceived as obstructing EU support for a Tunisian human rights group (Yacoubian, 2004: 8).

14 Political rights score of 7 and civil liberties score of 6, where 1 is the highest rating and 7 is the lowest (Freedom House, 2010).

15 The EU IT on DGHR, co-chaired by Germany and Portugal, involved 13 EU member states, the Council Secretariat, the EU Delegation at the AU and the European Commission. It completed a concept paper by July 2008. The African IT was chaired by Egypt, with the involvement of 11 African countries and the AU Commission (Africa–EU Ministerial Troika, 2008).

16 The full list of African participants is: Zambia, Egypt, Ethiopia, Ghana, Nigeria, Senegal, Burkina Faso, Kenya, Uganda, Burundi, Algeria, South Africa, Zimbabwe, plus the AU Commission.

17 Examples are: Senegal (0.0 per cent), Burundi (2.7 per cent), Kenya (3.0 per cent), Zambia (3.3 per cent), and Uganda (3.4 per cent).

18 The sparse DGHR roadmap contrasts with the Energy Partnership, for example, which comprises 31 pages of discussion and analysis.

19 This paradox emerged with the 'Mugabe' affair, originally causing the postponement of the second EU–Africa Summit in 2003, and continuing to overshadow the event up to and during its eventual occurrence in December 2007, where various African leaders rallied to President Mugabe's defence (*New York Times*, 9 December 2007).

References

Africa–EU Ministerial Troika (2008) 'Joint Progress Report on the Implementation of the Africa-EU Joint Strategy and its first Action Plan

(2008–2010)', Addis Ababa, 20–21 November, available at: www.africa-eu-partnership.org/documents/documents_en.htm (accessed 30 June 2012).

Africa–EU Ministerial Troika (2009a) 'Draft Joint Roadmaps for the Implementation of the 1st Action Plan (2008–2010) of the Joint Africa–EU Strategy', Luxembourg, 28 April, available at: www.africa-eu-partnership. org/documents/documents_en.htm (accessed 30 June 2012).

Africa–EU Ministerial Troika (2009b) 'Assessment Report', 9 October, final, 13th Africa–EU Ministerial Troika, 14 October, available at: www.africa-eu-partnership.org/pdf/assessment_report_101009.pdf (accessed 30 June 2012).

Africa–EU Strategic Partnership (2007a) 'A Joint Africa–EU Strategy', Lisbon, 9 December 2007, available at: www.africa-eu-partnership.org/documents/ documents_en.htm (accessed 30 June 2012).

Africa–EU Strategic Partnership (2007b) 'First Action Plan (2008–10) for the Implementation of the Africa–EU Strategic Partnership', available at: www. africa-eu-partnership.org/documents/documents_en.htm (accessed 30 June 2012).

Amirah-Fernández, H. and Menéndez, I. (2009) 'Reform in Comparative Perspective: US and EU Strategies of Democracy Promotion in the MENA Region after 9/11', *Journal of Contemporary European Studies*, 17 (3): 325–38.

Bicchi, F. (2006) '"Our Size Fits All": Normative Power Europe and the Mediterranean', *Journal of European Public Policy*, 13 (2): 286–303.

Bicchi, F. (2010) 'Dilemmas of Implementation: EU Democracy Assistance in the Mediterranean', *Democratization*, 17 (5): 979–96.

Carbone, M. (2010) 'The European Union, Good Governance and Aid Co-ordination', *Third World Quarterly*, 31 (1): 13–29.

Carothers, T. (2006) 'The Backlash Against Democracy Promotion', *Foreign Affairs*, 85 (2).

Cavatorta, F., Chari, S.C., Kritzinger, S. and Gomez Arana, A. (2008) 'EU External Policy-Making: "Realistically" Dealing with Authoritarianism? The Case of Morocco', *European Foreign Affairs Review*, 13: 357–76.

Council of the European Union (2003) *European Security Strategy: A Secure Europe in a Better World*, adopted by the European Council meeting in Brussels on 12 December 2003.

Council of the European Union (2007), *The Africa–EU Strategic Partnership: A Joint Africa–EU Strategy*, Lisbon, 9 December, 16344/07 (Presse 291).

Crawford, G. (1996) 'Whither Lomé? The Mid-Term Review and the Decline of Partnership', *Journal of Modern African Studies*, 34 (3): 503–18.

Crawford, G. (1998) 'Human Rights and Democracy in EU Development Co-operation: Towards Fair and Equal Treatment', in M.R. Lister (ed.), *European Development Policy*, Basingstoke: Macmillan, pp. 131–78.

David, D. (2000) 'Forty Years of Europe–ACP Relationship', *The Courier*, special issue on the Cotonou Agreement, September.

Del Biondo, K. (2011) 'EU Aid Conditionality in ACP Countries: Explaining

Inconsistency in EU Sanctions Practice', *Journal of Contemporary European Research*, 7 (3): 380–95.

Emerson, M., Noutcheva, G. and Popescu, N. (2007) *European Neighbourhood Policy Two Years On: Time Indeed for an 'ENP Plus'*, CEPS Policy Brief No. 126, Brussels: Centre for European Policy Studies.

European Commission (2001) *The European Union's Role in Promoting Human Rights and Democratisation in Third Countries*, COM (2001) 252, 8 May.

European Commission (2004), *European Neighbourhood Policy – Strategy Paper*, COM (2004) 373, 12 May.

European Commission (2008) *Annual Report 2008 on the European Community's Development and External Assistance Policies and Their Implementation in 2007*, Luxembourg: Office for Official Publications of the European Communities.

European Commission (2009a) *Annual Report 2009 on the European Community's Development and External Assistance Policies and Their Implementation in 2008*, Luxembourg: Office for Official Publications of the European Communities.

European Commission (2009b) *Implementation of the Joint Africa–EU Strategy and its First Action Plan (2008–2010): Input into the Mid-Term Progress-Report*, SEC (2009) 1064, 20 July.

European Commission (2010) *Annual Report 2010 on the European Community's Development and External Assistance Policies and Their Implementation in 2009*, Luxembourg: Office for Official Publications of the European Communities.

European Union (2000) 'Partnership Agreement between the Members of the African, Caribbean and Pacific Group of States of the One Part, and the European Community and its Member States, of the Other Part, Signed in Cotonou on 23 June 2000', *Official Journal of the European Communities*, L 317/3, 15 December.

Freedom House (2010) *Freedom in the World 2010*, available at: www.freedomhouse.org/template.cfm?page=363&year=2010 (accessed 30 June 2012).

Habermas, J. (1990) *Moral Consciousness and Communicative Action*, Cambridge: Polity Press.

Hyde-Price, A. (2006) 'Normative Power Europe: A Realist Critique', *Journal of European Public Policy*, 13 (2): 217–34.

Joint AUC–EC Task Force (2009) 'Report of the 9th meeting of the Joint AUC–EC Task Force', 8–9 October, Addis Ababa, available at: www.africa-eu-partnership.org/documents/documents_en.htm (accessed 30 June 2012).

Lightfoot, S. and Burchell, J. (2005) 'The EU and the World Summit on Sustainable Development', *Journal of Common Market Studies*, 43 (1): 75–95.

Manners, I. (2002) 'Normative Power Europe: A Contradiction in Terms?', *Journal of Common Market Studies*, 40 (2): 235–58.

Molenaers, N. and Nijs, L. (2009) 'From the Theory of Aid Effectiveness to the Practice: The European Commission's Governance Incentive Tranche', *Development Policy Review*, 27 (5): 561–80.

Pace, M. (2009) 'Paradoxes and Contradictions in EU Democracy Promotion in the Mediterranean: The Limits of EU Normative Power', *Democratization*, 16 (1): 39–58.

Reynaert, V. and Del Biondo K., (2009) 'The EU and Democratic Governance in Africa: A Comparative Analysis of the EU's Policies towards North and Sub-Saharan Africa', paper presented at the UACES conference, Angers, 3–5 September.

Scheipers, S. and Sicurelli, D. (2008) 'Empowering Africa: Normative Power in EU–Africa Relations', *Journal of European Public Policy*, 15 (4): 607–23.

Seeberg, P. (2009) 'The EU as a Realist Actor in Normative Clothes: EU Democracy Promotion in Lebanon and the European Neighbourhood Policy', *Democratization*, 16 (1): 3–19.

Sicurelli, D. (2010) *The European Union's Africa Policies: Norms, Interests and Impact*, Farnham: Ashgate.

Sjursen, H. (2006a) 'What Kind of Power?', *Journal of European Public Policy*, 13 (2): 169–81.

Sjursen, H. (2006b) 'The EU as a "Normative" Power: How Can This Be?', *Journal of European Public Policy*, 13 (2): 235–51.

Yacoubian, M. (2004) 'Promoting Middle East Democracy: European Initiatives', United States Institute of Peace, Special Report 127, October 2004, available at: www.usip.org (accessed 30 June 2012).

Youngs, R. (2002) 'The European Union and Democracy Promotion in the Mediterranean: A New or Disingenuous Strategy?', *Democratization*, 9 (1): 40–62.

Youngs, R. (2004) 'Normative Dynamics and Strategic Interests in the EU's External Identity', *Journal of Common Market Studies*, 42 (2): 415–35.

Youngs, R. (2008) *Is European Democracy Promotion on the Wane?* CEPS Working Document No. 292, Brussels: Centre for European Policy Studies.

9

Economic Partnership Agreements and Africa: losing friends and failing to influence

Christopher Stevens

Both the Euro-Africa Summit of December 2007 in Lisbon and its successor in Tripoli of November 2010 illustrate Europe's difficulty in marrying its rhetorical goal of a strategic partnership with Africa and its trade policy towards the continent. The lofty aims of the Lisbon Summit were lost in a bad tempered row over Economic Partnership Agreements (EPAs), given that it took place one month before what the EU billed as the 'ultimate deadline' for interim agreements. Three years later, with only a quarter of sub-Saharan African states having signed interim EPAs, the subject still loomed over the discussions.

There have been numerous analyses of EPAs but this chapter addresses one particular aspect: the light that the EPA saga throws on the ability of the EU to conduct a coherent negotiation that advances its objective interests whether these are defined in terms of values or of *realpolitik*. There are two strands to the argument. One is that over time Europe's response to a particular problem in the World Trade Organisation (WTO) focused increasingly narrowly on a single solution. The other is that this solution involved substantial risks and few rewards, a feature that has been obscured by the apparent strength of the EU in the negotiations.

Many of the existing analyses focus on the anticipated economic effects of the EPAs (Fontagne *et al.*, 2008; Hallaert, 2010; Karingi *et al.*, 2005; Keck and Piermartini, 2007; ODI, 2006; Perez, 2006; PWC, 2007). Others use EPAs as a lens through which to view changing European political interests (Babarinde and Faber, 2004; Farrell, 2005; Goodison, 2007; Manners, 2009; Nwobike, 2006; Storey, 2006). Since most of these pre-date initialling of the eight interim or full EPAs they are based on assumptions about the detailed terms and are necessarily blind to the subtleties of what has actually been agreed. These details are now available and are reflected by the more recent items in the substantial literature that has monitored progress (from economic, political science and legal perspectives) during the decade-spanning negotiations and examined the implications of the detailed texts as they have emerged (Bilal and Stevens, 2009; Desta, 2006; Forwood, 2001; Khumalo and Mullet, 2010; Matambalya and Wolf, 2001; Meyn, 2008; ODI, 2007; Stevens *et al.*, 2009; Thallinger, 2007).

The first section analyses the evolution of the EPA, and its place within broader European trade policy, from the early 1990s. The second section examines the arguments for and against the EU's position and asks how far EPAs have advanced European interests. Whilst asking a somewhat different question, they follow the approach of two articles (Elgström and Pilegaard, 2008; Elgström and Frenhoff Larsén, 2010) that use the EPAs to shine a spotlight on the modalities of internal decision-making within the EU.

The road to the Economic Partnership Agreements

EU trade preferences

This chapter is organised mainly chronologically to analyse the origin, history and position setting of the negotiating parties. But the reader must be able to place this within the context of the trade and economic issues at stake through an understanding of the EU's complex hierarchy of trade policies and their commercial importance for some African countries.

The EU has always had multiple import regimes offering different levels of market access to different countries. The original six inherited colonial trade preference schemes from France, Belgium and Italy, and the UK extended these when it joined. African states were the principal beneficiaries of this system until 1975 when they were joined by some of the UK's former colonies and the African, Caribbean and Pacific (ACP) group was created (Lister, 1988; Lister, 1997). The access of ACP exports to the EC market was determined by the provisions of the Lomé Convention which was signed in 1975 and renegotiated three times until it was superseded in 2000 by the Cotonou Partnership Agreement (CPA) (European Union, 2000). In its early years, Lomé was at the apex of the EC's 'pyramid of privilege' in the sense that a wider range of its actual and potential exports entered the European market at lower tariffs than did those of any other trade partner.

Apart from Lomé and a set of bilateral preferential trade agreements with most Mediterranean countries, the main provision in the EU's trade regime for developing countries was the Generalised System of Preferences (GSP). Following an initiative in UNCTAD and justified in the GATT and WTO by the 1979 'Enabling Clause', all the developed countries have created a GSP that offers lower tariffs on some of their imports from developing countries. All of the GSPs are 'autonomous policies' of the developed countries concerned (that is, they have been designed and agreed voluntarily by each state acting alone) and so all are different in their details.

The GSP began life as the poor relation of Europe's trade preferences for developing countries (Stevens, 1981). It covered fewer products than either Lomé or the Mediterranean accords and often imposed higher tariffs. Consequently, whilst all developing countries are eligible, it was used only by those that did not have superior access under another regime. But during the 1990s the GSP began to develop internal differentiation. Initially, there was a scheme to offer extra preferences to Central American and Andean states which was wider than 'the Standard GSP' (that is, more products were covered) and deeper (that is, tariffs were lower). Then, in 2001, the EU launched the Everything But Arms (EBA) scheme under the GSP. This provides duty-free and quota-free (DFQF) access for all exports from least developed countries (LDCs), albeit with a transition period for bananas (to 2006), rice and sugar (to 2009). Finally, in 2005, the special tranche for the Central American/Andean states was transformed into a broader GSP+ regime available to a wider group of developing countries but still excluding Mercosur, and most of South, South East and East Asia (Stevens, 2007a). The GSP+ has an important place in the EPA story, explained below, both in its provisions and in the stance of the Commission.

The result was a more complex pattern of trade preferences. Lomé/Cotonou was joined at the apex by EBA, with GSP+ close behind. Despite the differences, one common feature of all the regimes is that they were non-reciprocal (that is the beneficiaries were not required to offer any special regime to European exports in return).

But in a parallel development Europe began to negotiate reciprocal free trade agreements (FTAs) with its close neighbours, with South Africa, Chile and Mexico. It also did so with its Mediterranean partners to replace the non-reciprocal accords. The result was, in broad terms, a three band inverted pyramid of privilege. At or near the apex were a large number of states that had liberal access to the European market for their goods exports under either a non-reciprocal preference agreement or an FTA; in the middle were those developing countries eligible only for the Standard GSP; whilst the base comprised the small number of OECD states not covered by an FTA that exported to the EU under the misnomer of the GATT/WTO 'most favoured nation' (MFN) regime and paid the highest tariffs.

1992–2000: the origins of EPAs

It is against this background that the journey from Lomé to EPAs took place. There was nothing inevitable about the journey. Several North–South non-reciprocal preferential trade agreements – such as the African Growth Opportunity Act (AGOA), adopted by the USA in 2000 – are still justified by the parties in the same way that Europe justified Lomé, that is by simple assertion that they are compatible with GATT/WTO principles. But in the case of Lomé this claim was put to the test in multilateral dispute settlement and ruled to be invalid. So a successor regime had to be found.

Influences on the Commission
The road to the multilateral dispute that was eventually to result in EPAs began in the early 1990s. The trigger setting the process in motion was a 1993 GATT judgement on the legality of the European banana trade regime which was, in turn, 'collateral damage' from the completion of the Single European Market (Stevens, 2000). Prior to 1992, the EU did not have a common trade regime for bananas. Separate regimes existed (largely reflecting colonial legacies) which resulted in markedly different prices in member states. They were made possible by a provision in the Treaty of Rome (Article 115) that permitted member states to restrict imports from their partners of goods that originated outside the Community. Coupled with the oligopolistic nature of the trade in bananas, this allowed the UK, France and Italy to restrict imports of cheaper Latin American bananas (both directly and via another member state) until their markets had absorbed all of the more expensive fruit exported by ex-colonies (and overseas *Départements*).

Because the Single Market allowed cheaper fruit imported into other EU states to be transhipped to UK/France in competition with the 'colonial fruit', the old regime had to be replaced with one that would impose sufficient restrictions on imports from the most efficient suppliers (in Latin America) to remove the incentive for producers to take over the market share of the former colonies. The result was a three-tier tariff (zero for the ex-colonies, low for Latin America within a quota and high outside the quota) which became the subject of a series of GATT and WTO disputes, all of which the EU lost.

In the first of the adverse judgements, the GATT panel concluded not only that the banana regime was contrary to the rules but also that the same applied to the entire Lomé Convention. A temporary fix was provided in 1994 when the EU was granted a waiver for Lomé until 2000, but the search was on to find a regime that would be less open to multilateral challenge.[1]

Although a response to the GATT ruling was inevitable, EPAs were not. One alternative was to seek a succession of waivers which, under both GATT and WTO rules, Members are allowed to approve. Up to that point, waivers had been the 'solution of choice' for North–South trade agreements such as the USA's Caribbean Basin Initiative (CBI) and Canada's Caribcan regime.

The other alternative was to adapt the EU's GSP in a way that would continue to provide the ACP with equivalent access without also providing it to large, low cost developing country competitors. This was technically feasible in the sense that the GSP could be altered unilaterally by the EU and already contained different tranches offering different levels of preference to different groups of countries. Since these tranches have changed in form and coverage during the period in which the EPAs were negotiated, there were a number of opportunities for the EU, had it wished to do so, to have tweaked the scheme to offer an EPA alternative to most if not all ACP. The question of whether such a GSP-based regime would have been challenged in the WTO is a matter of contention, and the issues are examined in more detail below.

From the outset, however, it was clear that EPAs were the Commission's preference. Although the Commission initially set out four options in a Green Paper (circulated in 1996 and published in 1997), only those with the key features of an EPA were described in such a way as to make them appear feasible. The four options described in the Green Paper were: maintenance of the *status quo*, integration of Lomé into the GSP and two variations on regimes under which the ACP would offer trade reciprocity, that is, the removal of restrictions to imports from Europe (European Commission, 1997). The first was ruled out in the Green Paper as infeasible and the second would have resulted in a sharp deterioration in market access for non-LDCs. Consequently, it was the two reciprocity options (that differed mainly on membership geometry) that were in the frame.

Despite this initial preference, the Commission's proposals at this point differed in one very important respect from what was subsequently offered. A key feature of the two reciprocity options is that they would cover only border measures on goods; the idea of including trade-related areas and an agreement on services was suggested only as 'an enhanced variant' (European Commission, 1997: 65).

The Cotonou negotiating mandate

The Commission's autumn 1997 proposal to the Council for a negotiating mandate was based closely on the reciprocity options in the Green Paper (European Commission, 1997). The Commission suggested that, after an initial round of negotiations on a Framework Agreement setting out the principles and an extension of the WTO waiver to 2005, ACP states be invited to negotiate reciprocal trade agreements with, as the only alternative, the GSP upgraded only for LDCs.

The mandate that was approved was a subtly different compromise between two clear viewpoints in the Council. One, with Germany as a prominent exponent, was to support the Commission's proposals as a way of fostering a liberal trade policy in ACP states and integrating the Lomé Convention into the 'mainline' provisions of the WTO. The other, with the UK (then in the Presidency) in the driving seat, was to reduce the contrast between the post-2005 options by improving the Standard GSP both to reduce friction in the negotiations and to promote a wider liberalism in EU trade policy (since enhancements to the GSP would tend to benefit all developing countries).

The mandate agreed in the dying hours of the UK Presidency at the end of June 1998 combined both approaches. On the one hand, it committed the EU to 'offer a process to establish free trade areas ... [to] be negotiated during the five years following the expiry of the current convention (2000–2005)' (European Commission, 1998). On the other hand, a footnote provided that the Council and the Commission will examine all alternative possibilities to provide a new framework for trade between non-LDCs ACPs and the European Union 'which is equivalent to their existing situation under the Lomé Convention and in conformity with WTO rules. In particular the Council and the Commission will take into account their interests in the review of the GSP in 2004, making use of the differentiation permitted by WTO rules' (European Council, 1998: 18, footnote 1).

2000–2006: the Commission's approach narrows

The CPA that succeeded Lomé reflected this compromise (Forwood, 2001). Although the deadline for a new regime was deferred until 2007 (rather than the 2005 proposed by the Commission) with the negotiations launched by a two-year 'period of reflection', the essence of the Commission's proposals remained intact. At the same time, though, the CPA also provided for a fundamental review in 2004 (European Union, 2000, Article 37: 5) and the possibility of alternative trade regimes for 'ACP countries which [do not] consider themselves in a position' to enter an EPA that would be 'equivalent to their existing situation' (European Union, 2000, Article 37: 6).

EPAs take centre stage

The CPA represented a double compromise: between the European member states (Elgström and Pilegaard, 2008) and between the EU and the ACP. The latter were able to use the very broad language in the CPA to claim support for positions that were very far removed from those of the EU (Matambalya and Wolf, 2001). The CPA, for example, defined EPAs (Article 36.1) merely as 'WTO compatible trading arrangements, removing progressively barriers to trade ... and enhancing cooperation in all areas relevant to trade'. This phraseology was shown by events to be open to very wide interpretation over the proportion of trade to be liberalised and the duration of the implementation period (Bilal and Stevens, 2009). The CPA also contained commitments (Article 35: 3) to 'maintaining special treatment for ACP LDCs and to taking due account of the vulnerability of small, landlocked and island countries'.

By the time the Commission obtained its mandate for the negotiations in June 2002, positions had hardened (European Council, 2002). There was no initial offer of significantly improved market access for ACP exports – an early proposal to extend DFQF to all ACP states had been blocked within the Commission – just a non-binding declaration by Sweden supported by the UK and Denmark attached to the mandate arguing in favour (Elgström and Pilegaard, 2008). The only mention of improved rules of origin (RoO) was that the EU would 'assess any specific request for change ... presented by the ACP' despite early drafts of the mandate having included more concrete initiatives (European Council, 2002). No mention was made of any alternative outcomes: the mandate recognised no regime other than an EPA, emphasised the centrality of reciprocity and stated that the 'objective' of EPAs is 'establishing free trade areas' and the liberalisation of services and investment as well as goods.

The year 2004 came and went without any serious examination of options other than EPAs. The official explanation was that the negotiations had not yet progressed sufficiently far for ACP states to decide whether or not they wished to proceed with EPAs. But whilst accurate (since none of the regional EPA negotiating groups[2] had got to grips with the details that are at the heart of any trade agreement) the lack of progress reflected the strong resistance in almost all parts of the ACP. The Caribbean Regional Negotiating Machinery (CRNM), which negotiated on behalf of the group, was virtually alone in arguing that appropriately worded EPAs could be helpful.[3] Even had a serious review been undertaken later, the only conclusion it could have reached (as explained below) would have been that the EU did not have in its complex array of trade regimes one that was 'equivalent' to Lomé/Cotonou. Instead, it was put about that the Standard GSP (to which all developing countries, including the ACP states, are eligible) was close enough, which was demonstrably untrue (Stevens *et al.*, 2007).

The alternatives foregone

This exclusive focus on EPAs was a deliberate choice not an unavoidable necessity. The Commission claimed throughout that there were no feasible alternatives (Curran, 2007; Curran *et al.*, 2008), but the counter argument is that it was 'not the absence of alternatives, but rather the Commission's unwillingness to consider alternatives, that ... tended to characterise the EPA debate' (Storey, 2006: 342). Although the alternative of seeking waivers appeared decreasingly viable as the decade progressed, that of utilising the GSP became more feasible, though its WTO legitimacy was contested.

In the more litigious atmosphere of the WTO, waivers that had been approved largely without contention in the GATT became the subject of hard bargaining – and pay offs. It took until 2002 to obtain the extension of the waiver foreseen in the CPA (and then only as part of the deal to launch the Doha Development Round) and even then the EU was required to 'compensate' Thailand, the Philippines and Indonesia by offering them improved fisheries access. With the banana dispute rumbling on, DG (Directorate General for) Trade was very reluctant either to seek a new waiver or to allow the existing one to expire in 2007 without EPAs being in place. Technically, no new policy had to be in place until mid 2009 at the earliest since this would be the soonest that a WTO complaint, even if launched on 1 January 2008, could have resulted in a judgment by the Appellate Body. But, whether tactically in order to maintain pressure on the ACP or because of a genuine concern to be seen as WTO compliant, the Commission refused publically to consider this option and let it be known during 2007 that Cotonou tariff rates would simply disappear from European customs computers on 1 January 2008. From that date onwards, imports from ACP states would be covered by the 'next most favourable tariff regime' for which they were eligible; in the absence of an EPA, this would be EBA for LDCs but the Standard GSP for the others.

The alternative of accommodating the ACP through a special tranche of the GSP became more feasible when, in 2005, the EU introduced the GSP+. The Commission strongly contests this, arguing that only the Standard GSP (which offers much less favourable preferences) would be WTO compatible (Curran *et al.*, 2008). Since no special GSP-based regime was introduced, it is impossible to know whether or not it would have been challenged. But the details of the GSP+ and the circumstances of its introduction show a much greater willingness on the part of the EU to 'push the WTO boundaries' in relation to the countries of Latin America than in relation to the ACP.

The GSP+ was introduced in the wake of a successful WTO challenge by India against the former EU regime for Central America and the Andean countries, which the EU had justified as supporting the fight against narcotics, and which had been extended to Pakistan. In delivering its verdict, the WTO Appellate Body ruled that it was legal to differentiate between developing countries within the GSP but only if the countries receiving more favourable treatment have common characteristics that the extra preferences address. The beneficiaries of EBA satisfy this criterion since LDCs are recognised in the WTO treaties as a special group. It is less certain that the same applies to the beneficiaries of GSP+.

There are two eligibility rules for GSP+. One is that countries must have ratified and be implementing a raft of international social, environmental and human rights conventions, and 14 of the 15 countries that applied for GSP+ were accepted by the EU as fulfilling this criterion. The other is that they must be 'vulnerable' according to a formula, which rules out from eligibility countries such as India, Brazil, Thailand and the Philippines.

Unless and until there is a WTO challenge to the GSP+, its conformity with the Appellate Body ruling cannot be known for certain, but it has certainly been questioned (Bartels, 2007). The fact that no other country recognises as a special group the GSP+ beneficiaries, and that the EU fails to do so in any arena other than the GSP+ suggests that, at the least, the regime's legitimacy is uncertain. Nonetheless, the EU was willing to 'push the WTO boundaries' in support of the Central American and Andean states that would otherwise have faced new tariffs on their exports. Moreover, it agreed to extend the GSP+ regime to these states on an interim basis *before* some had ratified all of the required conventions (Stevens, 2007b).

Had the EU been willing to have accorded GSP+ status *temporarily* to all non-LDC ACP states from January 2008, it would have been doing no more than had already been done for Central America and the Andean states. If, more ambitiously, it had created a sub-tranche especially for the ACP this would not have been wholly out of line with its approach to the creation of GSP+. Whilst not a complete alternative to the CPA preferences, this would have significantly removed the pressure of the 2007 deadline on most non-LDC ACP states and could have offered many states a permanent alternative to EPAs.[4]

2007 onwards: negotiations continue

As the deadline approached, it became increasingly clear that it was technically not possible for it to be met. FTAs are extremely complex agreements that, in addition to setting out broad principles and a set of overarching rules, must specify exactly what will happen to thousands of separately identifiable goods (and, in the case of the EPAs, also what would be done in relation to a large number of services, to investment and to other trade-related issues). This work was insufficiently advanced.

In recognition, the European Commission agreed in November 2007 to split the negotiations into two stages: only 'interim agreements' needed to be initialled before the end of 2007; these had to include complete provisions on goods, but services, investment and some other contentious issues could be deferred until 2008. When it became clear that even this timetable was too ambitious, the EU adopted a regime that would extend temporarily DFQF[5] on a non-reciprocal basis to all countries that had initialled interim or full EPAs by the end of 2007 (European Council, 2007). This avoided the political embarrassment of ACP states that were actively negotiating EPAs having their exports to the EU disrupted. But, by the same token, it weakened greatly the EU's negotiating leverage.

This deadline was met by 35 of the 76 negotiating states and Zambia followed suit during 2008. Only the CARIFORUM states initialled a full EPA; all of the others were interim agreements. Altogether 19 African states initialled, though Namibia did so only 'provisionally'. The negotiations continued, but in an effort to bring the process to a close the EU decided in 2011 to withdraw by 2014 the temporary DFQF to non-signatories. The CARIFORUM states signed a full EPA in the third quarter of 2008 and several of the other initiallers signed interim EPAs in 2009. But there were notable exceptions. By mid 2012, the countries of the Southern African Customs Union, Zambia and the EAC states had still not concluded their negotiations – and no other states had initialled an interim EPA.[6] Moreover, even by then, only preliminary moves were being taken, and only by some signatory states to implement the changes to tariff and trade policy that they were required to make.

Analysis of the EU's position

Losing friends

So far so uncontroversial: the EU had an underlying strategy which it pursued single-mindedly, using the threat that CPA preferences would disappear on 1 January 2008 to exert additional pressure during the final months. Compromises may have been made on the way, but it is tempting to see a guiding strategy whereby the Commission would slowly tighten its grip on the negotiations as 2007 approached so that all options other than a deep and demanding EPA were removed from contention. There were some questioning voices among the European member states but there were 'many factors [that] spoke against member state action' (Elgström and Pilegaard, 2008: 374).

Because of this, the EPA negotiations have been a favoured subject for analysis of whether they reflect the exercise of normative power or *real-politik* (Farrell, 2005; Manners, 2009; Storey, 2006). But this focuses on the Commission's stated objectives and its partial success in obtaining signatures from some ACP states. It does not also cover the likelihood of the commitment made in the EPAs being fulfilled.

This chapter argues that there is a low likelihood of measures favoured by the Commission actually being implemented fully by governments that have agreed EPAs under duress and remain unconvinced of the benefits. Until December 2007, the negotiating advantage lay with the EU (since it could impose tariffs on imports from some recalcitrant ACP states); from January 2008 onwards, a plausible case can be made that the balance shifts to the ACP signatories (that have some latitude to ignore provisions they do not wish to implement without provoking the imposition of EU tariffs). Since this turning point was predictable, one relevant question is why it appears to have been disregarded by the EU. Another, even more interesting question is why the Commission weakened its own position during the negotiations. These lead in turn to the overarching question of why the Commission chose to 'lose friends' without being able to 'influence people'.

The Commission pressed hard a strategy that failed to offer a fallback position not only for the ACP but also for Europe, and, in doing so, stirred up opposition that was very strongly publicised by civil society organisations (CSOs) in the EU and the ACP (Bilal and Stevens, 2009) as well as scepticism from the European and some national parliaments. Moreover, the policy has also created major problems for African regional integration (Khumalo and Mulleta, 2010; Meyn, 2008) a cause that the EU's declaratory policy supports (Farrell, 2005; Forwood, 2001; Storey, 2006).

One answer could be that it was the 'right thing to do' either narrowly for commercial advantage or for ACP development. This is examined in the next section. The evidence is insufficiently strong to explain why the EU embarked upon a course that would inevitably produce a stand-off in December 2007 unless it backed down. And it is hard not to interpret as backing down the decision to offer unilaterally DFQF to any country that 'initialled' (or, as in Namibia's case, 'provisionally initialled') an interim EPA.

It is also notable that, in pushing so hard for FTAs, the Commission necessarily backpedalled on other objectives such as enforceable environmental, social and labour provisions. Such provisions were 'inserted with greater vigour' into the CPA compared with Lomé (Farrell, 2005: 271). And they are found extensively in the GSP (Manners, 2009). Indeed, they provide the critical trigger for GSP+, the principal practical alternative to EPAs for non-LDC ACP states. The EU has the absolute right unilaterally to withdraw GSP+ from countries that fail to implement a raft of international conventions, and has done so for Sri Lanka. But the WTO requirements for FTAs contain no such provisions. Only the CARIFORUM EPA has enforceable provisions in this area, but it is hard to see any parts of the relevant text (CARIFORUM-ECEPA, 2008, Title IV, Chapters 4 and 5) that are expressed in sufficiently precise terms for an infraction to be easily identifiable. Moreover, the range of actions that the EU could take in cases where treaty obligations are not being fulfilled explicitly excludes trade remedies (that is, the imposition of tariffs).

Failure to win over the ACP

An obvious first question is why the Commission thought it could obtain agreement on EPAs by 2007 when the CPA compromise suggested that it had failed to do so in 2000. Its negotiating cards in 2000 were not likely to become stronger over the next eight years. They included one strong 'stick', a modest 'carrot' and a 'joker'.

The stick was the one actually used – the threat to remove preferences from ACP states that failed to agree the new deal. The carrot was the opportunity to offer improved market access for exports from any state that did sign an EPA. The joker was 'ideological conversion' by an ACP government to the economic ideas underpinning EPAs, as had worked in some cases a decade earlier with structural adjustment.

The weakness of the stick was that it had force only for a relatively small number of ACP states. The 76 ACP states that took part in the EPA negotiations fall into one of three categories in terms of their vulnerability to the loss of CPA preferences:

- those that stood to lose in a very tangible way because they exported to the EU goods that would face higher (and in some cases punitive) tariffs if the CPA regime were not replaced by something equivalent – this applied to Kenya's horticulture and processed tropical fruit exports, and to Botswana's beef, which risked the imposition of EU import taxes equivalent to 80 per cent of export value (Meyn, 2007);

- those with 'a good alternative' to CPA consisting primarily of LDCs (the option of DFQF access under EBA) plus South Africa (which had a bilateral Trade, Development and Co-operation Agreement giving it access to the EU market that is preferential, although not as good as that available to other ACP states); and
- those with 'non-sensitive exports' facing EU standard tariffs that are either zero or very low (such as the oil exporters Nigeria, Gabon and Congo).

Table 9.1, column 4, shows that almost all the states in the first category have signed whilst only a small minority in the other categories have done so; all but three of the African negotiating states that failed to initial an interim EPA are LDCs.[7] In other words, the stick appears to have worked but only on those countries that had vulnerable exports (of which there were just ten in Africa). The creation of EBA *after* the CPA was signed profoundly weakened the EU's stick – and casts doubt on the idea that the Commission was always committed to forcing EPAs on unwilling ACP states (Faber and Orbie, 2009).[8]

The potential carrot (of improved access to EPA signatories) was never a strong one since the scope to improve upon the CPA regime was limited. The few residual tariffs that were in place under the CPA have been removed for signatories, but the gains from this are small. Big gains would require a change to the RoO which determine where a product containing imported inputs has been 'produced' (and, hence, whether it is entitled to a preference because it is has been 'produced' in the beneficiary state or is ineligible because it been produced in the country that supplied the imported inputs). There have been some small improvements in the EPA rules compared to the CPA, but any substantial change has been deferred. This is partly because the DG Taxud has taken an initiative to press for a radical change to the basis for setting the origin rules (Stevens, 2006). Again, actions by the Commission weakened its scope to win countries over to supporting EPAs.

Ideological conversion may well have been a goal, with statements from DG Trade having a strong evangelical flavour, urging the merits of a more liberal economic policy (Elgström and Pilegaard, 2008). But there is little evidence that it produced any converts among the ACP. It lacked three key features that had characterised the earlier successful efforts to win support for structural adjustment by the Bretton Woods Institutions (BWI), the IMF and World Bank. These were: a strong theoretical basis (since the case for liberalising towards just one trade partner is much weaker than that for multilateral liberalisation); analytical capacity (noticeably weaker in DG Trade than in the BWI); and inducements

Table 9.1. Overview of African interim EPA initialling states

	Members	Initialling/ Signatory states [a]	Countries falling into EBA/ standard GSP	Proportion of signatory countries %	*Number of liberalis- ation schedules*
ESA IEPA	Comoros Djibouti Eritrea Ethiopia Madagascar Malawi Mauritius Seychelles Sudan Zambia Zimbabwe	*Comoros Madagascar* Mauritius Seychelles *Zambia* Zimbabwe	Djibouti Eritrea Ethiopia Malawi Sudan	55	*6*
EAC IEPA	Burundi Kenya Rwanda Tanzania Uganda	*Burundi* Kenya *Rwanda Tanzania Uganda*	—	100	*1*
SADC IEPA	Angola Botswana Lesotho Mozambique Namibia South Africa Swaziland	Botswana *Lesotho Mozambique* (Namibia) Swaziland	Angola	71	*2*
CEMAC IEPA	Cameroon Chad Cent. African Rep. Congo DR Congo Eq. Guinea Gabon S. Tomé/ Principe	Cameroon	Chad Cent. African Rep. <u>Congo</u> DR Congo Eq. Guinea <u>Gabon</u> S. Tomé/ Principe	12.5	*1*
ECOWAS IEPA	*Benin Burkina Faso Cape Verde Côte d'Ivoire Gambia Ghana Guinea Bissau Liberia Mali Mauritania Niger Nigeria Senegal Sierra Leone Togo*	*Côte d'Ivoire Ghana*	*Benin Burkina Faso Cape Verde [b] Gambia Guinea Bissau Liberia Mali Mauritania Niger <u>Nigeria</u> Senegal Sierra Leone Togo*	13	*2*

Notes:
(a) Countries in italics are classified as LDCs.
(b) Cape Verde has been classified as non-LDC since January 2008 but will be able to export to the EU under the EBA initiative for a transitional period of three years.

Source: http://ec.europa.eu/trade/wider-agenda/development/economic-partnerships.

(with the offer to improve rather than remove trade preferences being less immediately persuasive than a structural adjustment loan).

Influences on the Commission

Part of the reason for the CPA compromise that delayed the date for EPAs by two years may have been a desire to kick a tricky issue into the long grass from where it would have to be retrieved at a later date by different officials. But this does not explain the tactics subsequently employed by these new officials (or by the two Commissioners for Trade – Pascal Lamy and Peter Mandelson – both of whom came from left of centre parties). Partly it may also have reflected a belief that some ACP states that had been reluctant during the CPA negotiations could be brought on board at a later date, and that their conversion would increase pressure on the others – and this did happen but not sufficiently to resolve the problem.

If it is hard to fathom why the Commission thought that it could avoid a confrontation in 2007, it is easier to see why it embraced the EPA concept as strongly as it did for at least three reasons. The first is that it was influenced by the negotiation of a free trade area with South Africa (Stevens, 2000).

The South Africa FTA may have led to confidence that a reciprocal trade agreement could be sold to the ACP. The South Africa negotiations were conducted by the Directorate General for Development (DG8), which produced the Green Paper that had launched the EPA concept. But there were several key differences (Stevens, 2000). Not least was the institutional change within the Commission. DG8 was responsible for all aspects of relations with the ACP, aid and trade. The existence of the ACP was its *raison d'être*. By contrast, for DG Trade which was created in 2000 and was responsible for all the post-CPA negotiations, the ACP were a peripheral concern alongside relations with the OECD and newly industrialising countries and with the WTO (Elgström and Pilegaard, 2008; Farrell, 2005).

Moreover, the EU's negotiating position with respect to South Africa was much stronger than it was with the ACP, and even then the deal nearly came to grief. Unlike the ACP, South Africa started the negotiations from a low position in the EU's hierarchy of preferences and so faced significant commercial obstacles in relation to highly efficient competitors such as Israel. Signing a deal would substantially improve its access to the European market, especially for agricultural goods.

Even so, a final agreement was in the balance until the very end, as 'deadlines' for the end of the negotiations came and went. The break-through at Commission level came in February 1999, but obtaining member state approval proved to be an additional obstacle and it was not until the European heads of government Berlin Summit on 25 March 1999 (concerned more with Kosovo than with South Africa) that approval was given – some five years after the negotiations began.

A second influence on the Commission was a growing disillusionment (shared in at least some member states) with the 'hands off' philosophy of the Lomé Convention, particularly by contrast with the more inter-ventionist policy-based lending of the BWI. Although a reaction to struc-tural adjustment had already set in by the end of the 1990s (not least in relation to the actual influence that donors can exert on recalcitrant governments), there was a strong view in Brussels that the Commission needed a new legal basis for its relations with the ACP so that it could influence their trade policy (Babarinde and Faber, 2004; Faber and Orbie, 2008).

A third change, which became more apparent in the following decade, was that the creation of the World Trade Organisation (WTO) had altered both the legal basis and, more importantly, attitudes to multilateral rule-making and enforcement (Thallinger, 2007). The WTO acquired 'binding' dispute resolution powers that had not applied to the GATT, and, related to this, WTO Members began to act in a much more litigious manner. As the decade turned, the EU found itself embroiled in several disputes relating to its hierarchy of trade policies. Two brought by India (on the GSP and on sugar) added to the pressure created by the banana disputes (which rumbled on) to make EU trade policy less open to challenge and, crucially, to avoid having to seek favours from other Members. Given the centrality of the WTO to DG Trade's mandate compared to the peripheral position of the ACP, it became clear that it was reluctant to jeopardise the first to favour the second.

Failing to influence

The Commission may have been influenced by an ideological conversion (that led it to press EPAs on unwelcoming ACP states either to advance Europe's commercial interests or because it was 'good' for ACP develop-ment) as well as by a pragmatic desire to tread the path of least resist-ance in the WTO. But the desired effects will occur only to the extent that the EPA policies are implemented, and it is far from clear whether this will be the case precisely because there were few ACP 'converts' to the liberal trade philosophy underpinning EPAs.

The impact on non-EPA members and their regional partners

Evidently, the countries that have not signed EPAs will not be removing barriers to trade with the EU. Less obviously, they will not now be likely to remove barriers to trade with any regional partner that has signed an EPA. One of the goals for EPAs set out in the CPA was to reinforce regionalism in the ACP. But the result has been the opposite and confirms the 2001 comment that 'To all but the most optimistic, Cotonou heralds the *de facto* break-up of the ACP group, as divisions emerge between the countries able to negotiate REPAs [as EPAs were then called], and others having to opt for alternative arrangements' (Forwood, 2001: 439).

As indicated in Table 9.1 column 5, apart from 100 per cent memberships in EAC (and CARIFORUM) the proportion of countries in each negotiating group that have signed varies between 13 per cent (in ECOWAS and CEMAC) and either 57 per cent or 71 per cent in SADC (depending on whether Namibia eventually signs). Members of a trade integration scheme need to have similar trade policies. By definition, customs union signatories must have one, common set of tariffs on most imports. States belonging only to a regional FTA have more latitude because they retain separate and different external tariff regimes, but if their tariff schedules are not harmonised they will have an incentive to keep rigorous border controls (for example to avoid tariff-free EU goods entering an EPA state then being transhipped to its neighbour, which still maintains a tariff on direct imports from Europe). Evidence suggests that it is the paraphernalia associated with physical barriers at land borders that are the most serious constraints to intra-regional trade, rather than differences in trade policy *per se*, although the latter do underpin the former (Charalambides, 2005; Hess, 2000; Visser and Hartzenburg, 2004).

So, although there is no intrinsic reason why a barrier should be created between neighbours that join and do not join an EPA, in practical terms the goal of the latter to avoid reciprocity would be undermined if this did not happen. Similar considerations apply even between countries that have signed EPAs. Only the EAC EPA has a single tariff liberalisation schedule applicable to all members (see Table 9.1 column 6). In all of the others, different countries are liberalising different products at different speeds, and have a renewed interest in maintaining intra-regional border restrictions to prevent their own schedules being undermined. 'Most regions submitted hastily drawn up liberalisation schedules which were the outcome of neither a national nor a regional consensus' (Meyn, 2008: 525).

The impact on EPA members

How far will EPA members actually go in liberalising towards the EU and adopting the other rules to which they have agreed? As in all trade agreements, not only is 'the devil in the detail' but also in the enforcement. EPAs will only limit a signatory's policy space if any requirement to take actions that the country would not otherwise choose is enforced.

Dispute settlement provisions underpin the enforceability of any agreement. Even if never activated, the combination of measurable obligations with a strong system for imposing penalties for non-compliance is likely to have an impact on implementation. If they are weak or lacking, the agreement is no more than the expression of good intentions. The EPAs contain only moderate enforcement provisions and some ambiguity over what needs to be done.

In terms of transparency, the EPA commitments fall into two groups: those where it is very clear what is to be done and when; and those where there exists a degree of uncertainty and ambiguity. Tariff liberalisation by the ACP signatories is the most prominent example of the first category, but for many signatories it will not be put to the test for a decade or more since they have managed to exclude from any liberalisation most of their very sensitive goods and to defer for a decade or more a significant reduction of high tariffs (Bilal and Stevens, 2009). Moreover, ACP states are allowed to impose without penalty 'safeguards' (under which tariffs are raised) in a range of quite broadly defined circumstances in most cases for up to eight years at a time. By contrast both the EU's FTAs with Mexico and Korea require the country imposing non-agricultural safeguards to pay compensation, whilst the FTA with Chile has very restrictive provisions on safeguards.

So any early infractions are likely to be on less transparent commitments such as the 'MFN clause' which requires an ACP state to extend to the EU any 'more favourable treatment' it subsequently offers to a 'major trading country' with which it agrees a 'free trade agreement'. The initial onus will be on any ACP state that negotiates another FTA autonomously to decide whether or not it meets these criteria. But they are subject to some degree of judgement: what counts as 'more favourable treatment' in an agreement that may have some parts that are less good and some that are better; is a trade agreement between developing countries[9] necessarily a 'free trade agreement'? Even the apparently precise definition of 'major trading country' given in the EPAs is open to some fudging.

Only if the Commission notes what it believes to be an infraction, takes the matter to dispute settlement and wins will an EPA party be faced with a hard choice. Whilst the provisions vary between EPAs, the norm is that the EU appoints one arbiter, the ACP appoints another and these two select a third who is not a national of either party who will chair the panel. So it is not certain that the EU will always win any cases that it brings. Even then, neither the EAC nor the ESA interim EPAs contain any provision for the complaining party to take 'temporary remedies' in case of non-compliance with a ruling. In the other cases, the complaining party is entitled to 'adopt appropriate measures' to increase the pressure on the complained about state to alter its policy. They include an EU commitment to exercise due restraint when applying temporary remedies and to select measures that are best aimed at bringing about compliance.

The ultimate 'risk' that any ACP government runs in forming a narrow interpretation of opaque provisions (or of failing to reduce its tariffs as agreed) is that it may be required to reverse this position at a later date on pain of losing some or all trade preferences. In other words, the stark choice facing a non-implementing ACP state if it is taken to dispute settlement and loses is similar to the one it faced in 2007 – with the penalties being the same or, probably, smaller. The reason why they may be smaller is that as the EU extends low tariffs to an increasing number of countries (e.g. through new FTAs) the commercial value of the ACP's preferences declines – a process known as 'preference erosion'. In many cases, therefore, the influence of the EU over the policies of EPA signatories may decline over time.

Conclusion

It is hard not to conclude that the increasing focus on EPAs as the only solution to the WTO challenge to Lomé preferences and the widening of their scope to include many policy areas that are not directly relevant to that challenge reflect the Commission's exercise of its 'hard power' over some ACP states. What is puzzling is why the pitfalls were overlooked by the Commission. Since alternatives existed and could have been chosen at points during the decade, the explanation of 'path dependency' is not sufficient.

The pitfalls indicate the limit to the EU's 'hard power' even in negotiations with a group of weak developing countries some of which are heavily dependent on Europe for trade preferences. These limits were tightened by DG Trade itself with the EBA initiative. No African country has agreed a full EPA and few of those not facing export disruption when the CPA goods trade regime expired have signed what are described as 'interim EPAs' but are, in effect, 'goods-only EPAs'. It remains to be seen how far the signatories will use their 'soft power' of foot dragging to avoid implementing even the interim EPAs.

One possible answer to the puzzle is to be found in personalities. The Commissioners and senior officials involved in the negotiations have moved on, claiming to have successfully completed the negotiations. Public attention wanes after signature. Some of the most vocal CSOs, such as Oxfam, have ended their EPA campaigns; implementation is simply not so newsworthy.

Regardless of the validity of this explanation, the EPA story adds evidence to the argument that the EU has built-in difficulties pursuing a coherent external policy, even in an area that is clearly a Union-level responsibility. DG Trade's policy was clearly 'consistent', at least until the end of 2007 when it backed down by extending for an indeterminate period DFQF to all states that were continuing to negotiate. But 'coherence' requires more than 'consistency'; it implies an outcome that advances an identifiable European interest, however defined. It is hard to discern what interest is served by an outcome in which the great majority of African states have declined to join EPAs, the few that have may not greatly change their policies, and a shadow has been cast over other areas of Euro-Africa policy.

Notes

1 In principle requests for a waiver had to be approved by 75 per cent of members, but in practice the GATT worked on the basis of consensus (as does the WTO) and so in reality all members must at least acquiesce in any request for it to go forward.

2 At this point there were six negotiating groups: CARIFORUM, ECOWAS, CEMAC, Eastern and Southern Africa (ESA), SADC and the Pacific. In 2007, there was a final realignment with the EAC countries withdrawing from ESA and (in the case of Tanzania) SADC to form a separate group.

3 The coda is that the CRNM was perceived by some regional governments as having exceeded its mandate and its existence as a separate institution was brought to an end in 2009 when it was made part of the Caricom Secretariat.

4 The Commission's rebuttals of the GSP as an option necessarily failed to address these issues. This is understandable since the EU cannot admit to the possibility that GSP+'s WTO legitimacy is questionable.

5 Apart from sugar and rice that continued to be treated as they had been under the CPA.

6 The members of the Southern African Customs Union are Botswana, Lesotho, Namibia, South Africa and Swaziland. South Africa, which had declined to initial in 2007, had also neither initialled nor signed, though it continued to negotiate.

7 The African countries that have acted non-predictably are all LDCs that initialled: Burundi, Rwanda, Tanzania and Uganda (to retain the EAC); Lesotho (for improved clothing rules of origin); Madagascar (to allow joint production with Mauritian firms), Mozambique (to reduce dependence on South Africa) and Zambia (to obtain a higher sugar quota).

8 Although DiCaprio and Trommer (2010) argue the opposite – that EPAs risk the dilution of special, non-reciprocal treatment for LDCs.

9 Notified to the WTO under Part IV, which deals with special and differential treatment for developing countries, rather than under Article 24, which covers FTAs and customs unions.

References

Babarinde, O. and Faber, G. (2004) 'From Lomé to Cotonou: Business as Usual?', *European Foreign Affairs Review*, 9: 27–47.

Bartels, L. (2007) 'The WTO Legality of the EU's GSP+ Arrangement', *Journal of International Economic Law*, 10: 869–86.

Bilal, S. and Stevens, C. (eds) (2009) 'The Interim Economic Partnership Agreements between the EU and African States', Policy Management Report 17, Maastricht: European Centre for Development Policy Management.

CARIFORUM-EC EPA (2008), 'Economic Partnership Agreement between the Cariforum States, of the one part, and the European Community and its member states of the other part', signed on 15 October 2008 in Barbados.

Charalambides, N. (2005) 'The Private Sector's Perspective, Priorities and Role in Regional Integration and Implications for Regional Trade Arrangements', Discussion paper 56, Maastricht: European Centre for Development Policy Management.

Curran, L. (2007) 'Response to the article "R. Perez, Are the Economic Partnership Agreements a First-best Optimum for the African Caribbean Pacific Countries?"', *Journal of World Trade*, 41 (1): 243–4.

Curran, L. Nilsson, L. and Brew, D. (2008) 'The Economic Partnership Agreements: Rationale, Misperceptions and Non-Trade Aspects', *Development Policy Review*, 26 (5): 529–53.

Desta, M.G. (2006) 'EC-ACP Economic Partnership Agreements and WTO Compatibility: An Experiment in North–South Inter-Regional Agreements?', *Common Market Law Review*, 43: 1343–79.

DiCaprio, A. and Trommer, S. (2010) 'Bilateral Graduation: The Impact of EPAs on LDC Trade Space', *Journal of Development Studies*, 46 (9): 1607–27.

European Commission (1997) *Green Paper on Relations between the European Union and the ACP Countries on the Eve of the 21st Century: Challenges and Options for a New Partnership*, Luxembourg: Office for Official Publications of the European Communities.

European Council (1998) 'Negotiating Directives for the Negotiation of a

Development Partnership Agreement with the ACP Counties', Information Note 10017/98, Brussels, 30 June.

European Council (2002) 'Recommendation Authorising the Commission to Negotiate Economic Partnership Agreements with the ACP Countries and Regions', Brussels, mimeo.

European Council (2007) 'Applying the Arrangements for Products Originating in Certain States which Are Part of the African, Caribbean and Pacific (ACP) Group of States Provided for in Agreements Establishing, or Leading to the Establishment of, Economic Partnership Agreements', Council Regulation (EC), No. 1528/2007, 20 December.

European Union (2000), 'Partnership Agreement between the Members of the African, Caribbean and Pacific Group of States of the One Part, and the European Community and its Member States, of the Other Part', signed in Cotonou on 23 June 2000, *Official Journal of the European Communities*, L 317/3, 15 December.

Elgström, O. and Frennhoff Larsén, M. (2010) 'Free to Trade? Commission Autonomy in the Economic Partnership Agreement Negotiations', *Journal of European Public Policy*, 17 (2): 205–23.

Elgström, O. and Pilegaard, J. (2008) 'Imposed Coherence: Negotiating Economic Partnership Agreements', *Journal of European Integration*, 30 (3): 363–80.

Faber, G. and Orbie, J. (2008) 'The New Trade and Development Agenda of the European Union', *Perspectives on European Politics and Society*, 9 (2): 192–207.

Faber, G. and Orbie, J. (2009) 'Everything But Arms: Much More than Appears at First Sight', *Journal of Common Market Studies*, 47 (4): 767–87.

Farrell, M. (2005) 'A Triumph of Realism over Idealism? Cooperation between the European Union and Africa', *Journal of European Integration*, 27 (3): 263–83.

Fontagne, L., Laborde, D. and Mitaritonna, C. (2008) 'An Impact Study of the EU–ACP Economic Partnership Agreements (EPAs) in the Six ACP Regions', 04 April 2008, Centre d'Etudes Prospectives et d'Informations Internationales (CEPII), Paris.

Forwood, G. (2001) 'The Road to Cotonou: Negotiating a Successor to Lomé', *Journal of Common Market Studies*, 39 (3): 423–42.

Goodison, P. (2007) 'The Future of Africa's Trade with Europe: "New" EU Trade Policy', *Review of African Political Economy*, 111: 139–51

Hallaert, J.-J. (2010) 'Economic Partnership Agreements: Tariff Cuts, Revenue Losses and Trade Diversion in Sub-Saharan Africa', *Journal of World Trade*, 44 (1): 223–50.

Hess, R. (2000) 'SADC: Towards a Free Trade Area', in T. Hartzenberg (ed.), *SADC–EU Trade Relations*, Harare: Sapes Books.

Karingi, S., Lang, R., Oulmane, N., Perez, R., Sadni Jallab, M. and Ben Hammouda, H. (2005) 'Economic and Welfare Impacts of the EU–Africa Economic Partnership Agreements', ATPC Work in Progress 10, March,

United Nations Economic Commission for Africa, African Trade Policy Centre.

Keck, A. and Piermartini, R. (2007) 'The Economic Impact of Economic Partnership Agreements (EPAs) in Countries of the Southern African Development Community (SADC)', *Journal of African Economies*, 17 (1): 85–130.

Khumalo, N. and Mullet, F. (2010) 'Economic Partnership Agreements: African–EU Negotiations Continue', *South African Journal of International Affairs*, 17 (2): 209–20.

Lister, M. (1988) *The European Community and the Developing World*, Aldershot: Avebury.

Lister, M. (1997) *The European Union and the South*, London: Routledge.

Manners, I (2009). 'The Social Dimension of EU Trade Policies: Reflections from a Normative Power Perspective', *European Foreign Affairs Review*, 14: 785–803.

Matambalya, F. and Wolf, S. (2001) 'The Cotonou Agreement and the Challenges of Making the New EU–ACP Trade Regime WTO Compatible', *Journal of World Trade*, 35 (1): 123–44.

Meyn, M. (2007) 'The End of Botswana Beef Exports to the European Union?', Project Briefing, Overseas Development Institute, London, September.

Meyn, M. (2008) 'Economic Partnership Agreements: A "Historic Step" towards a "Partnership of Equals"?', *Development Policy Review*, 26 (5): 515–28.

Nwobike, J. (2006) 'The Emerging Trade Regime under the Cotonou Partnership Agreement: Its Human Rights Implcations', *Journal of World Trade*, 40 (2): 291–314.

Overseas Development Institute (2006) 'The Potential Effects of Economic Partnership Agreements: What Quantitative Models Say', Briefing Paper 5, Overseas Development Institute, London.

Overseas Development Institute (2007) 'Economic Partnership Agreements: What Happens in 2008?', Briefing paper 23, Overseas Development Institute, London.

Perez, R. (2006) 'Are the Economic Partnership Agreements a First-Best Optimum for the African Caribbean Pacific Countries?', *Journal of World Trade*, 40 (6): 999–1019.

PWC (2007) *Sustainability Impact Assessment (SIA) of the EU–ACP Economic Partnership Agreements: Agro-Industry in West Africa, Tourism Services in the Caribbean, Fisheries in the Pacific*, Revised final report, Paris: PricewaterhouseCoopers.

Stevens, C. (ed.) (1981) *The EEC and the Third World: A Survey*, Sevenoaks: Hodder & Stoughton.

Stevens, C. (2000) 'Trade with Developing Countries: Banana Skins and Turf Wars', in H. Wallace and W. Wallace (eds), *Policy Making in the European Union*, Oxford: Oxford University Press.

Stevens, C. (2006) 'Creating Development Friendly Rules of Origin in the EU', Briefing Paper 12, Overseas Development Institute, London, November.

Stevens, C. (2007a) 'Creating a Development-Friendly EU Trade Policy', in A. Mold (ed.), *EU Development Policy in a Changing World*, Amsterdam: Amsterdam University Press, pp. 221–36.

Stevens, C. (2007b) 'Economic Partnership Agreements: What Happens in 2008?', Briefing Paper 23, Overseas Development Institute, London, June.

Stevens, C., Meyn, M. and Kennan, J. (2007) *The Costs to the ACP of Exporting to the EU under the GSP*, report prepared for the Netherlands Ministry of Foreign Affairs, London: Overseas Development Institute.

Stevens, C. Kennan, J. and Meyn, M. (2009) 'The CARIFORUM and Pacific ACP Economic Partnership Agreements', Economic Paper 87, Commonwealth Secretariat, London.

Storey, A. (2006) 'Normative Power Europe? Economic Partnership Agreements and Africa', *Journal of Contemporary African Studies*, 24 (3): 331–46.

Thallinger, G. (2007) 'From Apology to Utopia: EU–ACP Economic Partnership Agreements Oscillating between WTO Conformity and Sustainability', *European Foreign Affairs Review*, 12: 499–516.

Visser, M. and Hartzenberg, T. (2004) 'Trade Liberalisation and Regional Integration in SADC', Working paper 3, Trade Law Centre for Southern Africa, Stellenbosch.

10

Unfulfilled expectations? The EU's agricultural and fisheries policies and Africa

Alan Matthews

Isn't Europe actually turning its back on our countries? Isn't there a kind of falling out of love – I don't want to talk about divorce – but a kind of falling out of love with Africa? Luc Magloire Mbarga Atangana, Cameroon's Trade Minister, 29 April 2010.[1]

This chapter examines the changing dynamics of EU–African relations in agri-food trade. The EU takes 50 per cent of Africa's food exports and supplies just under 30 per cent of its food imports.[2] It is a trade relationship which has been heavily influenced by policy interventions on both sides. In the Joint Africa–EU Strategy (JAES) (European Union, 2000) adopted at the second EU–Africa Summit in Lisbon in 2007, the EU along with Africa committed in the area of agriculture to promoting policy coherence for development, food security, food safety and food quality. Policy coherence for development means that the EU, in pursuing its domestic policies, will seek to avoid creating barriers to African development and, where possible, will actively use these policies to facilitate African development (Carbone, 2008). A new EU–African Union partnership on agricultural development was promised to support Africa's agricultural agenda as set out in the Comprehensive Africa Agricultural Development Programme (CADDP). The EU and Africa also jointly committed to improving policy coherence for development in fisheries, notably with respect to fisheries access arrangements, trade and controlling illegal, unregulated and unreported fishing (Council of the European Union, 2007). The question we address in this chapter is whether the changes in EU agricultural and fisheries policies over the past decade have contributed to greater policy coherence for development, as promised in the JAES, or whether Africa is worse off as a result.

Africa's agricultural sector has long had a greater external orientation than other developing country regions.[3] Reflecting its experience of colonial economic development, Africa developed a significant dependence on the export of tropical commodities and other agricultural raw materials largely to European export markets as the former colonial powers. Even in the 2000s, over 50 per cent of Africa's agricultural and fisheries exports were destined for the EU, and these accounted for around 20 per cent of its total exports to the EU, with much of the remainder made up of oil and other minerals. Africa's agri-food exports to the EU are dominated by a range of tropical products that could not readily be produced within Europe, including cocoa, coffee and tea. Also important are fish, fruits and vegetables (influenced strongly by banana exports), cut flowers, sugar and tobacco. In 2007, Africa accounted for 12 per cent of the EU's food imports, although the share for some products was significantly higher. Africa supplies three-quarters of Europe's cocoa imports and 40 per cent of its cut flowers. Products heavily protected by the EU's Common Agricultural Policy (CAP), such as meats, dairy products and rice, with the partial exception of sugar, hardly figure at all in Africa's exports to the EU.

Agri-food products make up around 11 per cent of all EU exports to Africa, and Africa absorbs a relatively small percentage of EU agri-food exports at around 9 per cent of the total. The most important EU export is cereals (including milling products and cereal preparations), followed by beverages, dairy products and fish. African fish imports consist mainly of tuna from the EU fleet used as raw material by the tuna canning industry located in African countries. In addition, some West African countries are major importers from the EU of herring and mackerel, which play an important role in guaranteeing adequate fish supplies for domestic consumption. Importantly, the EU is not the main supplier of Africa's agri-food imports, accounting for just under 30 per cent of the total in 2007.

Policy interventions both on the EU and African side in addition to differences in climate and other resources have helped to shape these trade flows. Within Africa, economic development following independence in the 1960s took the form of import-substitution industrialisation which led to heavy discrimination against the agricultural sector (Anderson and Masters, 2009). As a result, Africa has become increasingly dependent on imports of foodstuffs and particularly cereals over time. The support given to EU agricultural production contributed to the decline in the level of world market prices, particularly for cereals, sugar and animal products. While this provided a benefit in the form of cheap food imports for African countries, critics argued that low prices undermined local strategies to increase food production and contributed to the dependence on food imports. African countries could have captured the terms of trade gain in the form of higher tariff revenue while at the same time protecting local producers, so it is more plausible to argue that cheaper food imports suited African governments bent on pursuing industry-led development strategies. However, the subsidised production of competing crops in the EU, such as sugar, cotton, tobacco and fruits and vegetables, undoubtedly damaged African producers of similar crops. At the same time, Africa's traditional agricultural and fish exports benefited from preferential access to the EU market, which privileged them relative to third country exporters (often other developing countries) while, in some cases, providing limited access to the high-priced protected markets for CAP commodities. What emerges is a complex picture of EU–African food trade relations where EU policy interventions have resulted in both benefits and losses to African exporters.

Recent EU policy changes risk squeezing Africa from two sides, leading to rising costs for Africa's cereals imports but lower prices and volumes for its agricultural exports (Stevens, 2003). Stevens' argument focused particularly on preference erosion resulting from EU agricultural policy reform. During the past decade, the value of traditional African preferences has been further eroded but, at the same time, the EU has extended significant new preferences in the form of duty-free quota-free access to its agri-food market to most African countries. Have these policy changes helped to improve Africa's position on EU markets in a way that will facilitate Africa in developing these sectors as engines of economic growth and poverty reduction? Or do they provide further evidence of the marginalisation of African interests in EU policy-making as the EU faces a widening number of challenges in its external economic relationships, as the quotation at the head of the chapter implies? Have the changes in EU policy fulfilled the expectations of those seeking greater coherence between the EU's agricultural and fisheries policies and African development?

In addressing these questions, we develop the argument in three steps. Firstly, we evaluate the 'traditional' policy interventions under the EU's CAP and its Common Fisheries Policy (CFP) up to around the year 2000, though these continued to be influential also in the following decade. Secondly, we describe the positive aspects of the policy activism which characterised the first decade of the twenty-first century, and which contributed to widening market access for African exporters to the EU market. Thirdly, we examine the negative downside of some recent and projected policy changes that are leading to the erosion of the benefits of preferential access, which African countries traditionally enjoyed. The final section concludes by summarising the state of coherence between EU agricultural and fisheries policies and African development.

EU agri-food and fisheries policies and Africa

Export subsidies

The most controversial impact of EU agricultural policy on Africa has been the use of export subsidies, also called export refunds. Export subsidies were an integral part of the CAP, particularly when high support prices ensured that EU production moved beyond the point of self-sufficiency and resulted in a surplus which had to be exported to world markets. Given that EU prices were often well above world market prices, exports could only take place by paying exporting firms the difference between the EU and the world price in the form of export refunds. Indeed, export refunds were managed in such a way as to ensure internal market balance at the high support prices, leading to the accusation that the EU was dumping its surpluses on world markets. When world prices were low, the EU simply increased its export subsidies to ensure its surplus was removed.

Development NGOs highlighted many instances where subsidised EU food imports undercut local producers and damaged local agricultural production: pigmeat in West Africa, tomato paste in Ghana, milk powder in Kenya and Cameroon, cheese in South Africa. However, it can be difficult to assess to what extent export subsidies can be held responsible for the inability of domestic farmers to develop local production, given the multitude of other handicaps under which they operate; for importing countries, they did reduce the cost of imports, although it is undeniable that the subsidies are unfair competition for other potential suppliers. At their peak, EU expenditure on export subsidies rose to over 10 billion euro in the early 1990s, although expenditure has fallen significantly in recent years and is now less than half a billion euro.[4]

Since 1992, the EU has significantly reformed its agricultural policy, lowering its support prices while compensating farmers for the loss of income, first through coupled direct payments and, more recently, through decoupled direct payments. EU self-sufficiency in major food products has fallen as a result, and the EU is once again a net importer of important commodities such as sugar, beef and poultrymeat where once it was a major exporter (Matthews, 2010a). The EU has offered to eliminate the use of export subsidies in the context of a WTO Doha Round Agreement, provided that similar disciplines are placed on other forms of export support such as food aid, export credits and state monopoly export boards, which are more commonly used by other major exporters. Although successive rounds of CAP reform have made the EU less dependent on export subsidies, the EU resorted again to subsidising exports of dairy products and pigmeat in 2009 and 2010 in order to lift low internal market prices.

Export subsidies are just the visible manifestation, and not necessarily the most important aspect, of the general incentive to higher EU production resulting from the CAP policy of supporting market prices and providing direct support to EU farmers. Any incentive to production, or disincentive to consumption, means that the EU is adding more to global supplies than it would do in the absence of support. World prices are depressed as a result relative to what they otherwise would be. This general world market price effect, because it affects all exports and imports of African countries of commodities protected by the CAP or their close substitutes, is a more important channel whereby the CAP influences African food production than export subsidies per se. Because of successive CAP reforms, the EU now distorts world food markets to a much smaller extent than before.

Preferential access and the commodity protocols

Offsetting this downward pressure on the prices of African exports of commodities protected by the CAP was the fact that, for many of these products as well as for non-CAP food and agricultural exports, African exporters benefited from preferential access to the EU market under the transitional trade provisions of the Cotonou Agreement up to the end of 2007 and, before that, the Lomé Conventions. One estimate of the average tariff applied by the EU in the agricultural sector put it at 16.7 per cent in 2001 (Bouët *et al.*, 2004). However, while some Latin American developing countries faced an average tariff of 18.3 per cent and South Asian countries faced an average tariff of 14.4 per cent, the tariff facing Sub-Saharan Africa agri-food exports was less than half these values at 6.7 per cent. The value of these preferences differed greatly across commodities, depending on the value of the most favoured nation (MFN) tariff and the degree of preference awarded. For most CAP products, protection took the form of an *ad valorem* (percentage) tariff and a specific (fixed) amount. The Cotonou preference for African, Caribbean and Pacific (ACP) countries was limited to the elimination of the *ad valorem* duty, leaving the specific duty (usually by far the biggest element of the total tariff) in place.

However, special trade protocols for four commodities; sugar, bananas, beef and veal, and rum, were introduced as part of the Lomé Conventions. In the sugar protocol, the EU undertook to purchase and import at guaranteed prices 1.3 million tonnes of cane sugar originating in specific ACP states (as well as India) for an indefinite period. The guaranteed price was to be negotiated annually within the price range obtaining within the Community. In practice, the price paid to ACP exporters was linked to the intervention price for beet sugar within the EU. When the single banana market was introduced in 1992, the external trade regime was modified to allow a tariff-free quota for traditional ACP banana exports, a tariff quota at a low tariff for traditional quantities of Latin American bananas and prohibitive tariffs on out-of-quota imports from both sources. This arrangement provided significant protection to ACP exporters. The banana protocol introduced in the 4th Lomé Agreement in 1990 provided that, in respect of banana exports to the EU market, no ACP state 'shall be placed as regards access to its traditional markets and its advantages on these markets, in a less favourable position than in the past or the present'. Under the beef protocol, certain ACP states were allowed to export specified quantities of boneless meat to the EU at import duties reduced by 90 per cent. The Lomé Conventions also provided for an annual quota within which rum could be imported duty-free. However, ACP rum was granted unrestricted duty-free access to the EU market in 2000, and the rum protocol was eliminated in the Cotonou Agreement.

These preferences were of substantial value to African exporters. Perez and Jallab (2009) calculated that sub-Saharan African agri-food exports to the EU were boosted by 1.2 billion euro annually as a result of preferences, and that sub-Saharan Africa overall was better off in welfare terms by 537 million euro as a result of these preferences. Milner *et al.* (2010) calculated that the preferential rent arising on African exports to the EU was equivalent to about 4 per cent of their total value of exports. They remarked on the high concentration of the rents both among beneficiaries and among commodities. Sixty per cent of the preferential rent on African exports went to just five countries, while 80 per cent accrued to ten countries. Similarly, one sector (sugar) accounted for 31 per cent of the total rent, and just three sectors accounted for 56 per cent. Mauritius and Swaziland were the main African beneficiaries of the sugar protocol, while Cameroon and Cote d'Ivoire were the main beneficiaries of the banana protocol. The beef protocol was mainly beneficial to Southern African exporters.

The EU Common Fisheries Policy

EU fisheries policy is implemented through the Common Fisheries Policy (CFP) established in 1983. The creation of the CFP was stimulated by the decisions in 1976 by member states, along with many other coastal states, to create Economic Exclusion Zones (EEZs) that extended control of marine resources from 12 to 200 nautical miles (Bretherton and Vogler, 2008). The CFP was intended to achieve market integration within the EU while also securing access for the European distant-water fleet to external fishing grounds from which they were excluded by the creation of EEZs. In 2002, EU member states agreed a reform of the CFP in the light of general agreement that it was not effective in delivering the sustainable exploitation of fisheries resources.

As fish stocks are at critically low levels in EU waters, resulting in decreasing levels of landings, the EU faces a growing supply deficit to meet its demand for fish. The supply deficit has been solved through the increased purchase of fish caught by non-EU vessels outside EU waters and by the rise in effort of EU fleets in distant water fisheries (particularly along the African coast) (Ponte *et al.*, 2007). EU tariff policy favours the import of fish from ACP sources through a system of discriminatory tariff preferences. EU MFN tariffs (in per cent *ad valorem* equivalent) on imports stood at 11.8 per cent for unprocessed fish and 15 per cent for semi-processed fish, giving a significant competitive advantage to African exporters with duty-free entry (Campling, 2008).

Bilateral fisheries agreements were the backbone of EU fisheries collaboration with African countries. As of mid-2005, the EU was party to 22 bilateral access agreements of which 16 were with African countries (Ponte *et al.*, 2007). The cost to the EU to enter these agreements was 145 million euro per annum in the late 1990s, which was paid to ACP countries as rent for obtaining fisheries access. European fishing companies paid an additional 30 million euro in access fees.

These agreements were criticised on a number of grounds. While their legal basis was that a harvestable surplus of fish must exist within the EEZs of partner countries, the extent to which surplus fish resources existed was questioned for some countries. Despite formal access rules governing fishing locations, fish stocks targeted and restrictions on access to local waters reserved for local artisan fleets, there were reports that the activities of foreign industrial trawlers led to the depletion of the artisan fisheries on which local fishermen depend. The development impact of the licence fee income received from the EU was questioned where the money was not used to further develop the fisheries sector or for investment in other parts of the economy.

In response to these criticisms, the EU proposed in 2004 as part of the reform of the CFP that access agreements with third countries should be replaced by Fisheries Partnership Agreements (FPAs) intended to provide a legally binding framework through which policy dialogue about sustainability issues would be promoted (Carbone, 2009). The FPAs were intended to demonstrate the EU's commitment 'both to sustainable and responsible fisheries policy and to poverty reduction in developing countries' (Bretherton and Vogler, 2008: 412). For example, the process of jointly agreeing on the use of the financial contribution was intended to ensure that most of these funds would be used for the conservation and sustainable management of fisheries resources. Despite the attempt to address some of the criticisms, it appears that problems remain. Although the prevention of over fishing, in particular for stocks of importance to local people, is a key concern of the FPAs, various types of leakage are possible in the implementation of these agreements.

Other restrictions on African market access

Rules of origin can potentially limit the value of preferences. For example, an origin rule which specifies that a minimum share of the selling price must be attributed to value added in the preference recipient may set the share at a level which makes it impossible for a developing country with a low level of industrialisation to comply.[5] Rules of origin are, in general, not a major hindrance to exploiting preferences in the agri-food sector. The evidence suggests that the utilisation rate of agri-food preferences is very high (Bureau *et al.*, 2007). However, the rules of origin determining eligibility for preferences for fisheries products have a significant practical and even perverse effect. To benefit from Cotonou preferences fish had to be caught using ACP or EU vessels and with a majority of crew who were ACP or EU nationals, and in the case of processing firms 50 per cent of the capital must be provided by either ACP or EU owners. This means that if the ACP states lack the fishing capacity themselves (in the case of tuna processing, for example), then their processing sectors must buy from EU vessels rather than potentially more competitive sources in order to benefit from preferential access to the EU market. This creates an incentive for ACP countries to grant EU vessels preferential access to their EEZs to ensure that their tuna canneries are supplied with 'originating' tuna. Some critics have argued that the preferential access offered to the ACP countries for the processed and canned tuna they export to the EU should be considered as a form of upstream subsidy to EU vessels rather than a trade concession to ACP countries (Ponte *et al.*, 2007). The EU has recently reformed its rules of origin for EPAs and its Generalised System of Preferences (GSP) scheme to exclude the officer and crew requirement.

Food safety and animal and plant health regulations (sanitary and phytosanitary standards) are another area which can create effective non-tariff barriers to African agri-food exports. Standards are needed to protect consumer, plant and animal health, but there is a suspicion that they may be implemented for protectionist purposes or enforced in a discriminatory manner. Campling (2008), for example, reports that ACP countries believe that their fish product exports are inspected more frequently than those of their major competitors. Ensuring and showing compliance with EU standards is also costly and may exclude some exporters from the market. On the other hand, standards can also help to improve access to the EU market by enhancing the reputation of African countries' exports. There is also evidence that standards may have a catalytic effect, in that investments, for example to meet hygiene requirements, may also lead to increased productivity and lower costs.

Improvements in Africa's access to the EU agri-food market

Since the turn of the new century, a number of initiatives have improved market access for African exporters of agri-food products to the EU market. This began in 1999 with the conclusion of the EU–South Africa Trade and Development Cooperation Agreement (TDCA), which entered fully into force in May 2004 (Council of the European Union, 2004). The Agreement provides for the liberalisation of 95 per cent of the EU's imports from South Africa within ten years, although the exclusions mainly cover agricultural products. This was followed in 2001 by the Everything but Arms (EBA) scheme introduced as a special arrangement under the EU's Generalised System of Preferences. This extended duty-free and quota-free access to all least developed countries (LDCs) to the EU market, with transition periods for three sensitive products rice, bananas and sugar. Most recently, from 1 January 2008, the Cotonou Agreement trade preferences were extended to duty-free quota-free access for those African countries which signed interim Economic Partnership Agreements (EPAs) with the EU.[6] Their main effect was to extend the tariff-free access enjoyed by the least developed African countries to the non-least developed African countries willing to sign, although all signatories may benefit in addition from somewhat more generous rules of origin.[7]

Increased access to the EU market was not the only policy change introduced by EPAs. As foreseen under the Cotonou Agreement, EPAs mark a fundamental change in Europe's trade relations with Africa by reintroducing the notion of reciprocity in trade concessions. Among the concerns expressed in relation to this was that opening African markets to EU imports could undermine food security in the EPA states. However, it appears that the ACP countries have made use of the flexibilities in the EPA negotiating process to exclude from liberalisation those products that are of importance from the point of view of food security (Matthews, 2010b). Where tariff elimination has been scheduled on such imports from the EU, this elimination has been delayed for more than a decade in many cases. There is also some evidence that, where tariffs are scheduled to go to zero, the initial tariff rate is relatively low. In those cases where initial tariff levels are higher, the EU may be a relatively minor source of imports, such that the domestic market price will continue to be set by the price of imports (including the MFN tariff) from the major supplier. It is thus hard to argue that EPA tariff liberalisation is going to open the floodgates to cheap imports that will undermine ACP food security (Matthews, 2010b).

However, the EPA provisions on other border measures commit the ACP partners to disciplines that go well beyond what the WTO requires and could, potentially, damage their food security. The puzzling feature is that, in some of the individual agreements, the EU has agreed to clauses that are more favourable from the ACP point of view, but these are not replicated in the other agreements. Some observers are sceptical of the efficacy of some of the measures now barred to ACP governments as a means to improve food security (for example, World Bank, 2008). But until ACP governments themselves are convinced by the merits of these arguments, it sows the seeds of distrust for the EU to insist that it knows best.

From preferences to preference erosion

At the same time as the *scope* of preferences for African agri-food exports on the EU market has been extended, the *value* of these preferences is being eroded. Preference erosion comes in several forms.

One channel is when preferences are generalised to a larger number of countries so that African countries are not the only beneficiaries. For example, the granting of GSP+ preferences (which now increasingly cover agri-food products) to vulnerable countries which have signed up to sustainable development and good governance or the granting of duty-free quota-free access to non-African LDCs under the EBA scheme increase the degree of competition in the EU market and reduce the value of African preferences.

A second channel is through reductions in MFN tariffs resulting from WTO agreements. African countries gain from these reductions in export markets where they do not currently enjoy preferential access, or in markets where the final MFN tariffs are lower than the initial preferential tariffs. However, they suffer from increased competition from other countries on those markets (principally the EU) where they do enjoy preferences currently. The growing number of free-trade agreements which the EU has signed results in a similar process of preference erosion, even if concessions on agricultural trade are usually limited in these agreements.

A third channel is when the EU unilaterally alters the support mechanisms for market prices under the CAP, thus also affecting the price paid for imports. Support prices for some commodities, such as cereals and beef, were reduced already in the MacSharry CAP reform in 1992, with EU farmers being compensated by the introduction of direct payments. These price reductions were deepened and extended to dairy products in the Agenda 2000 reform. The Fischler reform in 2003 (the Luxembourg Agreement) initiated a further wave of price reductions. Such changes have been particularly important for the commodity protocol products of bananas, sugar and beef.

With the scheduled entry into force of EPAs at the beginning of 2008, the three commodity protocols came to an end, albeit with transitional arrangements for sugar. In September 2007, the EU announced that it was invoking its right under the Cotonou Agreement to renounce the Sugar Protocol (SP) with effect from 1 October 2009. The Banana Protocol had already been weakened in the Cotonou Agreement compared to the Lomé Conventions, and the gains to beneficiaries were further reduced by the reform of the EU's banana import regime in 2006 and the Geneva Agreement on Trade in Bananas in 2009. The Beef Protocol was much less significant with Namibia the last remaining beneficiary, and successive CAP reforms have also reduced its value to that exporter.

The termination of the SP was the most contested of the three, because its provisions guaranteed a market at a guaranteed price to the SP signatories. The ending of the guaranteed minimum price for ACP exports after 2010, together with the increased import supply of cane sugar, was expected to lead to a fall in the value of African sugar exports. In principle, low-cost producers can now compensate for lower prices by expanding exports beyond their SP quotas. However, in the period up to 2015 a 'dual safeguard' operates to limit any increase and could even force a decrease in the current export volumes from non-LDC exporters if exports from LDCs were to increase sufficiently rapidly. In practice, the volume of ACP sugar delivered to the EU has fallen in recent years, but this has been due more to unusually high world sugar prices as well as buoyant domestic demand rather than lack of available markets in the EU.

The value of banana preferences was considerably eroded in 2006 when the EU complied with the terms of the Doha Cotonou waiver and introduced a tariff-only regime. Further preference erosion will take place as part of the settlement of the WTO dispute (the 'banana wars') with Latin American exporters (the Geneva Agreement) which will see the tariff on Latin American bananas fall from euro 176/t to 114/t over a time period that depends on when and if a Doha Round Agreement is reached (*Bridges*, 2 June 2010).

The Doha Round and preference erosion

The value of African preferences for agri-food products on the EU market will be further eroded if there is a successful conclusion to the Doha Round negotiations. The mandate for these negotiations called for the 'fullest liberalisation' of trade in tropical products, but also recognised that action should be taken to guard against the erosion of long-standing preferences. The problem is that many of the products that could be considered 'tropical' are among those for which African countries have preferential access to the EU. This led to a tense stand-off between African countries (represented by the ACP group) and Latin American exporters over the products that should be classified as 'tropical' (and therefore subject to deeper and faster tariff cuts), and those that should fall into the 'preference erosion' category (subject to standard tariff reductions over a longer implementation period) (Perry, 2008; Stevens and Kennan, 2006). The draft modalities in December 2008 produced by the Chairman of the agriculture negotiating group reflected these tensions with widely different draft texts proposed for the tariff reduction modalities for tropical and preference erosion products (WTO, 2008).

The Geneva Agreement between the EU and Latin American exporters on bananas also extended to proposed modalities for both tropical and preference erosion products. The agreement covers both the definition of the two lists and the treatment of the products on each list (*Bridges*, 13 January 2010). This proposal was communicated to the WTO Director General in December 2009. Although noted by the WTO General Council and implicitly supported by the United States, reservations were expressed by India and Pakistan with regard to the proposed extent of tropical product tariff reductions and the length of the implementation period for cutting duties on sensitive products on the preference erosion list. African countries acquiesced to the agreement, although expressing unhappiness about it, but would benefit from acceptance that sugar and cut flowers would be treated as preference erosion rather than tropical products.

Preference erosion in fisheries

A similar process of preference erosion is at work in the fisheries sector, driven again by the extension of preferences to non-African suppliers and by reductions in MFN tariffs as a result of WTO negotiations. For example, an agreement between the EU and Thailand, the Philippines and Indonesia created an additional quota of 25,000 tonnes of canned tuna at a preferential tariff for these three countries following a complaint by Thailand and the Philippines against the discriminatory Cotonou preferences on canned tuna. Reductions in fisheries tariffs are being discussed in the negotiating group on 'Non-Agricultural Market Access' (NAMA) in the Doha Round. The NAMA negotiations have considered two alternative approaches. One is the 'critical mass approach', which would require that a critical mass of major fish-producing, importing and exporting countries establish a sector-specific agreement to liberalise fish trade. The other is the 'formula approach', which would require agreement on a formula to be applied to current tariff regimes, so as to reduce them over time, and which is the EU's preferred option. The Doha negotiations are now suspended, which postpones the risk that preferential access to EU markets accorded to fish and fishery products from African counties will gradually by eroded by the general liberalisation of international fish trade.

Fisheries subsidies were also a focus for discussion for the first time in the Doha Round. The Declaration launching the Round committed Ministers to 'clarify and improve WTO disciplines on fisheries subsidies, taking into account the importance of this sector to developing countries' (World Trade Organisation, 2001). Fisheries agreements between the EU and ACP countries were a particular source of concern as many payments nominally for access rights or to aid the development of local fisheries have been construed as subsidies to the EU fleet.

The Chair of the Negotiating Group on Rules circulated his first draft text on possible new disciplines on fisheries subsidies in November 2007 (WTO, 2007). The main concern of African countries was to ensure that fisheries access fees should be exempted from any new disciplines on fisheries subsidies. The draft text explicitly states that access fees shall not be deemed to be subsidies within the meaning of that agreement, subject to certain conditions. The text faced considerable criticism, including from African countries which sought greater flexibility for the strategic use of subsidies to attract domestic and foreign investors and to support the fishing industry in times of need. However, the Chair had not circulated a revised text by the time the Doha Round negotiations broke down in July 2008.

The Africa Group restated its position in September 2009 in a paper which reiterated that any new rules should explicitly not include fishing access agreements within the definition of subsidies.[8] However, the paper drew attention to concerns over the treatment of 'secondary' transactions where a government having purchased access rights transfers these to its private fishing fleet for less than the full amount it paid to the coastal or island government, which could be construed as a subsidy to the private fishing fleet. The Chair's November 2007 draft text includes a carve-out for the onward transfer of the access rights from a distant water fishing nation to its private fleet provided that the fishery in question is within the range of the EEZ of a developing country. In the context of the imminent revision of the Common Fisheries Policy, a key question is whether EU fishing vessels should cover all the costs of their fishing activities in third country waters (Agritrade, 2009).

Adjustment assistance

The fact that the benefits of EU preference schemes are heavily concentrated on a small number of countries and commodities makes it relatively easy to identify the losers from preference erosion and, if desired, to provide financial assistance for adjustment and amelioration of adjustment costs. The EU has provided special aid packages for ACP sugar and banana exporters to help them to cope with adjustment to lower preferences, from which African countries have also benefited. The Accompanying Measures for Sugar Protocol countries introduced as part of the 2006 sugar reform support the necessary adjustment processes in 18 ACP sugar producing countries. An indicative budget of 1.28 billion euro was provided for the period 2006–13. The EU provided over 450 million euro in the period 1994–2008 (not all funds have been disbursed and activities will continue for another three years) to ACP banana exporting countries in order to diversify away from banana exports or to improve competitiveness. A further 190 million euro support package for ACP banana exporters was agreed in 2009 as part of the Geneva Agreement on Trade in Bananas. In addition, the EU–Africa Partnership on Cotton has mobilised more than 300 million euro in support to cotton from EU member states and European Development Fund (EDF) resources. This adjustment assistance was criticised in the past for using financial resources to improve competitiveness in some countries where the sector had little or no potential to be competitive, and for long delays in fund transfers from Brussels (Hubbard *et al.*, 2000). However, it appears that lessons have been learned and that more recent experiences with adjustment assistance have been more positive.

Conclusion

The EU has modified its agricultural and fisheries policies during the past decade in ways that have brought important benefits to Africa and particularly African farmers. EU agricultural policy reforms have lowered the effective support price (including coupled direct payments) paid to European farmers, particularly for crops in direct competition with Africa's exports such as cotton, tobacco and sugar. Support prices for other CAP products, including cereals, meats and dairy products, have also been reduced. While this has put upward pressure on the price paid by African consumers, it has reduced the degree of competition with African farmers.[9] The EU use of export subsidies has been greatly reduced. Market access has been significantly improved, starting with the TCDA with South Africa in 1999, then with the grant of duty-free quota-free access to all African least developed countries under the EBA in 2001 (with transition periods for the three most sensitive products rice, bananas and sugar which have now ended) and finally with the offer of similar duty-free quota-free access to all African economies prepared to sign an EPA. This unrestricted access to the EU market, combined with continued high EU protection of agri-food products against third country exporters, creates a valuable preferential rent for African exporters to the EU market which must be set against the price-depressing effect of greater competition from EU supplies, or supplies displaced from the EU market, on domestic and third country markets. The rent from leasing access rights for fisheries to the EU fleet also provides a valuable source of income for some coastal and island states.

Incoherencies undoubtedly remain. The EU continues to resort to export subsidies, and reserves the right to do so even in the case of countries which have signed EPAs. Some product-specific support linked to production and/or prices still exists, albeit at reduced levels, for competing products such as rice and cotton. The very size of the decoupled payments made to EU farmers (accounting on average for around 40 per cent of EU net farm income) is criticised for stimulating EU production at the expense of competitive imports, although their main impact is more likely on the structure of EU production rather than its overall level. The TDCA with South Africa, which is by far the region's most competitive exporter, excludes important agri-food products from the EU market access offer. South Africa is eligible to negotiate EPA access as part of the SADC grouping, but so far has balked at signing an interim EPA because it believes the cost in terms of policy autonomy foregone in other areas would be too high. Rules of origin in EPAs, particularly in the case of fisheries, remain almost as restrictive as under Cotonou (Campling, 2008). There is scepticism whether the Fisheries Partnership Agreements in fact live up to the management and conservation objectives they have set.

However, the main African concern is that some of what the EU has given in the form of improved preferences has been taken away by other policy changes which have eroded the value of these preferences, most notably the disappearance of the commodity protocols. Another concern is that the quid for pro for the extension of market access has been to concede reciprocity in the form of EPAs, with fears that this would undermine their food security. Detailed analysis of the tariff liberalisation schedules of African EPA states suggests that these fears are overstated, although some of the other EPA provisions appear to go beyond what African countries have accepted as part of WTO disciplines. African ACP countries have been slow to ratify full EPAs. The European Commission has now threatened to end its duty-free market access offer to those countries which have not completed negotiations and either signed or ratified EPAs by 1 January 2014, although the view of the European Parliament is that this deadline should be extended by two years to give the negotiations a reasonable chance of success.

Commodity protocols were an example of trade-related aid. Because the quantities benefiting were limited by quotas, the main gain to African exporters was the additional rents they earned on protocol exports. The Protocols helped to sustain high cost production which would not otherwise have been viable. However, the existence of the Protocols, particularly the Sugar Protocol with its guaranteed market at very favourable prices, may have helped to induce some of the high production costs in the first place. While the protocols might have been successful in maintaining exports, they provided little incentive to improve the competitiveness of African production. Thus, from a dynamic perspective, some of the short-term gains may have had a longer-term cost. The implicit aid transfer was also generated at a very high cost. The value of the protocols depended on continued high protection for these commodities on the EU market, which imposed a significant cost on EU consumers and on other developing country exporters. The commodity protocols were a relic of a period of dependent development and had outlived their usefulness. EPAs provide some measure of preference but in a more market-oriented way which should help to avoid some of the worst side effects of the protocols (Matthews, 2009). The more general concern with the erosion of African preferences also needs to be placed in context. Preferences are useful when they provide some temporary protection to enable an export industry to gain a foothold in world markets; they are not desirable as a permanent crutch to uncompetitive sectors. The EU has accepted the need to provide adjustment assistance where preferences are being phased out even if African negotiators will always argue that more money should be made available.

On balance, the EU appears to have decisively improved the coherence of its agricultural and fisheries policies with African development. However, Africa continues to face a massive challenge to reduce the extent of undernutrition and hunger, and to reverse its growing dependence on food imports. As a result of initiatives taken over the past decade, African exporters now have broadly tariff-free access to the EU market for unlimited volumes of agri-food exports. Exporting food products requires that exporters meet a wide variety of requirements (for example, with respect to food safety) set by public authorities and, increasingly, private supply chains. Many argue that these non-tariff measures have substituted for tariffs as barriers to increasing developing country exports. However, unless these measures are implemented in a deliberately discriminatory way, they are now part of the general costs of doing business in export markets. They underline the challenge to ensure that Africa can take advantage of its favourable market access by overcoming supply-side constraints and facilitating the integration of its predominantly smallholder agriculture into global supply chains.

This requires, primarily, that African governments implement their commitments under the agreed African Union Comprehensive African Agricultural Development Programme (CAADP), including increasing expenditure on agricultural development to 10 per cent of their budgets. However, the EU also has a responsibility under the JAES to provide adequate complementary resources. Recently, the EU mobilised 1 billion euro through the new EU Food Facility, of which approximately 560 million euro is destined to Africa, as well as committing other funds to support food security and agricultural development initiatives. However, it will be important that the EU and member states reverse the decline in the share of their aid disbursements allocated to African agriculture if the expectations created by the new market access opportunities are to be fulfilled.

Acknowledgement

This chapter draws on research conducted under the Framework Agreement on Policy Coherence project funded by the Advisory Board for Irish Aid whose financial support is gratefully acknowledged. The paper is also an output of the 'New Issues in Agricultural, Food and Bio-energy Trade (AGFOODTRADE)', Small and Medium-scale Focused Research Project, Grant Agreement no. 212036, funded by the

European Commission. The opinions expressed in the chapter are those of the author and should not be taken to represent the views of either Irish Aid or the European Commission. Helpful research assistance was provided by Patrick Wustmann.

Notes

1 Luc Magloire Mbarga Atangana, Cameroon's Trade Minister, said this while he was lobbying the European Parliament for additional adjustment aid for ACP banana producers, 29 April 2010 (*Europolitics*, 29 April 2010).

2 Data for the trade shares quoted in this introduction are from the BACI database (Gaulier and Zignago, 2010). The BACI database uses COMTRADE as its unique source of information. Since countries report both their imports and their exports, the same trade flow is reported twice. BACI takes advantage of the double information on each trade flow to fill out the matrix of bilateral world trade providing a unique value (and quantity) for each flow reported at least by one of the partners. The cost of the improved trade coverage in BACI is that there is a greater delay in making the figures available. The 2007 data quoted in this chapter are the latest available at the time of writing.

3 Throughout this chapter we use Africa to stand for sub-Saharan Africa. While the relationship between the EU and North Africa shares many of the characteristics described in this chapter, it developed within a different set of institutional arrangements (the Euro-Mediterranean Agreements) and there is not space to provide an adequate treatment of these differences here.

4 See graph of CAP expenditure over time on the Directorate General for Agriculture and Rural Development website, available at: http://ec.europa.eu/agriculture/cap-post-2013/graphs/graph2_en.pdf (accessed 1 December 2010).

5 The value added criteria to qualify for preferences are 30 per cent in the Everything But Arms (EBA) scheme and 45 per cent in the Generalised System of Preferences (GSP), whereas beneficiary countries argue that a more realistic target is 10–15 per cent.

6 Interim EPAs only cover trade in goods, while full EPAs are intended to also cover other issues such as services, public procurement, intellectual property, competition and investment rules.

7 All non-LDC African countries except Nigeria, Gabon and Congo signed interim EPAs by the end of 2007.

8 African Group paper on fisheries subsidies, Experts Group Meeting on enhancing Africa's participation in the WTO negotiation process, UNECA, 7–8 September 2009. http://www.uneca.org/atpc/egm0909/AfricanGrouppaperon%20FisheriesSubsidies.pdf

9 We should note that production incentive effect of high EU milk prices was held in check by a milk quota system which is scheduled for elimination in 2015. Simulations indicate that EU milk production will increase as a result, even at the lower support prices. Similarly, the production incentive effect

of high sugar prices is also abated by quotas. The EU Commission has highlighted the need to bring an end to sugar quotas in a non-disruptive way in its November 2010 Communication on the post-2013 CAP (Matthews, 2010a).

References

Agritrade (2009) 'WTO and Fisheries Subsidies on the African Group's Agenda', available at: http://agritrade.cta.int/en/content/view/full/4823 (accessed 30 June 2012).

Anderson, K. and Masters, W. (eds) (2009) *Distortions to Agricultural Incentives in Africa*, Washington, DC: World Bank.

Bouët, A., Decreux, Y., Fontagne L., Jean, S. and Laborde Debucquet, D. (2004) 'A Consistent, Ad-valorem Equivalent Measure of Applied Protection Across the World: The MAcMap-HS6 Database', Working Paper 2004–22, Centre d'études prospectives et d'informations internationals, Paris.

Bretherton, C. and Vogler, J. (2008) 'The European Union as a Sustainable Development Actor: The Case of External Fisheries Policy', *Journal of European Integration*, 30 (3): 401–17.

Bureau, J., Chakir, R. and Gallezot, J. (2007) 'The Utilisation of Trade Preferences for Developing Countries in the Agri-Food Sector', *Journal of Agricultural Economics*, 58 (2): 175–98.

Campling, L. (2008) 'Fisheries Aspects of ACP–EU Interim Economic Partnership Agreements: Trade and Sustainable Development Implications', Issue Paper No. 6, International Centre for Trade and Sustainable Development, Geneva.

Carbone, M. (2008) 'Mission Impossible: The European Union and Policy Coherence for Development', *Journal of European Integration*, 30 (3): 323–42.

Carbone, M. (2009) 'Beyond Purely Commercial Interests: The EU's Fisheries Policy and Sustainable Development in Africa', in G. Faber and J. Orbie (eds), *Beyond Market Access for Economic Development: EU–Africa Relations in Transition*, London: Routledge.

Council of the European Union (2004) 'Council Decision of 26 April 2004 Concerning the Conclusion of the Trade, Development and Cooperation Agreement between the European Community and Its Member States, on the One Part, and the Republic of South Africa, on the Other Part (2004/441/EC)', *Official Journal of the European Union*, L 127/109, 27 April.

European Union (2000), 'Partnership Agreement between the Members of the African, Caribbean and Pacific Group of States of the One Part, and the European Community and its Member States, of the Other Part', signed in Cotonou on 23 June 2000, *Official Journal of the European Communities*, L 317/3, 15 December.

Gaulier, G. and Zignago, S. (2010) 'BACI: International Trade Database at the Product-Level, the 1994–2007 Version', CEPII Working Paper No. 2010–23, CEPII, Paris.

Hubbard, M., Herbert, A. and Roumain de la Touche, Y. (2000) *Evaluation of*

EU Assistance to ACP Banana Producers, Final Report, Denmark: Nordic Consulting Group.

Matthews, A. (2009) 'EPAs and the Demise of the Commodity Protocols', in G. Faber and J. Orbie (eds), *Beyond Market Access for Economic Development: EU–Africa Relations in Transition*, London: Routledge.

Matthews, A. (2010a) *How Might the EU's Common Agricultural Policy Affect Trade and Development After 2013? An Analysis of the European Commission's November 2010 Communication*, Geneva: International Centre for Trade and Sustainable Development.

Matthews, A. (2010b) 'Economic Partnership Agreements and Food Security', in O. Morrissey (ed.), *Assessing Prospective Trade Policy: Methods Applied to EU–ACP Economic Partnership Agreements*, London: Routledge.

Milner, C., Zgovu, E. and Morrissey, O. (2010) *Policy Responses to Trade Preference Erosion: Options for Developing Countries*, London: Commonwealth Secretariat.

Perez, R. and Jallab, M. (2009) 'Preference Erosion and Market Access Liberalization: The African Dilemma in Multilateral Negotiations on Agriculture', *Review of World Economics*, 145 (2): 277–92.

Perry, S. (2008) 'Tropical and Diversification Products: Strategic Options for Developing Countries', Issue Paper No. 11, International Centre for Trade and Sustainable Development, Geneva.

Ponte, S., Raakjær, J. and Campling, L. (2007) 'Swimming Upstream: Market Access for African Fish Exports in the Context of WTO and EU Negotiations and Regulation', *Development Policy Review*, 25 (1): 113–38.

Stevens, C. (2003) 'Food Trade and Food Policy in Sub-Saharan Africa: Old Myths and New Challenges', *Development Policy Review*, 21 (5–6): 669–81.

Stevens, C. and Kennan, J. (2006) *Tropical Products under Doha: Balancing Liberalisation and the Avoidance of Preference Erosion*, Geneva: International Centre for Trade and Sustainable Development.

World Bank (2008) *World Development Report 2008: Agriculture for Development*, Washington, DC: World Bank.

World Trade Organisation (WTO) (2001) *Doha Ministerial Declaration*, 14 November.

World Trade Organisation (WTO) (2007) 'Chairman's Introduction to the Draft NAMA Modalities', JOB(07)/126, Geneva.

World Trade Organisation (WTO) (2008) 'Revised Draft Modalities for Agriculture', TN/AG/W/4/Rev.4, Geneva.

11

Out of Africa: the energy–development nexus

Amelia Hadfield

Energy security is something of a policy newcomer. It emerged as a tacit component of EU foreign policy in 2000, and a strategic issue after a series of gas stoppages between Gazprom and European markets that have occurred periodically since 2004. A composite term, energy security, associates national security needs with the use of various natural resources for the consumption of energy. Consequently, energy powers contemporary states; underwrites society; constitutes markets, trade and investment patterns; and provides a standard of living that is drastically different for individuals, based on their physical access to the available energy supply. Energy security policies operate at public, national and transnational levels, and involve public and private actors, from small companies to the interests of the entire EU. Given the EU's current dependence on various forms of imported gas and oil, energy security policy now dominates contemporary agendas of both EU member states and the EU itself, touching on issues of competition, trade and the security and reliability of key exporters, all of which effectively counts as foreign policy. EU energy security policy, however, remains an unclear dimension of EU foreign policy. While the EU's dependence on imported oil and gas has highlighted the role of established exporters and new producers in North and sub-Saharan Africa, its energy security policy continues to be defined along a disconcertingly wide spectrum, ranging from the high politics of tough security stances with Russia to soft power tools of building state capacity in developing states as part of its EU–Africa Strategic Partnership.

Having existed as an original feature of Community external affairs since 1957, development policy has a far more venerable legacy. Development has also enjoyed a relatively clear ethos of preferential trade and aid-based assistance, despite changes in the emphasis of poverty-reduction and security since the 2005 amendments to the Cotonou Partnership Agreement (European Union, 2000), the 2005 European Consensus on Development (European Union, 2006) and the

2011 European Commission Communication entitled *Increasing the Impact of EU Development Policy: an Agenda for Change* (European Commission, 2011a). Development policy generally aims at reducing and eventually eliminating poverty in those countries that cannot provide a living standard above the poverty level for the majority of their citizens. Development policy has a relatively clearly defined ethos of political reform driven by economic assistance, which operates alongside social and security issues that have become increasingly visible in European development policy since 2005.

What, therefore, do energy security and development have in common? More importantly, can an 'energy-development nexus' of synergistic policies be said to exist in the EU's dealings with sub-Saharan Africa?[1] This chapter suggests that energy sectors in developing countries, and development policy in general, have two key features in common, but possibly in opposition to each other:

- Capacity building: endowing a state with the infrastructure, trading power and investment pull to increase its ability to provide public goods to its citizens, tackle poverty reduction based on increased amounts of export-based revenue. Development policy effectively lays out a methodology by which to accomplish such progress; the export energy sector (oil, gas, minerals) is for most states the most effective sector by which to kick-start that process.

- Sustainability: Development policy strives to permanently transform states and their various governmental modes from a condition of underpowered dependence to one of regionally integrated inter-dependence featuring good governance and economic expansion. Fossil fuel exports are ultimately non-sustainable commodities, and within the broad ethos of climate change, energy security as a 'green commodity' focused on enhancing the provision of public services such as electricity via sustainable methods of solar and wind power, nuclear power in some cases, and biofuels that neither exacerbate food shortages (ethanol crops) nor produce additional carbon (wood).

The dilemma is clear. Developing states in Africa are under pressure from major development funders like the EU to use their national resources such as oil, gas (natural and unconventional), metals such as copper, or minerals such as uranium and convert them to a live and lucrative export industry. On the basis of fossil fuel exploitation, more than a third of African states could reap the economic benefits of some aspect of energy export. States who successfully join the exporter band-wagon ultimately expand the number of gas and oil exporters from which the EU can choose, thus lessening its dependence upon Russian and Gulf suppliers and enhancing its overall energy security.

However petro-states, or at least the 'resource curse' of heavy dependence upon a single commodity, twinned with poor governance that enables corruption, is a serious threat. Weak governance standards are the norm, not strong state capacity. Equally, encouraging use, and even reliance on the use of hydrocarbons, flies in the face of the EU's hard-won international environmental credibility and its own commitment to become carbon-neutral. Promoting sustainable use of renewable and alternative energy resources fits more credibly within the EU's overall energy security/climate change ethos, and since 2009 has been a major pillar of the EU development ethos. In material terms however, a 'green energy Africa', rather than a 'hydrocarbon based Africa', will certainly worsen its own energy security prospects, at least in the short and medium term.

And yet, the EU appears dedicated to finding policies that work towards positive returns in both energy and development. Three areas of overlap, and one major implication arise from such an ambition. Firstly, energy and development are both market norms. Development is a neo-liberal norm based on the ethics of assistance to the developing world driven by patron-led trade, aid and investment. Energy operates within the EU as a neo-liberal market, based on the ethics of competition and deregulation and driven by a few major exporters, to which African states could feasibly be added. Secondly, both energy and development are potentially sustainable in the long run. Thirdly, they both possess implications of societal stability, and in some cases outright state security. The implication is the need for a convincingly clear consensus on what precisely an 'energy-development policy nexus' comprises. While aspects of the three commonalities are an increasingly regular feature in EU policies, the establishment of a full-fledged nexus is still uneven and unpersuasive.

This chapter contributes to the burgeoning interest in the area by, firstly, assessing the role of energy in Africa (subsistence, climate change and export potential); secondly, by looking at the various steps that the EU has taken in the past decade to construct an EU–Africa partnership, then looking in detail at the extant energy–development policy nexus within EU–ACP (African, Caribbean and Pacific) development strategies such as the Joint Africa–EU Strategy (JAES), and the role of Economic Partnership Agreements (EPAs). The chapter concludes by considering the overall impact that development and energy security have upon the coherence and consistency of EU foreign policy. In broad terms, are energy policies fused holistically and logically to development objectives for sub-Saharan Africa, or are they annexed hastily in an *ad hoc* effort to generate policy consistency? At the heart of the following analysis therefore is a broad, value-added question: namely, what if any sustainable benefits have arisen for the EU and ACP states alike from the addition of energy to the EU–Africa Strategic Partnership? And, equally, how effectively has the energy–development nexus contributed to the coherence and consistency of overall EU foreign policy?

Competences and policy implications

According to Articles 3–6 of the Treaty of Lisbon, energy, development and foreign policy operate according to different levels of competence. In its market aspects, energy is a core shared competence (like the internal market), where the EU implements market mechanisms to encourage liberal trade and enhanced competition. However, EU member states still retain exclusive sovereign competence in the most key aspect of energy security, that is, overall choice about the type, the supplier and contractual obligations relating to the energy resources used within their national market. This makes for difficult EU–Member States' relations regarding exporter choice, and is also no guarantee of a completed liberalised gas or electricity market within the EU. Development is a parallel competence, meaning that the EU and its MS have traditionally undertaken development objectives together, but also separately. A variety of recent EU documents however (e.g. the *Agenda for Change*, European Commission 2011a) have put much stress on the need for streamlining EU development efforts, preventing possibly duplication. Lastly, EU foreign policy is the outcome of inter-governmental Council agreements that define and implement a common foreign and security policy where needed, but limited firmly to the provisions of the Treaty on European Union.

In terms of constructing synergies between energy security and development, it is clear that the national interests of the EU member states are never far away, for neither policy has been deemed to fall within the purview of an exclusive competence. This has not however prevented the European Commission in particular from keeping the upper hand in managing a vast range of development programmes that it has traditionally enjoyed, or taking on – through treaty or technocracy – a widening range of energy security responsibilities, particularly those that touch on climate change obligations.

With inherent third-party implications vividly displayed in both energy security and development in the EU, the question therefore is twofold. Firstly, to what extent do both policies operate as a form of EU foreign

policy? EU development policy has allowed the EU to move from a pur-
veyor of market-based values to a template of political norms including
democracy, rule of law, human rights and myriad forms of good governance simply by tying such values and norms directly into the nature of
assistance on offer. Equally, energy security as conceptualised by the EU
is a neo-liberal market norm that encourages competitive and integrated
markets at the macro-economic level, requires dialogue with designated
strategic partner-suppliers at the political level and fosters a variety of
sustainable practices at the social level, from capacity building to public–
private partnerships. The EU has yet to concretise this into a working template however, firstly because of the sheer volume of issues that fall within
energy security (including climate change), secondly because of the very
different views held by member states and even inter-institutional factionalism and thirdly because the target audience is generally far less homog-
enous in their energy profiles than they are in their development needs.

Secondly, and relatedly, what precisely is the 'synergistic message'
that is being sent by the EU to key sub-Saharan African states in the con-
scious twinning of these two policy areas? The obvious message is that
development and energy security are natural partners in the fight against
poverty reduction and that a temporary use of fossil fuels – if sustainably
managed – is a win-win situation. The tougher message is that the two are
good partners in some sectors, for some states only, and that sequencing
is key. Basic state capacity and eradicating poverty and corruption are a
prerequisite to promises of export market construction, refining capacity and supply chain construction. The danger is that neither side finds
its way out of the 'state–capacity–sustainability' paradox, with African
states being regarded solely as new energy partners, while development
objectives slowly give way to EU interests found in energy policy. The
positives are that both sides make sustainable use of the market aspects
of trade and investment inherent in both development and energy policy,
establishing a series of harmonious – rather than harmonised – norms.

The role of energy in Africa

Itemising the role of energy in Africa is an exhaustive process, and
one best managed via an abbreviated literature review that highlights
writings on the issues that make up this varied topic.

Resources and resource wars

The material substance of energy products as naturally occurring
national resources is the baseline for exploring energy in Africa. Divided

into hydrocarbons (oil and gas), precious metals (gold, copper, tin) and minerals (uranium), African geography is unusually bountiful; a cruel irony considering its near-universal subsistence scarcity (Klare, 2002). The vast majority of all three occur in West Africa, though there are significant outcroppings in other states. Producing regions do have a history of conflict, but not always for obvious reasons (Hiscock, 2012). Ancestral rites and expectations frequently contribute to symbolic reasons for strife, alongside spats over land ownership and use. The Democratic Republic of Congo (DRC) is the bloodiest recent example of resource wars, whilst Ghana appears to be a model case for resource-driven good governance (Shah, 2010). The DRC bloodshed has cost 5.4 million lives and an additional million displaced people. The nine-year UN presence has failed to prevent a variety of rebel forces from terrorising DRC citizens, themselves funded by tin, tantalum and tungsten.

Elsewhere, violence in the Gulf of Guinea, chiefly Nigeria, is directly related to the rich onshore and offshore oil deposits that have been regularly retrieved and refined by a variety of European energy companies. Royal Dutch Shell in particular stands accused of contributing to tensions amongst the minority ethnic groups inhabiting the Niger Delta (primarily the Ogoni and Ijaw) by worsening the economic and health consequences for the people in the region through slow reactions to oil spills and allegations of human rights violations from 1990 until 2007 (Human Rights Watch, 1999). Whilst Nigeria is nominally democratic (an outcome of the Obasanjo government), competition for oil wealth has visibly fuelled violence between various ethnic groupings, which in turn has prompted the militarisation of the region by ethnic militia groups, the Nigerian police, the military and a range of assorted environmental activists (such as MEND or the Movement for the Emancipation of the Niger Delta). As 90 per cent of Nigeria's GDP is derived from its hydrocarbon wealth, the one-third cut in Nigerian oil production caused by this regional resource war now threatens to destabilise the state itself (Ojakorotu and Gilbert, 2010). Ghana however, with lucrative gold and offshore oil holdings, exemplifies a state steeped in norms of democracy, good enough to qualify in 2009 for US President Obama's first visit to sub-Saharan Africa, indicating that the resource curse can be successfully thwarted with the right combination of state tools and political will.

Energy actors: suppliers and buyers[2]

Until the Arab Spring of 2011, Libya and Algeria were the key North African oil export actors, and Nigeria and Angola, as the leading sub-Saharan African suppliers, produced between 2.62 per cent and 2.31 per cent of world oil production. Libyan exports have dropped by more than half during 2012, but forecasts are reasonably buoyant about a 25 per cent upswing by 2014. In terms of natural gas, Algeria and Egypt continue to dominate the Maghreb, whilst Nigeria, Cameroon, Congo-Brazzaville and now Angola hold sway in the south. A range of established and newcomer exporters, these meet the demands of European and North American markets using a combination of established gas (export) pipelines, liquefied natural gas trains and ports and crude oil tankers. In 2010, European markets imported roughly 15 per cent of its oil and gas from Africa, making Algeria the third largest exporter of gas to the EU after Russia and Norway.

The Gulf of Guinea oil producers in West Africa, which includes established exporters of Nigeria and Angola, are particularly key for the EU. These two states, along with smaller and more recent exporters of Gabon, Equatorial Guinea, Sudan and Cameroon also feed the North American, European and now Asian markets. West Africa's exports were once almost exclusively oil, and the dominant role played by Nigerian exports thus renders it 'an important partner in the EU's diversification efforts' (Tarradellas, 2008). However, the West African gas industry has advanced in recent years, though key projects have underperformed, including the West African Gas Pipeline, with a history of sub-capacity operation, despite an abundance of regional gas.

Between North and West Africa is the Sahel region, comprising Niger (key for uranium), Mauritania and Mali (transit routes for pipelines). While East Africa does not produce oil or gas in significant quantities, the region has been identified as a productive future region, particularly Mozambique, Namibia and Tanzania that possess significant natural gas reserves. Such finds 'provide a welcome opportunity to diversify and include gas-fired power generation' in an area with very little current output; but the real question is the extent to which 'East African gas be commercialised for power production' (Pago, 2011).

In terms of use, a key question is why Africa, north and sub-Saharan south alike, is so tilted towards exports rather than internal use of its own energy assets As argued by Enfield (2009: 158):

> lower levels of economic growth and industrialization hold back energy demand, and energy infrastructure remains underdeveloped in most parts of the continent. Moreover the preference of many oil-producing governments has been to secure immediate revenue flows through exports, rather than to utilize the domestic hydrocarbon as part of a strategic industrialization policy.

The outcome contributes to the problem at the heart of the energy–security nexus: namely, the tension between focused, capacity building or longer term prospects for sustainability. Securing immediate revenue flows without the infrastructure and governance structures to contain them sees vanishing revenue either through poorly negotiated export contracts or domestic corruption. This virtually guarantees that petro-states remain dependent on development aid when they could and should be self-sustaining. Equally however, a lack of commitment to a post-fossil fuels future means that, as far as energy is concerned, Africa is very clearly open for business.

Due to its readiness to work with foreign companies (whether national or international), Africa's foreign investment climate is generally conducive to present and future initiatives. This contrasts with frostier attitudes to foreign direct investment and a more pragmatic view of energy security found especially in Russia and some areas of the Middle East. Because most African producers are not yet consumers, this has two key effects, both with major implications for relations with Europe. Firstly, an unusual producer–consumer relationship in which a generally wider range of access and options to higher quality oil is available for European energy companies regarding onshore and offshore extraction, with little or no energy nationalism featuring in the form of burdensome or unanticipated obstacles, and negligible domestic consumption requirements. Secondly, and as a corollary of this, an unorthodox understanding of energy security in which security entails for sub-Saharan African states simply security of demand for its own upstream (explorative and exploitative) energy industry.

For North African states, security of demand is a little different in that there are some features of nationalisation in the energy sectors of some states, and less of an openness about unlimited upstream exploration. Taken together, however, the continent as a whole does not yet place any major emphasis upon security of supply of its own (or imported) energy resources. It has virtually none of the dependence anxieties, not only because its own natural resources have not been industrialised into a mature energy market or infrastructure, but also because of the overwhelming concentration on biomass as a source of energy, rather than the mainstreaming of gas or oil. Is this good news for Europe?

Certainly such an investment climate and the offshore reserves still to be tapped in the Gulf of Guinea alone promise sound yields. The EU may not in fact want African states to move in the traditional direction of increased reliance on its indigenous hydrocarbons. Not only will this gradually reduce the production amount on offer for key European export markets, but it works against the entire sustainable philosophy of the EU itself and its relations with third parties. Energy efficiency is a key requirement and can usefully be applied to wasteful upstream and downstream practices found in some states (and thus the responsibility of both governments and private investors), but this does not solve the wider problem of how to modernise developing African states without doing harm to the environment, undermining the investment potential of its current markets, and inadvertently increasing the security of supply dilemma faced by Europe.

Energy and climate change

As indicated above, energy security has serious consequences for climate change. For developing states, with a premium on growth, accessing the natural resources of one's state not only makes economic and social sense. However, with South African trends indicating medium-term developments (Wolde-Rufael, 2005; Menyah and Wolde-Rufael, 2010) economic growth that directly fuels carbon emissions augurs potential difficulties for African states, and highlights a key area for the EU energy–development policy nexus. In this area, the EU has perhaps played a role equal to its development status, in highlighting the long-term view and a variety of practical requirements that could be put in place within wide-ranging development instruments to kickstart more sustainable attitudes.

Energy and development

This area focuses upon the burgeoning role played by energy policies in broad development programmes, as propounded by donors such as the EU, the USA and the World Bank. In addition, the most notorious 'bottlenecks to development in Africa', including cultural transitions and chronic poverty, are described by leading African commentators (Maathai, 2010: 47), highlighting the individual requirements of states and actors in terms of accountability and responsibility.[3]

The first observation is the assumption that energy products are necessary public provisions and beneficial commodities. As Amirat and Bouri (n.d.) argue, the connection between energy and economic growth is established on a variety of factors, including 'the causal relationship between the per capita energy consumption and the per capita GDP' generated from annual data, usually including capital and labor variables to chart 'the energy-growth nexus'. Interestingly however, petro-states are the exception rather than the rule. In a study on Algeria for instance, analysis revealed 'that energy was no more than a minor contributing factor to output growth and certainly not the most important one when compared to capital and labor' (Amirat and Bouri, n.d.). Stemming from these and other studies is an enduring reality, namely that states in need of development aid need to prioritise capital and labour, critical state infrastructure and state capacity before investing in an extractive industry like oil or gas.

The main observation is that chronic state capacity problems hamper African states from both providing basic energy security to their own citizens, as a form of public good such as electricity or water, and enhancing their structural ability to make use – directly or indirectly – of energy resources to enhance their welfare. Not only do the majority of African states fall below global levels of economic and political development, ranking seventh out of seven categories of the UN Human Development Index (United Nation Development Programme, 2011), they do so despite prolific regional energy assets and staggering revenues. Nigeria and Angola, along with Sudan, represent African examples of the 'resource curse', the phenomenon recognised by economist Jeffrey Sachs, among others, in which some countries rich in natural resources operate with far slower economic growth than countries without such resources (Humphreys *et al.*, 2007). The reasons are many, and their relationship complex, but the problem is generally a distributive one. In the more traditional scenario, revenues simply do not get reinvested back into the state to provide basic social services either because of a lack of institutional or governmental transparency and accountability, or explicit corruption, or outbreaks of violence.

The knock-on effect for governance is substantial: democracy is weakened, institutional capacity and human resource development is undermined or non-existent and there are increased levels of conflict. This profile typifies much of the Gulf of Guinea (International Crisis Group, 2006). The other scenario is equally devastating, but for different reasons; states like Gabon and Equatorial Guinea with good reinvestment strategies are still heavily dependent upon the revenue generated by oil sales; price drops like those during the 1980s and 1990s force such states to borrow heavily despite a well-functioning energy sector, producing crippling levels of debt. Even Africa's considerable investment and market openness in the energy sector does not prevent external competition for its resources. Europe, and indeed the EU itself, has been decidedly skittish with the appearance of Russian, Chinese and Indian buyers on the African scene. Thus, as Dannreuther (2007) points out, 'it is this mix of geopolitical ambition among the major external powers and the internal instability of many oil-producing states which makes oil a significant contribution to conflict and global insecurity'.

One could point out that chronic poverty is the vehicle that most aids and abets such insecurities in Africa.

Are there ways out? Development policy can play a part, witnessed by American efforts at institutionalising reforms such as the Extractive Industries Transparency Disclosure Act or state-specific conflict acts which require extractive companies listed on the Securities and Exchange Commission (SEC) to disclose further information in their financial reporting, ensuring that the sale of extractive products do not support the conflict. As argued by McHaney and Veit (2009), '[t]he Extractive Industries Transparency Disclosure Act would help ensure that some of our close allies in the continent – Africa's new petro-states – use their oil riches in ways that promote development, not civil unrest and conflict'.

Constructing the energy–development nexus: key stages

To commence this section, a brief review of the changes in EU development policy is required. The EU's 50-year commitment to poverty reduction has undergone a threefold reorientation. Firstly, WTO rulings regarding the phasing out of preferential trade and a shift in global aid consensus has forced the EU to rework the very heart of its development philosophy. Preferential trade will come to a gradual end for all ACP states by 2020, and aid will continue to operate on a conditional basis. The conditions themselves however have now been subjected to an overhaul, chiefly by becoming more securitised. Secondly, whilst ostensibly remaining committed to poverty reduction as part of its goal to achieve the UN Millennium Development Goals, the rise of a securitised narrative by which to view both developing and fragile states has encouraged the EU to put equal, if not greater emphasis, upon state security over and above poverty reduction. With the rise of energy imports as a key part of the EU's own security, energy security has gradually been added as a development objective in its own right. Thirdly, the rise of a 'foreign policy portfolio' that designates chief regional allies as strategic partners has pushed the EU to amalgamate the sub-Saharan component of the ACP with its Mediterranean neighbourhood of North Africa in an uneasy attempt to reclassify the continent as a whole as a strategic partner. However, whilst North African states remain firmly embedded within the ENP, ACP states still funded by the European Development Fund (EDF) and the JAES now a component of the proposed 2014–20 Development Cooperation Instrument (DCI), the EU has created untold confusion in its attempt to deal clearly with both the development and energy security needs of the continent.

Contemporary developments: 2002–2012

The critical energy security elements in developing African countries range from chronically low energy capacity for both urban and rural areas, lack of access to the sporadic energy (generally electricity) that is available and inefficient, potentially harmful, use of limited energy types which results in only low productivity of energy use across a community or region. Even before the major initiatives of 2002 onwards, the EU's largest share of energy-oriented financial commitments between 1997 and 2001 went to developing countries in South and Central Asia (25 per cent) and sub-Saharan Africa (24 per cent).

The first major step towards fusing energy security objectives within the established EU development framework was the 2002 Commission Communication on 'Energy cooperation with the developing countries', which focuses on the idea of encouraging various forms of energy cooperation with developing countries (European Commission, 2002). Here, the development goal of poverty reduction was placed alongside the chief European energy goal of security of supply, as well as environmental protection. The EU suggests two demand-side methods to ensure cheap, widely available (and frequently rural) electricity needs: energy efficiency in current use and renewable energy sources for future uses.

In addition to the basic mantra of efficiency and renewables, the EU then turned to the increasingly cross-cutting nature of energy, specifically its role in either hindering or fostering wider societal development. By far the most notable initiative is the EU Energy Initiative for Poverty Eradication and Sustainable Development (EUEI), launched at the 2002 World Summit on Sustainable Development (WSSD) (European Commission, 2003).

A joint effort of the Commission and Member States, the EUEI is a demand-side project in which the development ethos of assistance and poverty-eradication underwrite the provision of adequate, affordable, sustainable energy services to the poor. Taking as its starting point the MDG goal to lift people out of poverty and hunger, the EUEI solution is to begin with the precondition of reliable access to affordable energy used in a sustainable manner. Further reasons in which the provision of basic energy sources is concomitant with the dynamics of poverty reduction include:

- The limited access to energy services and a heavy reliance on traditional biomass (wood, agricultural residues and dung) for cooking and heating. Biomass reliance is a hallmark of poverty in developing countries (2.4 billion people as of 2003).
- Better access to a wider, more sustainable range of energy serv-

ices helps to develop businesses and income-generating activities, including homes, schools, health centres, city and rural lighting, communication, water supply, heating and cooling.

Limited access and the catalytic effect that such access can provide to communities and regions highlight the importance of energy for poverty reduction and sustainable development. The EU's role is thus to fill the gap between traditional development provisions, underpowered developing markets and the practical needs of providing basic energy services. In other words, because 'current market and aid mechanisms are unlikely to bring modern energy services to the poor in the foreseeable future', the EU's development–energy nexus is to broker key partnerships between governmental or regional public authorities, the private sector, its own EU institutions and member state contributions and the local population (European Commission, 2003). Such partnerships generally take four forms: top-down legislation and regulatory frameworks; specific energy policies and strategies; broad institutional support; and specific and technical forms of capacity building (European Commission, 2003: 6).

The EUEI thus sets out the development–energy nexus logic in five stages:

- Raise political awareness among high level decision-makers of the important role energy can play in poverty reduction.
- Clarify the need for energy services for poverty reduction and sustainable development.
- Make apparent the need for energy services in national and regional development strategies.
- Encourage the coherence and synergy of energy-related activities.
- Stimulate new resources (capital, technology, human resources) from the private sector, financial institutions, civil society and end-users (European Commission, 2003: 5).

With this strategy, the EU suggests that development as a poverty-reduction vehicle must arm itself with a variety of categories, including energy. The EUEI provides this focal point which uses both institutional and technical methods to broker partnerships which in law, regulation, capacity building or institutional linkages all foster a critical awareness of the basic need to solve energy security as defined by the developing world, namely procuring access to adequate affordable sustainable energy services in both rural and urban areas.

The EUEI is therefore the policy vehicle which most extensively first incorporated energy cooperation into the EU's engagement with ACP states, and which subsequently led to the inclusion of energy as an autonomous priority area in the European Consensus on Development. The EUEI also kick-started key foreign energy policy instruments, including the ACP–EU Energy Facility, the COOPENER Programme that focuses on energy services for poverty alleviation in developing countries and includes a number of projects in sub-Saharan Africa and the member state funded Partnership Dialogue Facility (PDF), which focuses specifically on the integration of energy policy in the development strategy for African countries, as well as the Johannesburg Renewable Energy Coalition (JREC),[4] all of which attempt to complement other EU-led energy partnerships in general, and the EUEI in particular.

The EUEI was followed in 2005 by the European Consensus on Development (European Union, 2006). In addition to the two main commitments to reduce poverty (via a focus on the MDGs), and ensuring that development spring from Europe's own democratic template of human rights, democracy and other fundamental freedoms, the Consensus emphasises again the idea that poverty eradication itself is a holistic, 'multidimensional' policy. It does so by arguing that poverty affects people in a number of ways from food security to limited public services. Consequently, the Consensus emphasises the variety of 'associated areas' still to be addressed in order to genuinely overcome poverty, including tackling HIV/Aids, chronic economic under-development and the inefficient/irresponsible use of natural resources.

The combination of consumptive practices and natural resources takes up the theme first established by the EUEI, lodging energy firmly within the EU's development strategy, as outlined in section 82 (under the heading 'Water and energy'):

> Large sectors of the population in developing countries have no access to modern energy services and rely on inefficient and costly household energy systems. Community policy therefore is focused on supporting a sound institutional and financial environment, awareness raising, capacity building, and fundraising in order to improve access to modern, affordable, sustainable, efficient, clean (including renewable) energy services through the EU Energy Initiative, and other international and national initiatives. Efforts will also be made to support technology leapfrogging in areas like energy and transport. (European Union, 2006: 13)

In Africa, energy is an absence, not a presence. It is a problem, not a catalyst. Lack of (reliable) electricity, the preponderance of women reliant on wood and coal for household heating and cooking are the main issues at hand, and targeting capacity building as a way of modernising energy supply and encouraging its efficient use appears as part of the answer. The resource wealth of sub-Saharan Africa is not of course mentioned, only passing references under the aegis of 'the sustainable management of natural resources'.

The 2005 Consensus focused on policy coherence and suggested for the first time that all Union activities should be coordinated to ensure a coherent and consistent approach in development policy. This was further strengthened by the 2007 *Report on Policy Coherence for Development* in which the EU explicitly 'seeks to build synergies between policies other than development cooperation that have a strong impact on developing countries, for the benefit of overseas development' (European Commission, 2007a).[5]

What is interesting are the four classifications afforded to the EU–Africa balance as a result of introducing energy into the relationship. As identified by the 2007 *Policy Coherence* document (ibid.):

1. 'A technical partnership' to ensure enhanced electrification by lessening Africa's traditional reliance on biomass fuels.
2. 'An energy security partnership' in which the EU effectively defines itself and Africa as both in need of 'adequate, affordable and sustainable energy', with early provisions for securing energy through EU policy vehicles like the EUEI, the Africa–EU Infrastructure Partnership and the Africa–EU Energy Partnership.
3. 'A tacit importer–exporter relationship' that connects the commitment of the 'Energy Policy for Europe to deliver sustainable, secure and competitive energy by integrating Europe's energy and development policies in a "win-win game"'. Developing countries will 'grow' their nascent energy markets thanks to EU support and become increasingly self-sufficient, while those with substantial natural resources are targeted to become mid-term suppliers of conventional fossil fuels and renewable energy to the EU.
4. 'A holistic landmark': the intense cross-cutting between environment, climate change and the developing world's access to energy services is – if successful – a testament to the ability to integrate energy initiatives into development and foreign policy.

EU imports and African exports thus not only make good market sense, but can logically fuse 'Europe's energy and development policies in a win-win game'. From 2005 onward, these ideas and ambitions began to have a marked impact on the EU's relationship with the ACP group in general, and Africa in particular. The most central policy tool in this regard was the 2005 creation of the joint EU–ACP Agreement to create the ACP–EU Energy Facility, the sole focus of which is to grapple with securing energy for Africa. In other words, the Energy Facility is designed 'to improve access to sustainable, modern and affordable energy services, including energy efficiency and renewable energy', itself catalytic to strengthening economic growth and improving the social conditions of ACP states (European Commission, 2007a: 228). The Energy Facility is actually a mechanism for facilitating solutions: bringing together public and private sector, partners, but also acting as the portal for proposals and funding (198 million euro for co-financing the first call of 80 projects).

Just as energy sources and provision require infrastructure, the Energy Facility is designed to work alongside the 2006 EU–Africa Infrastructure Partnership, which facilitates investments necessary for generation, cross-border interconnections, grid extension and rural distribution (European Commission, 2006). The Infrastructure Partnership is funded by the EU Infrastructure Trust Fund for Africa, established with the European Investment Bank (EIB) to support transport, energy, water and ICT. Its 9th EDF budget was 60 million euro, to which the member states contributed 27 million euro, releasing a further 260 million euro in loan financing from the EIB. Again, the EU template is never far away. Public–private initiatives are favoured, combining private investment in energy infrastructures with government policies whose sole focus should be the construction of the much-vaunted 'favourable regulatory environment' across African regions and the continent as a whole.

By 2006 and 2007, the EU–Africa initiatives were becoming a feature of the growing institutionalisation of EU–ACP relations, which has been a result of the Cotonou Agreement of 2000 involving civil society actors and representatives of the private sector in the political and economic dialogue between the EU and the ACP. In other words, the energy–development nexus was becoming the centrepiece of high-level summits, including the October 2006 Joint Africa–EU Ministerial Troika, which called for a comprehensive Africa–EU Energy Partnership (AEEP) but which to some extent had already been quietly established since 2002 (Council of the European Union, 2006). A public pronouncement however transformed the energy partnership from a mid-level project to a flagship initiative able to stand as an autonomous policy area. This is precisely what happened at the EU–Africa Lisbon Summit in December 2007, where the energy–development nexus effectively created a framework for bilateral continental policy dialogue on security of supply, energy access and climate change by looking at catalysts such as investment, transparency and liberalisation to promote outcomes such as robust energy markets in chief African suppliers such as Nigeria and Angola.

From late 2007, the AEEP saw the development of a long-term but fairly feasible framework for 'structured political dialogue and cooperation between Africa and the EU on energy issues of strategic importance, reflecting African and European needs' (European Commission, 2009b: 11). Africa in other words was buying what the EU had to sell in institutional and development terms; and the EU had found a ready political partner in an already well-established commercial energy partner. The biggest step forward was the eight-part 2008–10 Africa–EU Partnership Action Plan, containing the blueprint for both a Partnership on Energy, and a Partnership on Climate Change. Funded by the EDF, the energy partnership is premised on a triple focus on financial support for energy development, infrastructure building and the sustainable management of resources, and aims to overhaul 'regional infrastructures and interconnections, supporting public and private investment, and improving access to energy services and energy efficiency' (Europa, 2012). The parallel climate change partnership is a multilateral initiative, with dialogue taking place within the Global Climate Change Alliance (GCCA), encouraging African states to strengthen their cooperation to international post-Kyoto agreement at the Copenhagen Summit in late 2009.

The years from 2009 to 2012 have witnessed further attempts by the EU to replace its tendency to rhetoric with policy substance. The 2009 Report on *Policy Coherence for Development* made clear the EU's commitment to placing energy issues firmly within the orbit of development policy, as a newly synergistic addition (European Commission 2009a). The benefits were clearly enumerated in Article 3.12:

- A natural link with the 2008 EU Climate Change and Energy Package, creating new opportunities for biofuel exports ... helping countries where there is good potential to develop biofuel production.
- Energy dialogues with developing countries, via thematic programmes to support the environment and the sustainable management of natural resources, including energy.
- Boosting economic development via the double goals of eradicating poverty through better access to energy, and keeping climate change within acceptable parameters (European Commission, 2009a).

The 2009 document effectively lays the groundwork by which energy security can help eradicate poverty (at least in the form of public goods), be linked to climate change to solidify a commitment to sustainable (possibly carbon free) approaches, including biofuels, via wide-ranging thematic programmes locked in place with specific partners. These aims move towards the long-term goal of increased and sustainable prosperity for African states, but must also be reconciled with the traditional virtues of development policy. These twin strategic objectives were explicitly stated in the 2011 European Commission communication entitled *Global Europe: A New Approach to Financing EU External Action* (European Commission, 2011b: 4).

Critical observations about poor quality development methodology used by the Commission produced a rather under-powered attempt to grapple with indicators by which such broad, synergistic goals could be measured in terms of their successful implementation. Unfortunately, the *Policy Coherence for Development Work Programme 2010–2013* provides few details as to how the EU can reasonably expect to practically implement development, energy and climate synergies in developing states. It offers only one key target, to 'identify the impacts of the renewable energy directive's biofuels provisions on developing countries exporting biofuels to the EU', with indicators suggesting how to measure biofuels production and impact (European Commission, 2010: 16). There is no reference to fossil fuel use for capacity building or poverty eradication, or even any alternatives to operate similarly apart from biofuels, merely a passing comment on supporting 'improved access to green technology for developing countries'. The only objective with any substance is the suggestion that the EU reduce tariffs on low-carbon technologies 'to promote the growth of markets inside and outside Europe and increase the take-up of low carbon technology, making it easier for developing countries to embark on low-carbon development plans' (European Commission, 2010: 16–17).

EU–Africa relations and the role of energy

Development is 'the policy with the greatest potential to both absorb the values and ambitions of its donors and transmit them in the shape of political, security [and economic] objectives to its recipients' (Hadfield, 2007: 39). Nowhere is this more evident than the series of Lomé Conventions and the Cotonou Agreements (2000, 2005 and 2010). To some extent, traditional donor–beneficiary asymmetries still dominate, with the EU retaining its ability to alter its overall development programme within the general rules of the WTO, or adjusting key aspects of its trade and aid giving. Although Lomé and Cotonou attempted to keep the preferential, asymmetrical trade regime between ACP states and the EU going for as long as possible, the 2007 WTO deadline for new trade arrangements forced a sea-change. It is clear that for all but the 42 poorest states – still protected under the Generalised System of Preferences (GSP)[6] – entry into the world economy through a transition of liberalisation and free trade is inevitable. The EU's recipe is to make such a transition lengthy, regional in nature and sufficiently linked to the goals of poverty reduction to still count as development policy. In terms of energy, the EU's long-term goal may also be to steadily encourage the liberalisation of energy markets within Africa, promote the same inter-regional template for energy infrastructure, including regional interconnectors, and in the short term to seek tariff-free status on low-carbon technologies.

From the general framework provided by the Cotonou Agreement, all sub-Saharan African states enjoy both a bilateral relationship with the EU, funded by the EDF and structured on Country Strategy Papers and Mid-Term Reports, and a *regional relationship* based on the EPAs. As the previous section indicated, a range of development initiatives have been attempted to renew the Africa–EU partnership, working variously toward regional goals, energy security goals and climate change. EU–ACP/Cotonou relations has been given renewed impetus with the *2010 Renewed Africa–EU Partnership*. Based on the Commission initiative to consolidate EU–Africa relations, the proposal lays down the foundation for a synergistic policy of development and energy. With the Millennium Development Goals underwriting international development requirements, the EU template of governance, democracy, rules of law and human rights providing normative direction, the energy component is effectively subsumed under the rubric of 'combating climate change and environmental degradation, the sustainable management of natural resources, and disaster risk reduction'. At this point the two existing instruments – the ACP–EU Water Facility and the ACP–EU Energy Facility – are not yet extensive enough to provide a foundation for this synergy. Instead, there are three more likely candidates: the energy and environment theme within the Development Cooperation Instrument (DCI), EPAs and the current Africa–EU Strategic Partnership and its thematic partnerships (including energy and climate change) within the JAES.

Development Cooperation Instrument

Energy and climate change are by definition cross-cutting issues, and furthermore are issues touching on the global goods so the EU has unsurprisingly conceptualised them in broad terms applicable to a transborder world, as well as a useful bilateral benchmark. The Development Cooperation Instrument launched in 2007 contains both bilateral support for states outside the ACP and Neighbourhood framework and five thematic programmes for all developing states including ACP states, the second of which is entitled 'environment and sustainable management of natural resources including energy', or ENRTP (European Commission, 2011c). Its twin objectives are of real interest, because they suggest a more profound example of an energy–development policy nexus, although within a thematic, rather than country-specific or regional template:

- integrating environmental protection requirements into the Community's development and other external policies,
- helping promote the Community's environmental and energy policies abroad in the common interest of the Community and partner countries and regions.

Here, the EU's own template of energy and environment practices appears to predominate, possibly at the expense of a shared dialogue. More positively, improved domestic energy provisions and possible exporter revenue are cast as clear poverty-alleviation tools. ENRTP is designed to 'address the challenges having a deep effect on the lives of poor people (rapidly degrading key ecosystems, climate change, poor global environmental governance, inadequate access to and security of energy supply)' (DG Development and Cooperation, 2012). Its fifth priority is the only one that touches on energy, and is firmly tilted towards future provisions for a green Africa, by 'broadening the options as regards sustainable energy, in particular by developing a legislative and administrative framework which favours investments and businesses' (ibid.). As the DCI is the most extensive and well-funded EU development instrument, future finances for its strategic objectives (implemented via regional Annual Action Plans) are likely to remain high, which bodes well for the substantive support required in critical African states.

Economic Partnership Agreements

EPAs represent the 'new, WTO-compatible trade arrangements' launched in the first round of Cotonou in 2000 (European Commission, 2007b: 2). To be clear, for the purposes of Article XXIV of GATT, EPAs are an interim agreement to plan for the formation of a Free Trade Area.[7] EPAs divide up the former ACP grouping of 79 states into six groups: Caribbean, Pacific, SADC, East African Community (EAC), Eastern and Southern Africa (ESA) and West Africa (ECOWAS).[8] Negotiations, however, have proved difficult. Of the 15 ECOWAS members from West Africa, only the Ivory Coast has signed up to an interim EPA and Ghana initialled it; in Central Africa, only Cameroon signed an interim EPA (in 2009); and in the Pacific, both Papua New Guinea and Fiji signed interim EPAs (again in 2009).

EPAs have been heavily criticised for: their unclear objectives, methodology and timetables; the unequal patterns of trade competition that will establish among economically disadvantaged countries within the same area; their apparently exclusive focus on regionalism and integration amongst countries with little or no local history of political or economic cooperation; the lack of structural fail-safes built to prevent ongoing corruption; their lack of attention to key philosophies of sustainable development and energy security; and, more broadly, for the ongoing policy reorientation by which EU development policy appears inexorably to be moving away from the original commitment to poverty reduction.[9]

Arguably, opening up national markets includes liberalising energy markets, allowing for competition and inviting in foreign investment. The ideal situation is one where the benefits of a liberalised, competitive energy market take off, with trade in energy goods as a genuine feature of African markets – nationally and regionally – with the sector operating to employ sufficient numbers and generating sufficient revenue to contributing positively and permanently to the capacity building of the energy sector, institutional building of the state and the economic welfare of a large chunk of the populace. Currently however, the onus for such changes and reforms lies entirely with the African states to demonstrate their needs within the aegis of EPA goals (made doubly difficult for those who have not signed EPAs). The European Commission indicates that African states must state clearly within their own Country Strategy Papers funding needs in three main categories:

- Trade-related assistance (funds to deal with the costs of liberalisation, e.g. creation of an Energy Regulator, or at least to begin the process of unbundling/privatising energy sector ready for national and international competition).
- Infrastructure (funds/assistance/expertise for developing public institutions and building capacity for private actors, plus the 'hardware' of venues, roads and, connections, including grids, pipelines, inter-connectors, LNG ports, etc.)
- Regional integration (funds for linking with other states; again interconnectors, regional energy regulators, etc.)

African states therefore need to coordinate both internally and regionally with the new Regional Preparatory Task Forces to get the most out of the EU's proposed 'regionally owned financial mechanisms' – the new funding instrument for the EPAs. The financial provisions exist, but it appears rather *more* provisions are on offer for ACP states who sign up to the EPA framework.

The European Parliament's Committee on Development suggests rightly that FDI is crucial for market opening and energy sectors to be concretely developed, and where 'potential investors need to be offered the security of transparent and stable rules'. EPAs are as yet missing a chapter on investment so that states can capitalise on their mining, electricity, gas and oil sectors (European Parliament, 2008: 16). The Parliament sees EPAs less as a catalyst for trade and market opening, and more a genuine development instrument to 'promote sustainable development, regional integration and a reduction of poverty' (European Parliament, 2008: 7). The EP's also suggests that the adjustment costs present the EU with an opportunity rather than an expense. From the perspective of overhauling market structures and rehabilitating energy sectors, this may well be so in theory, but in practice ACP states will suffer serious (if temporary) financial impact in readjusting their markets, opening to competition and installing regulatory mechanisms necessary for growth and investment.

Africa–EU Strategic Partnership

Despite being caught between the expediencies of its own gas market and the ethical imperatives of its development relations with Africa, the EU continues to opt for a development-driven approach to cultivating an energy relationship with Africa. This differs radically from the instrumental approach to hydrocarbon exploration and exploitation in Africa undertaken in recent years by China, where there is solid support for overhauling underpowered local infrastructure, and a measure of state capacity building, but virtually nothing in the way of normative obligations. The EU, in other words, is doing it the hard way by incorporating a normative bedrock, and complicating the definition of energy to encompass poverty reduction and associated goals of capacity building. The upshot is that access to, and trade in, energy should contribute to poverty-reduction in Africa, while the pursuit of Africa energy security goals should be understood to coincide with the EU's own energy security objectives. EU development policy is therefore now persuasively presented as the principal means by which to improve energy security (Youngs, 2009).

The Africa–EU Strategic Partnership strives 'to bridge the development divide between Africa and Europe' in which enhanced political dialogue among partners will allow serious cooperation on eight key areas, including energy. Prompted by the fallout over EPAs, as well as the rise of China, the EU was led to assert that:

> most of Africa's development challenges require a response of a political nature. The EU can therefore no longer envisage nor limit its development

cooperation to a form of institutional charity deprived of a key political dimension ... [but] must build upon the qualitative improvements allowed for by the EU Strategy for Africa ... in order to better meet EU and African interests and needs ... Emerging challenges, such as climate change, energy security, migration, terrorism among others, require a new and adapted type of partnership between the two continents. (European Commission, 2007c: 6)

The Africa–EU Strategic Partnership must therefore be clearly based on a consensus of values, common interests and strategic objectives. It furthermore needs to be institutionalised via effective multilateralism and must create a workable environment (manageable sectoral reform with visible deliverables). In effect, the Partnership is the *political counterpart* to the EPAs, giving the EU a foundation 'to implement its aid effectiveness objectives towards Africa ... the necessary framework for ... an overarching political partnership ... a new inter-continental alliances' (ibid.).

The First Action Plan (2008–10) outlined eight themes, of which energy is number five. All eight are identified as 'strategic priorities' to be worked upon vigorously between 2008 and 2010, mainly because success here will 'have a positive impact on the daily lives of the citizens of Africa and Europe' (Council of the European Union, 2010a). As illustrated in table 11.1, achieving development and energy security objectives is a wholly interdependent process, a process which itself is a means to achieving the ends of the Strategic Partnership.

Now in its Second Action Plan (2011–13), this continent-to-continent partnership remains committed to growth, investment and job creation (Council of the European Union, 2010b). A brief look at energy (the fifth theme) reveals the increasingly pragmatic use of the capacity-building vs. sustainability dilemma. Mention of accessibility, affordability, regulatory and investment mechanisms is all drawn from the

Table 11.1. The JAES, development and energy security

JAES	Development	Energy security
Political impact	Dialogue on governance	Rule of law; fight against corruption; institutional development
Economic impact	Regional integration initiatives; trade integration	Supply side attention: Promotion of private sector development, supported by FDI; strengthening of physical infrastructure networks (transit of goods); trade integration
Environmental impact	Migration, deforestation, desertification, flood/drought	Energy efficiency, biodiversity, toxic waste, nuclear safety

EU lexicon of energy security, which can now be more practically guaranteed with enhanced EU–Africa electricity capacity and increased African gas exports. The Action Plan makes clear that EU energy security interests geared at accessing Africa's 'vast and untapped natural resources' are major drivers behind the energy partnership within the JAES, oriented along lines that will ensure genuine African growth. Yet the sustainability of this growth seems divided between fossil fuel reliance and use of renewables, and will rest upon the practical outputs of the new Renewable Energy Cooperation Programme.

Conclusion

The brief overview of the energy partnership within the JAES indicated that EU energy security interests are clearly at play, but tempered with an ongoing commitment to a sustainability ethos. This chapter suggests that the connection between development and energy policy, while relatively adolescent will swiftly become a critical issue for the EU. What is necessary is a ranking of needs and interests and implementation of the most needed and the most doable goals.

Africa–EU dialogue on energy access and energy security is urgent, in that it applies equally to all African states: that of communicating the need to strive for improved access to reliable, secure, affordable, climate friendly and sustainable energy services. Africa's radically uneven hydrocarbon distribution means that some areas and states will become prominent energy actors, and others will remain as importers, transit states. For African energy actors – particularly exporters in the Gulf of Guinea – the chief concern will be security of resource supply and consistent market demand. The EU should be able to capitalise on the latter; however, the historic framework of its development polices ties the EU into the stability of the former. Can EPAs, as the post-preference development template, successfully engender political reform, inter-state trade and the local and regional stability necessary to produce consistent access to known and unknown hydrocarbon fields? Is development as envisaged by the EU a substantial enough vehicle by which to use public sector tools to not only transform attitudes to governance but also to support medium- and long-term private sector ventures?

State capacity, however, following traditional definitions of consolidated infrastructure, specialised personnel and permanent institutions yielding an autonomous state, continues to be urgently required in the majority of African states. And in some senses, it is a natural precursor to subsequent negotiations on energy access and security. More likely, low-level provision for public services (electricity, water) will move in parallel with high level politics concerning energy security. The focus is certainly a local one, and the onus is upon the EU to establish development structures that can reliably underwrite municipal and national projects, from basic requirements of increasing electrification and access to water to more complex scenarios of moving to carbon-free economies via increased use of African-appropriate renewables and a renewed stress on the more efficient use of current hydrocarbon energy sources.

Two final questions therefore remain. The synergistic message that is being sent to Africa by the EU regarding an energy–development nexus, and the extent to which this nexus operates coherently and consistently within EU foreign policy. This chapter has laid out in some detail both the broad contours and particular objectives that the EU has used in the past decade to produce such a nexus. While its origins are arguably the EU's own climate change and internal energy policies, the nexus itself has clearly evolved from the AEEP to the JAES into a focused series of complementarities between growth and investment, between access and trade, between sustainability and renewables, between security of supply for the EU and emerging security of demand for African producers, between basic provisions and far-reaching climate change commitments. With the possible exception of the EPAs, the overall energy–development nexus is therefore unwieldy but relatively solid; the majority of initiatives capture effectively the progress that has occurred previously, and work to deepen it relative to Africa. Further progress can of course be made, most likely by further refining the core issues within energy security provisions that the EU views as directly linked to poverty reduction, translating them into ranked targets, with a decent budget and a concrete methodology for evaluating their implementation, located for instance within country and regional action plans.

As a self-standing connexion that exhibits degrees of complementarity, the energy–development policy nexus thus exists and possesses an emerging and inherent value for discovering the poverty reduction elements within energy security as part of the EU–Africa framework. This same nexus has not however worked well within two other aspects of EU development policy: namely the Development Cooperation Instrument and the European Neighbourhood Policy (ENP). The DCI applies to dozens of other underdeveloped states, and contains frequent mention of energy and environment objectives in both its bilateral and thematic components, with little concrete implementation or evaluation methodology. Here, energy is only one of many annexes to a virtually global development project; and will not be taken up precisely because the scope is so vast and the targets so ambiguous. The ENP, where virtually every state is a key hydrocarbon exporter or transit state, is not definitively classified as development (managed by the European External Action Service rather than DG DevCo). Second, the ENP does not contain a fully fledged neighbourhood energy tool at the regional level but rather one of many provisions operationalised within the bilateral ENP Action Plans, despite the robust albeit private-sector history of EU-North Africa energy trade relations. In addition, there is an uneasy division between ENP states in North Africa, the Sahel, sub-Sahel states and the state of South Africa, in addition to the different EPA groupings; some states are dealt with in separate categories, others together, some see the energy–development nexus as the centrepiece of their foreign relations with the EU, where for others it is less evident. Despite the early success of the energy–development nexus within EU–Africa relations, it has yet to make a toehold within the DCI and ENP in a manner demonstrating a genuine coherence of message or consistency of strategy within the broader remit of EU foreign policy.

Notes

1 This chapter concentrates upon energy-development relations of sub-Saharan Africa, that is all the African states south of, and including the Sahel, that fall within the grouping of African, Caribbean and Pacific (ACP) states that have traditionally been the focus of EU development until the creation of the Development Cooperation Instrument (DCI) in 2007, which broadened EU policy extensively to include other regions. North Africa falls under the European Neighbourhood Partnership, a separate development instrument geared to reform and modernisation that groups all North African states – together with Middle Eastern states – into a separate category, benchmarking progress via bilaterally agreed ENP Action Plans (each of which contain provisions on energy security).

2 Readers should be aware of the difference between *production*, which refers to barrels of oil/cubic meters of gas extracted daily, and *supply*, which refers to market availability of gas and oil products most refinery treatment.

3 Wangari Maathai was the 2004 Nobel Peace Prize winner and Green Belt Movement founder.

4 A forerunner of the agreements to follow, the JREC includes roughly 100 developing and developed governments committed to cooperate on the basis of both national and regional targets and timetables towards a visible increase in the share of renewable energy in the global energy mix.

5 The political framework for PCD incorporates the Communication on PCD in April 2005, the subsequent Council Conclusions in May 2005, the European Consensus on Development in December 2005, and the Council Conclusions on the EU PCD report in November 2007. See Carbone (2009).

6 Under the GSP, LDCs receive duty and quota-free access to EU markets via the Everything but Arms (EBA) arrangement. South Africa is the

exception, which benefits from the provisions of the Trade and Development Cooperation Agreement (TDCA) in the absence of any new trade regime.

7 FTAs exist when duties and other restrictive regulations of commerce are eliminated on substantially all the trade between parties.

8 SADC comprises Namibia, Botswana, Lesotho, Swaziland and Mozambique; with Angola and South Africa as envisaged partners; East African Community (EAC) includes Uganda, Kenya, Tanzania, Rwanda and Burundi; Eastern and Southern Africa (ESA) includes Zimbabwe, Seychelles, Mauritius, Comoros and Madagascar; West Africa (ECOWAS) includes Ivory Coast and Ghana, Nigeria (50 per cent of which is oil), Mauritania, and 12 more.

9 See DFID, *Ten Myths about EPA's*, available at: www.dfid.gov.uk See also Kotte (2011).

References

Amirat, A. and Bouri, A. (n.d.) *Energy and Economic Growth: The Algerien Case*, University of Economics and Management of Sfax-Tunisia, available at http://ps2d.net/media/Amina%20Amirat.pdf (accessed 30 June 2012).

Carbone, M. (2009) *Policy Coherence and EU Development Policy*, London: Routledge.

Council of the European Union (2006) *7th EU–Africa Ministerial Troika Meeting Brazzaville*, 10 October 2006 – Final Communiqué, 13823/06 (Presse 283), 11 October.

Council of the European Union (2010a) *First Action Plan (2008–2010) for the Implementation of the Africa–EU Strategic Partnership*, available at: www.africa-eu-partnership.org/sites/default/files/eas2007_action_plan_2008_2010_en_6.pdf (accessed 5 March 2012).

Council of the European Union (2010b) *Joint Africa–EU Strategy Action Plan 2011–2013*, available at: *http://europafrica.files.wordpress.com/2008/07/second-final.pdf* (accessed 5 March 2012).

Dannreuther, R. (2007) *International Security: The Contemporary Agenda*, Cambridge: Polity.

DG Development and Cooperation (2012) *Environment and Sustainable Management of Natural Resources*, available at: http://ec.europa.eu/europeaid/how/finance/dci/environment_en.htm (accessed 2 November 2012).

Enfield, M. (2009) 'Africa in the Context of Oil Supply Geopolitics', in A. Wenger, R. Orttung and J. Perovic (eds), *Energy and the Transformation of International Relations: Toward a New Producer–Consumer Framework*, Oxford: Oxford University Press.

Europa (2012) 'The Africa–EU partnership at work', available at: http://europa.eu/legislation_summaries/development/african_caribbean_pacific_states/rx0006_en.htm (accessed 5 March 2012).

European Commission (2002) *Energy Cooperation with the Developing Countries*, COM (2002) 408, 17 July.

European Commission (2003) *EU Energy Initiative for Poverty Eradication and Sustainable Development (EUEI)*, Luxembourg: Office for Official Publications of the European Communities.

European Commission (2006) *Interconnecting Africa: the EU–Africa Partnership on Infrastructure*, COM (2006) 376, 13 July.

European Commission (2007a) *EU Report on Policy Coherence for Development*, COM (2007) 545, 20 September.

European Commission (2007b) *Economic Partnership Agreements*, COM (2007) 635, 23 October.

European Commission (2007c) *Commission Staff Working Document Accompanying the Communication 'From Cairo To Lisbon – The New EU–Africa Strategic Partnership'*, SEC(2007) 855, 27 June

European Commission (2009a) *EU 2009 Report on Policy Coherence for Development*, COM (2009) 461, 17 September.

European Commission (2009b) *Assessment Report of the Joint Africa–EU Strategy*, 9 October, available at: http://ec.europa.eu/development/icenter/repository/ jaes_assessment_report_20091010_en.pdf (accessed 30 June 2012).

European Commission (2010) *Policy Coherence for Development Work Programme 2010–2013*, SEC (2010) 421, 21 April.

European Commission (2011) *Increasing the impact of EU Development Policy: an Agenda for Change*, COM (2011) 637, 13 October.

European Union (2000), 'Partnership Agreement between the Members of the African, Caribbean and Pacific Group of States of the One Part, and the European Community and its Member States, of the Other Part', signed in Cotonou on 23 June 2000, *Official Journal of the European Communities*, L 317/3, 15 December.

European Union (2006) 'Joint Statement by the Council and the Representatives of the Governments of the Member States Meeting within the Council, the European Parliament and the Commission on European Union Development Policy: "The European Consensus"', *Official Journal of the European Union*, C 46/1, 24 February.

European Commission (2011a) *Increasing the Impact of EU Development Policy: An Agenda for Change*, COM (2011) 637, 13 October.

European Commission (2011b) *Global Europe: A New Approach to Financing EU External Action*, COM (2011) 865, 7 December.

European Commission (2011c) *Proposal for a Regulation of the European Parliament and of the Council Establishing a Financing Instrument for Development Cooperation*, COM (2011) 840, 7 December.

European Parliament (2008) *Report on Development impact of Economic Partnership Agreements* (EPAs) (2008/2170/ (INI)), Committee on Development, A6-513/2008, 18 December.

Hadfield, A. (2007) 'Janus Advances? An Analysis of EC Development Policy and the 2005 Amended Cotonou Partnership Agreement', *European Foreign Affairs Review*, 12 (1): 39–66.

Hiscock, G. (2012) *Earth Wars: The Battle for Global Resources*, Singapore: John Wiley & Sons.

Human Rights Watch (1999) *The Price of Oil: Corporate Responsibility and Human Rights Violations in Nigeria's Oil Producing Communities*, New York, January 1999.

Humphreys, M., Sachs, J.D. and Stiglitz, J.E. (eds) (2007) *Escaping the Resource Curse*, New York: Columbia University Press.

International Crisis Group (2006) 'The Swamps of Insurgency: Nigeria's Delta Unrest', Africa Report No. 115, 3 August.

Klare, M. T. (2002) *Resource Wars: The New Landscape of Global Conflict*, New York: Owl Books.

Kotte, D. (2011) 'Getting Trade Right', *IP: Global Edition*, 12: 25–32.

Maathai, W. (2010) *The Challenge for Africa*, Anchor Books.

Menyah, K. and Wolde-Rufael, Y. (2010) 'Energy Consumption, Pollutant Emissions, and Economic Growth in South Africa', *Energy Economics*, 32: 1374–82.

McHaney, S. and Veit, P. (2009) 'Stopping the Resource Wars in Africa', World Resources Institute, 10 August, available at http://www.wri.org/stories/2009/08/stopping-resource-wars-africa.

Ojakorotu, V. and Gilbert, D.L (2010) 'Special Report: Checkmating the Resurgence of Oil Violence in the Niger Delta of Nigeria', *Journal of Energy Security*, 18.

Pago (2011) *Powering Africa: The Gas Option*, available at www.energynet.co.uk/pa/PAGO2011/ index.html (accessed 30 June 2012).

Shah, A. (2010) 'The Democratic Republic of Congo', *Global Issues*, available at www.globalissues.org/article/87/the-democratic-republic-of-congo (accessed 30 June 2012).

Tarradellas, F. (2008) 'EU, Africa Unveil "Ambitious" Energy Partnership', EurActive.com, 9 September.

United Nations Development Programme (2011) *Human Development Report 2011*, New York: UNDP.

Wolde-Rufael, Y. (2005) 'Energy Demand and Economic Growth: The African Experience', *Journal of Policy Modeling*, 27: 891–903.

Youngs, R. (2009) *Energy Security: Europe's New Foreign Policy Challenge*, Abingdon: Routledge.

12

Climate change and the Africa–EU Strategy: coherence, leadership and the 'greening' of development

Simon Lightfoot

At a 2007 African Union (AU) Summit, Uganda's president, Yoweri Museveni, declared climate change an 'act of aggression by the rich world against the poor one – and demanded compensation' (*The Economist*, 2007). As a continent, Africa is extremely vulnerable to climate change in its major economic sectors, with the vulnerability being 'exacerbated' by the development challenges faced on the continent (Boko *et al.*, 2007; Stern, 2006). As the European Union (EU) notes, climate change superimposes itself on existing vulnerabilities and will hit Africa earlier and harder than in many other parts of the world (European Commission, 2007). What this means is that even relatively modest rises in temperature could 'tip the balance' and lead to severe water shortages and reductions in crop yields (Stern, 2006). The 4th Assessment Report of the Intergovernmental Panel on Climate Change (IPCC) predicts, with high confidence, food and water insecurity, sea level rises, migration and new health problems across the continent as a consequence of climate change. The report clearly outlines that the worst affected populations are anticipated to be the poorest citizens, who have the least resources to adapt (Boko *et al.*, 2007).

This chapter sets out why relations between the EU and Africa around the issue of climate change are crucial, in particular for the developmental dimension. It then goes on to examine tackling climate change in Africa through the prism of the EU as a normative power. In particular, it examines EU claims to offer a global leadership role in the field of climate change, and analyses how policy incoherence has the potential to weaken support from African countries for its attempts to shape the post-Kyoto climate agreement. The chapter then sets the current Africa–EU Partnership on Climate Change in the context of previous discussions linking development and climate change, before outlining the elements of the partnership, such as the integration of climate change into Africa–

EU development cooperation, measures to prevent and reduce deforestation and attempts to integrate Africa into the global carbon market. The chapter highlights some of the main barriers to the potential success of the partnership, such as EU policy incoherence (trade, energy and agriculture), access to resources by African states and a potential lack of true dialogue. Overall, it concludes that despite strong rhetoric the climate change partnership risks failing due to insufficient EU funds. Finally, it points to the marginalisation of the EU at the Copenhagen Summit and the rise of China as an alternative development actor, challenging to the EU's normative leadership in this area.

Africa and climate change

Africa's climate is likely to be affected more severely than that of other regions. The level of poverty in many African states also means that they lack the resources to tackle climate-change-induced effects (Collier *et al.*, 2008). The quote from the Ugandan president highlights the classic problem with regard to international climate change politics: who is responsible and who pays? The question of equity is often explained using a peeling the onion metaphor: Parks and Roberts (2006: 332) argue that to understand climate injustice we must peel away ten layers of inequality and injustice to see how the benefits and costs of climate change are being distributed. Holland notes that, despite the connections, environmental sustainability has historically been under-addressed in development and poverty reduction strategies (Holland, 2008: 346). This in part may have reflected different priorities, as there is a literature that argues that environmental issues like climate change, despite their obvious impacts upon developing countries, are not the key concerns of developing countries (Redclift and Sage, 1998). For many years, climate change was often seen to be a 'northern issue' of concern only to the Northern nations that had created the problem (Paterson and Grubb, 1992). However, since the various IPCC reports and the Millennium Development Goals (MDGs), it is evident that development and climate change are clearly linked (Ayers and Dodman, 2010). Indeed the UNDP has produced a summary of how climate change affects each of the eight MDGs.

At the same time, we have seen the EU take a relatively clear line that the developed world is responsible for climate change and therefore needs to bear the overwhelming costs imposed by climate change mitigation and adaptation in the developing world. Since the signing of the United Nations Framework Convention on Climate Change (UNFCCC) at Rio in 1992, the EU has offered global leadership in this policy field (Oberthür and Roche Kelly, 2008). By 2005, it had a Strategy on Climate Change and Development, which informed its actions (European Commission, 2005a) and climate change formed an integral part of development cooperation (Lamin, 2004; Behrens, 2008). Indeed, it is argued that many states see Europe's climate change policy as inspiring their own approach to fighting global warming (Kilian and Elgström, 2010). However, it is also clear that climate change in Africa has the possibility to produce unfavourable challenges for the EU, such as increased illegal migration (Adger *et al.*, 2003), threats to security (Faust and Messner, 2005; Bagoyoko and Gibert, 2009) and increased conflict in the region (Raleigh, 2010). The issue of the extent to which the actions of the EU amount to global 'normative' leadership is the next question for this chapter.

EU global leadership and normative power[1]

The concept of normative power is most closely associated with the work of Ian Manners (2002). He argues that we should transfer the debate away from traditional concerns over whether the EU should be a civilian or military power, and focus more upon the ideational impact of the EU's international identity/role as representing normative power with the ability to shape conceptions of normal (Manners, 2002: 238–9). He suggests that the fact that the EU has been constructed on a normative basis 'predisposes it to act in a normative way in world politics' (Manners, 2002: 252). A key aspect of this process is that the EU should act to extend its norms into the international system. The expansion of the EU's normative basis on to the world stage represents a key process in providing greater legitimacy for the EU as an international power and 'allows the EU to present and legitimate itself as being more than the sum of its parts' (Manners, 2002: 244). Sustainable development is seen as a minor norm within the EU, in part because of the issues associated with its definition and operationalisation. Recent treaty developments, including Lisbon, have moved sustainable development and in particular the specific issue of climate change, into the core activities of the EU. All this poses several challenges.

The first is that environment and therefore climate change are a mixed competence in the EU between the European Commission and the member states. Whilst the Commission initiates many proposals, many national policies/strategies also exist and external action can also be state led (Egenhofer, 2006: 7). The treaty commitments mean that sustainable development must inform all Union policies, including external commitments. This was reinforced by the Barcelona European Council in 2002, which added an external dimension to the Sustainable Development

Strategy (SDS). This strategy was designed to overcome the problem of whether the EU could be regarded as a single entity in international affairs (Manners and Whitman, 2003: 397). Previous research has highlighted the lack of consistency in applying the environmental *acquis* within different directorates (see Bretherton and Vogler, 2006: 82). The last principle of the EU's sustainable development strategy, international responsibilities, is the principle which is of most consequence for both global environmental governance more broadly and theories of 'normative power Europe'. In a 2005 review of the SDS, the Commission stressed this point: 'The EU will use its considerable influence to bring more nations behind an ambitious sustainable development agenda. It will also use its own instruments, such as trade and cooperation agreements, to drive change and will factor the external dimension into its internal policymaking' (European Commission, 2005a: 6).

This statement links very well to the second challenge, that of policy coherence. Environmental issues are cross-cutting issues, something explicitly recognised by the European Consensus on Development (Lamin, 2004; Dearden, 2008: 190), which means they touch upon other bureaucratic competences, such as energy, agriculture and transport. Given the ambiguous position of energy policy within the EU (see Hadfield, 2006) and the complexity of development cooperation structures prior to Lisbon (see Carbone, 2007), the possibility of policy incoherence exists.

Policy coherence for development is a crucial aspect of EU development policy and an 'ambitious agenda' for the Africa–EU Strategy (Carbone, 2010: 248). The Maastricht Treaty introduced the notions of coordination and coherence to try and ensure all other EU policies did not undermine EU development policy (Peskett *et al.*, 2009). Carbone (2008: 330–1) argues that focusing attention on the process rather than the results means that 'development policy can be vulnerable to conflict with more powerful interests, notably those of trade, fisheries and agriculture'. Studies have shown that policy incoherence significantly weakens the ability of the EU to act as a normative power (Syzmanski and Smith, 2005; Storey, 2006; Holden, 2006), especially as the EU uses its development policy to shape values in developing countries (Bonaglia *et al.*, 2006). It therefore makes sense to use the concept of normative power in this chapter, especially as the concept has been applied to EU–Africa relations in a number of works (see Scheipers and Sicurelli, 2008; Storey, 2008; Sicurelli, 2010; Langan, 2012). The extent to which policy incoherence has affected the ability of the EU to act as a normative power in Africa is the topic to which this chapter now turns.

Climate change, Africa and the EU

For the EU to offer leadership on climate change is one thing but to achieve results the EU needs supporters. Part of the strategy of the EU as a normative power is to shape the behaviour of other states. Previous EU documents have demanded an increased level of commitment on the part of the African countries to push through institutional reforms and to implement policies in the economic, social and environmental arenas arising out of the undertakings agreed to at international conferences (Farrell, 2005). From this, we can identify a clear EU line on both sustainable development and climate change in Africa (see Peskett *et al.*, 2009). EU pressure can in part explain the first-ever meeting of ACP Ministers of the Environment in Brussels in 2004 and the 2006 'Joint ACP–EU Declaration on Climate Change and Development'. The African Union also drafted a climate change and development policy at the eighth ordinary session of the Assembly. As Scheipers and Sicurelli (2008: 617) argue 'these meetings show that climate change has become an increasingly relevant issue for SSA countries and that the EU considers the African, Caribbean and Pacific (ACP) group and the AU as interlocutors in drafting post-Kyoto commitments'.

The need for money is vital as one estimate is that OECD countries would have to invest roughly 1 per cent of their GDP in global environmental policy to put an end to the growing degradation of the world ecosystem (Faust and Messner, 2005). The EU has at its disposal a variety of funding streams (Peskett *et al.*, 2009). This allows the EU, as Brande (2008: 170) puts it, to either win or buy their souls. Therefore, the perception of the EU from the global south deserves particular attention (Fioramonti and Poletti, 2008: 617), especially when EU policies, 'such as development, poverty, multilateralism, trade and the environment', will have specific effects in developing countries. The EU has a very clear strategy to mainstream national climate change adaptation plans for African LDCs into broader sustainability frameworks (van der Grijp and Etty, 2010). This chapter now goes on to explore how well these sentiments were carried forward into both the Africa–EU Strategy and the debates about a post Kyoto regime. In doing so it sets the current EU–Africa Partnership on Climate Change in the context of previous discussions linking development and climate change before outlining the elements of the Partnership, such as the integration of climate change into Africa–EU development cooperation, measures to prevent and reduce deforestation and attempts to integrate Africa into the global carbon market.

The Africa–EU Partnership on Climate Change

In 2007, EU–Africa relations entered a new phase with the signing of the Joint Africa–EU Strategy (JAES), which aimed to provide an overarching long-term framework for EU–Africa relations. The issues addressed can be seen as a sequel to many that form part of the Cotonou Agreement, which identified 'common cause' between the EU and ACP on, amongst other issues, climate change (Bagoyoko and Gibert, 2009).[2] In relation to environmental degradation and climate change, it highlighted four key aspects (Harmeling *et al.*, 2007). The first was that 'Africa and the EU have a clear common interest to address environmental sustainability and climate change'.[3] Both the African states and the EU clearly identify the link between the negative effects of climate change on sustainable development aspirations, alongside other common goals, such as the MDGs. They also accept the basic tenet of the IPCC's conclusions outlined above that climate change 'will be most immediately and severely felt in the poorest and most vulnerable countries, which do not have the means and resources to adapt to the changes in their natural environment'. This importantly appears to provide evidence that the EU's normative power is the major factor in prompting a change in attitude amongst African states that climate change is no longer just a 'northern problem' and highlights the benefits of using the partnership approach in EU–Africa relations (Taylor, 2010; Carbone, 2010).

However, it is also clear that the myriad of different strategies and approaches on offer could lead to confusion (Olsen, 2004). Therefore, when we look at efforts to integrate development issues into other policy areas such as climate change, we see that where Directorate General for (DG) Development has been in the lead 'development implications have been better integrated' (Egenhofer, 2006: 18; Carbone, 2008: 328). A good example is seen to be the Communication on Climate Change in the Context of Development Cooperation. However, having DG Development in the lead may not always be politically acceptable to the member states (Egenhofer, 2006: 18). The risk of policy incoherence is therefore increased due to the large number of cross-cutting areas under scrutiny and the range of actors involved on the EU side.

The EU–Africa Partnership on Climate Change will be the main focus of the rest of this chapter. It is clear, though, that any discussion of climate change will also need to consider energy, particularly efforts to promote adaptation in the energy sector. The EU–Africa Partnership on Climate Change focuses mainly on building up Africa's capabilities to respond to climate change (Council of the European Union, 2008).

This Partnership includes two interlinked priority actions: (1) building a common agenda on climate change policies and cooperation, and (2) addressing land degradation and increasing aridity.

Priority action 1: build a common agenda on climate change policies and cooperation

The main objectives were for Africa and the EU to work together to push forward an ambitious post-2012 climate agreement and to strengthen capacities to adapt to climate change and to mitigate its negative effects. The creation of the Global Climate Change Alliance (GCCA) is an important aspect of the partnership (Ayers *et al.*, 2010). The GCCA aims to help the developing countries that are most vulnerable to climate change by funding actions in African LDCs and small island states to help these states tackle the effects of climate change (Sicurelli, 2010). The action anticipated 'systematic integration of climate change into African national and regional development strategies as well as into Africa–EU development cooperation' and 'reduced rates of deforestation and better preservation of forest ecosystems'. To do this, the strategy called for national/regional adaptation plans to climate change, as well as the need to support the implementation of Climate for Development in Africa (ClimDev Africa).[4] It also saw the need for greater African participation in the global carbon market and increased energy efficiency.

The EU also committed itself to a better geographical distribution of Clean Development Mechanism (CDM) projects. To do this, it has invested in capacity building projects raising the knowledge and awareness for CDM in Africa and other regions. These efforts, however, have not been very successful. Still less than 2 per cent of the CDM projects and investments are in Africa. Moreover, they tend to be small scale and getting CDM approval is often difficult and relatively expensive (European Commission, 2007; De Lopez *et al.*, 2009). The main issue appears to be that as these countries are less well integrated into the global economy than others and more aid-dependent, important support can be provided by other donors to enable clean energy transition (Newell *et al.*, 2009: 735). As a result, the majority of projects has tended to go to South East Asia or Central and Southern America, despite the need for the poorest nations to access funds for climate change adaptation (Brande, 2008: 170). In a clear issue of policy incoherence, it has been EU member states calls for environmental integrity that have narrowed the terms of additionality (Lovbrand *et al.*, 2009: 80).[5] These requirements have so far seemed to favour streamlined projects and resourceful project developers at the expense of projects designed and implemented by low income communities at low rates of return (Lovbrand *et al.*, 2009: 94). In demanding high environmental standards, the EU was actually undermining the aims of its policy and there appears little evidence that the EU is able to change its stance on this issue quickly (De Lopez, 2009; Wettestad, 2009).

Much of the Partnership on Climate Change focuses on monitoring and improving reporting techniques, as well as improving adaptation strategies. There is a strong focus on changing incentive structures to prevent deforestation and promote sustainable resource management. The strategy also takes a very clear technological line, with a commitment to share clean technology. Importantly, finance sources are identified including the 10th European Development Fund (EDF), bilateral contributions from EU member states and African states, UN sources and the Global Energy Efficiency and Renewable Energy Fund (GEEREF) (Holstenkamp, 2010). One proposal was that auctions from the Emission Trading System (ETS) could go into the GEEREF (Behrens, 2008). This would be an interesting development as it would shift the cost of mitigating climate change from states, who traditionally contribute to this fund, to polluting companies, but this has not happened. Part of the problem with the GEEREF was that their target market lacked geographical focus, much like the CDM. Whilst there was meant to be a focus on ACP states, investment decisions are made by project availability and financial performance. Therefore, it is likely that funds will go to South Asia, non-EU Europe, Russia and Central Asia and the Middle East and North Africa, especially give the challenges created by the global recession (Holstenkamp, 2010).

The need for finance is crucial as a common criticism of EU action is that the political message is there but the money is not. The GCCA has a budget of 60 million euro, which is seen as a 'drop in the ocean' to what is actually needed to fund real adaptation (Ayers *et al.*, 2010: 240). Therefore, whilst the GCCA might provide added value in political terms, such as helping to contribute to the joint Africa–EU declaration on climate change prepared for the Poznan Conference on the UN Framework Convention on Climate Change in December 2008, it is not clear what it provides in real financial terms (Peskett *et al.*, 2009). First of all, the money is not additional to ODA as it is meant to be. Second, the GCCA shows very little evidence of partnership (Ayers *et al.*, 2010). Finally, there are issues within the sub-groups of the AU, with the small islands and low-lying states having markedly different views on climate change from OPEC members, such as Angola and Nigeria (Sicurelli, 2010: 130).

Priority action 2: cooperate to address land degradation and increasing aridity

This priority builds upon a weakness of the initial strategy (see Harmeling, 2009) by attempting to combat desertification in the Saharan and Sahel zones of Africa, regions which have been identified as at high risk from the effects of climate change (Stern, 2006). A key aspect of this was the Green Wall for the Sahara Initiative, which plans to plant trees around the edge of the Sahara to prevent desert encroachment.[6]

However, Biopact highlighted the problem that a large proportion of the population in both regions depends upon unsustainably harvested wood fuel for domestic energy, despite the lack of these natural resources in the region. They also argued that the scheme would need to change farming and cultural practices. For example, in Madagascar, traditional slash-and-burn agriculture has wiped out about 90 per cent of the native tree cover. Therefore, the inclusion of land management issues in national development strategies is vital. The problem is ensuring that these strategies are joined up and that key stakeholders drive the project forward. Some critics claim that these types of projects are good at grabbing the headlines but that resources could be spent more effectively elsewhere. There is also tension between different experts. Climate change specialists tend to focus on greening energy sources, while development specialists argue that developing countries should be free in their choice of energy sources (Egenhofer, 2006: 54). A report in 2006 highlighted that a 'lack of awareness of climate change within the development community and limitations on resources for implementation are the most frequently cited reasons for difficulties with integrating climate change adaptation within development activity' (Harmeling *et al.*, 2007:57). A major issue is that deforestation remain outside the CDM, and afforestation/reforestation projects are not accepted by the EU Emissions Trading Scheme (EU ETS) – the world's largest carbon market (Collier *et al.*, 2008: 350).[7]

This leads nicely onto a discussion of the environmental aspects of the energy partnership. Energy is also a crucial aspect of development (Flint, 2008), as can be seen in the fact that the 'Joint Statement on the Implementation of the Africa–EU Energy Partnership' was the first such agreement signed after the Lisbon Summit. The main element is exploring Africa's renewable energy potential in a sustainable way, improving energy efficiency and reducing waste (European Commission, 2008). However, it is clear that EU energy decisions can have at best unintended and at worst contradictory consequences. For example, the decision to adopt biofuels as a means to reduce carbon reliance can be seen to create a market opportunity for some sub-Saharan African countries and therefore help adaptation rather than hinder it (Charles *et al.*, 2009). However, the rush for biofuels has also been identified as resulting in a loss of biodiversity and additional pressure on the food market, leading to higher food prices (Matthews, 2008).

There are also a range of interconnected tensions within the EU's agricultural policy that harm policy coherence in this policy field, especially with regards to adaptation to climate change. Matthews (2008) argues that despite recent reforms making the CAP less incoherent, there are still tensions. It is argued that EU subsidies have often led to overproduction, with the excess being 'dumped' on developing countries (Flint, 2008: 101). It is clear that in many countries the need to diversify livelihoods is vital, but that the CAP and trade agreements shape the choice of agricultural products in LDCs (Egenhofer, 2006: 55). It is also clear from the Cotonou Agreement that the EU favours environmentally friendly production processes, although Flint (2008: 89–92) argues that the increasingly stringent sanitary, phyto-sanitary and environmental standards imposed by EU could be seen as a 'green trade barrier'. The need to conform to the EU's environmental standards is clearly part of the EU's desire to shape 'normal' in developing countries. The final point is that the EU had a ban on the use of genetically modified crops. This policy prevented many African farmers from utilising genetic modification to enable agricultural adaptation in case their use harmed possible exports to the EU (Collier *et al.*, 2008: 347).

A clear connection between the Partnership on Climate Change and the Cotonou Agreement was the fact that Country Strategy Papers (CSPs) were meant to include an analysis of the environmental situation in each country (Hout, 2010: 6). A 2006 report by the European Court of Auditors (ECA) found that progress with environmental integration into the 2002–07 CSPs 'was mostly weak' and 'the environment had not been satisfactorily mainstreamed'. Climate change was only mentioned in one of the 46 CSPs of any of the African countries (van der Grijp and Etty, 2010: 196). The challenge of mainstreaming at the EU level is a constant challenge, particularly in cross-cutting themes such as the environment (see Geyer and Lightfoot, 2010). The reasons for this failure ranged from insufficient appreciation of Treaty commitments through to lack of data and adequately trained staff (Peskett *et al.*, 2009). Climate change considerations are meant to be 'systematically addressed' when preparing CSPs for 2013. However, of the four countries examined by DG Development, none had climate change integrated as priority in their CSP, and only two had climate change integrated in their national development plans (Peskett *et al.*, 2009). Whilst these examples arise out of EU–ACP Agreements, it shows that attempts by the EU to shape behaviour in African nations are mixed. To what extent this hinders the EU ability to shape the post-Kyoto framework is a crucial question.

Normative power Europe and post-Kyoto framework

The EU has set out to offer a global leadership role in the field of climate change, but, without support from African countries, its attempts to shape the post-Kyoto climate agreement will be significantly weakened. Previous research has shown how the EU promoted the Kyoto protocol to sub-Saharan African states 'by referring to them in a framework of solidarity and partnership' (Scheipers and Sicurelli, 2008: 607). This apparent partnership approach has been an integral part of EU–ACP relations in the past and is now crucial for the subsequent discussion of the Africa–EU Strategy below. The combination of money via the EDF and the diplomatic pressure to be part of the EU-led 'coalition of the willing' allowed the EU to construct a positive image of its climate change policy in the eyes of African representatives (Scheipers and Sicurelli, 2008: 616). This paper now turns to the efforts to construct an 'ambitious post-2012 climate framework'.

The 2009 joint ACP–EU declaration on climate change and development stated that the two parties would 'work together to mobilize political support for stronger action on climate change and to examine how the international framework could facilitate such action'. The declaration goes on to say that the future perspective must include 'perspectives on low-carbon development, climate-safe paths and enhanced climate resilience (ACP–EU, 2009). Shaping the post Kyoto global environmental governance structures is of such high importance to both African states and the EU that it deserves special attention. In previous summits from Rio to Johannesburg, the developing world was relatively marginalised. However, since the 2002 WSSD many of the bigger developing countries realised that their biggest bargaining power was their ability to obstruct (Parks and Roberts, 2006: 342). Therefore, if the EU was going to achieve its goals for an ambitious post-Kyoto system it needed the support of African and other states. The EU's position for Copenhagen was to get agreement on halving global emissions by 2050 with a view to keeping global warming below 2 degrees. It was willing to go to a 30 per cent reduction target if other countries made significant commitments to attain this goal. A meeting of environment Ministers from Africa and Europe was held in Nairobi in 2009 during the 25th session of the UNEP Governing Council to discuss climate change, international environmental governance and the EU–Africa Partnership. Whilst the meeting produced an understanding, there were two key issues unresolved. The first was the level of funding and the whether this funding was in addition to planned ODA (Sicurelli, 2010: 129). The EU was accussed of 'repeatedly dawdling on translating the rhetoric of responsibility into action on meeting its obligations to support adaptation in vulnerable developing countries' (Ayers *et al.*, 2010: 246). The other issue was that of the choice of EU instruments to achieve its norms, with many AU states critical of the market solutions to climate change, especially in relation to the CDM (Sicurelli, 2010: 129–30).

The issue of funding became a massive internal issue for the EU. It is estimated that developing countries will need up to 100 billion euro per year by 2020 to adapt to climate change and cut greenhouse gas emissions (European Commission, 2007). The EU proposed that between 22 billion euro and 50 billion euro per year could come from developed countries. The rest would come from carbon markets or from developing countries themselves. Despite rhetoric on responsibility, the then Environment Commissioner, Stavros Dimas, said that the EU would not sign a blank cheque to fight climate change in developing countries. The problem was that some of the poorer EU states, notably Poland, objected to the sums of money being suggested, especially during the recession. In particular, they objected to the idea that contributions should be based upon historic emissions (*Irish Times*, 10 October 2009). Poland feared that a past-emissions formula would leave it with an 'unfair' and 'excessive' amount to pay. To reduce tension within the EU, the European Commission produced a re-distribution formula to ensure that poorer EU countries would not be worse off as a result of the climate deal (Sicurelli, 2010: 138). However, the EU, wary of revealing its negotiating position early, did not specify an exact sum to help adaptation.

The extent to which the money to support adaptation would be 'new' money was another issue of contention. At a UN meeting in Bonn in 2001, the EU, Canada, Norway, Switzerland, Iceland and New Zealand said they would jointly pay developing countries 410 million dollars a year from 2005 to 2008. It is claimed that less than 10 per cent of that money has thus far been delivered. There was also the problem of previous commitments. Pallemaerts and Armstrong (2009: 16) argue that considering the importance of the funding issue in multilateral climate negotiations, it was surprising that there was no official document issued by the EU with data on the total level of financial support to developing countries for climate change mitigation and adaptation: in their view, such as 'lack of transparency is clearly inconsistent with the EU's claim to global leadership in the climate change process'. It also raises serious questions about the commitment of the EU to the norm of sustainable development and climate change reduction.

In the run up to Copenhagen, there were claims that the EU was back-tracking on pledges to provide additional money to finance climate change adaptation. Again a split within the EU weakened its position. Britain and the Netherlands argued strongly that funds be largely additional, but Germany, France and most small member states said they wanted existing aid to be used. What was interesting prior to Copenhagen was that a number of African states tried to coordinate their negotiating position. In previous international environmental negotiations, LDCs were disadvantaged due to their very limited capacities. They had also made limited headway getting the 'negotiating group of the G77 to constantly express their concerns' (Harmeling *et al.*, 2007: 82). The 2009 AU summit was recognition of past failures of the continent to make its voice heard on this issue. The objective was to agree a set of key ideas so that Africa could be represented by one delegation in Copenhagen. Whilst this did produce some positive outcomes, overall the verdict was that 'despite the long process of preparation to come forward with an African common position the united front collapsed when concessions had to be made through internal strife. The African Group was considered a possibly influential force to be dealt with before the negotiations but ended up as an outsider on the few accomplishments of Copenhagen' (Hoste, 2010: 8–9).

Copenhagen was also major setback for the EU, as it was widely described as having failed to play a leading role in the negotiations (Killien and Elgström, 2010). The outcome was driven by the 'BASIC' block of Brazil, South Africa, India and China, along with the USA. The Copenhagen Accord therefore had no real input from the EU either 'conceptually' or in terms of its substance (Curtin, 2010). As the EU was not involved in these talks, it was obliged to accept the deal (Dai and Diao, 2010). There were three main problems: the fact the EU was an incoherent and internally divided; the EU's pledges before the summit were not seen as credible; and the fact that the EU is a 'relatively minor power in terms of global emissions' (Kilian and Elgström, 2010: 267). A major concern for the EU is not only the fact that it was marginalised, but that an alliance was built between China, India and the developing countries of Africa, despite the EU pushing the attractive category of 'most vulnerable' states (Curtin, 2010: 9). As Bossuyt and Sherriff (2010: 8) note, 'Between Africa and Europe there may well be shared issues, but not necessarily shared interests.'

Conclusion

This chapter has argued that making use of the 'normative power' paradigm can add value to the discussion of climate change policy in

EU– African relations. However, due to the links between climate policy and development, we must also take into account the idea of policy coherence. Where norms and policy coherence have gone together, the EU can be seen to be 'ahead of the game' in relation to a policy discourse on promoting adaptation to climate change in developing countries (Ayers *et al.*, 2010: 246). It is clear that there are a few examples where it appears that the EU has been quite successful in getting climate change adaptation on the African agenda, and using its resources (financial aid, trade agreements) to achieve this, an indirect achievement of the EU's if nothing else (Sicurelli, 2010: 127). This statement is testament to the EU's desire to pursue an external dimension to its environmental policy. It also confirms that the EU's environmental influence extends well beyond the requirements it makes of its member states and potential applicants – it is willing to use its own instruments to promote the necessary collective action to combat environmental degradation.

However, the biggest problems are caused when normative power is undermined by policy incoherence. As Yamin (2005: 350) argues 'long-term European strategy on climate change post-2012 will require the EU, and other negotiating partners, to focus much more deeply on adaptation issues than has hitherto been the case. Advocating developing countries' mainstream climate adaptation will require, in turn, that the EU needs to pursue policy integration within the EU to ensure that environmental considerations, specifically climate protection and adaptation policies, are integrated into other policy areas – a process which is currently at an early stage. Issues such as biofuels, trade policy and the CAP can be seen to undermine EU actions in the field of development. A good example is that of EU lobbying to ensure the environmental integrity of CDM projects, which had the effect of reducing the attractiveness of projects in LDCs. Then there is the question of money. EU states are keen on rhetoric but less willing to fund measures to aid adaptation to climate change. Part of the problem lies with the fact that climate change involves so many EU actors, which means that a range of different documents are produced. We therefore have the Commission communication on climate change and development, the EU–Africa Climate Change partnership and the Global Climate Change Alliance. The last grouping, bringing together African LDCs and Small Island States has been relatively successfull politically, despite the differing consequences of climate change in each grouping. However, focus on the GCCA can be said to deflect attention and money from other initiatives, such as the Africa–EU Partnership on Climate Change.

The Copenhagen negotiations showed that whilst the 'EU can certainly set standards ... this does not necessarily mean that "everybody follows"' (Kilian and Elgström, 2010: 266). The need for coherence both in terms of actorness and policy is vital if the EU wishes to play a normative leadership role in this crucial policy field. Part of the problem stems from the EU's silence on the 'political challenges involved in reconciling the diverging interests of both continents on this dossier with strong North-South connotations' (Bossuyt and Sherriff, 2010: 5). Since the financial crisis hit the EU, this silence is now deafening. Having said that, it does appear that there is true dialogue between the AU and the EU on many issues associated with climate change, including both the need for the developing world to take some responsibility and the responsibility of the developed world for the major burden. Indeed, it is clear the EU has had influence over norms in African states. However, the EU can be seen as an incoherent actor by African states. These disagreements between African states and the EU over the actual amounts of aid, the rationale behind that aid and even the consistency of the EU positions within negotiations (see Sicurelli, 2010) weakens the relationship between the two, thereby allowing other actors to shape global climate change negotiations.

Acknowledgements

Earlier versions of this chapter were presented at the University of Kent Energy Analysis Group Workshop 2010 and UACES Annual Conference, Bruges 2010. Thanks to participants for comments and to Maurizio Carbone, Charlie Burns and two referees for their extremely useful suggestions. All errors remain my own.

Notes

1 This section draws upon Burchell and Lightfoot (2004).
2 One clear issue with EU–Africa relations is the large number of treaties and agreements in place. It is clear that on the issue of climate change, there is overlap between EU–ACP Agreements (which do not cover a number of North African states) and EU–Africa strategies.
3 See for example the AU Addis Ababa Declaration on Climate Change and Development in Africa which stresses many of the same issues as the EU document (Harmeling *et al.*, 2007).
4 The Climate for Development in Africa Programme is an integrated, multi-partner programme addressing climate observations, climate services, climate risk management and climate policy needs in Africa.
5 The complex methodological frameworks developed to ensure real, measurable and additional emission reductions is seen as a crucial hindrance to projects in LDCs (see Lovbrand *et al.*, 2009).

6 Tackling desertification are key elements of the Environment Initiative of the New Partnership for Africa's Development (NEPAD), the Johannesburg Plan of Implementation (JPOI), adopted by the World Summit on Sustainable Development (WSSD) in 2002 and the MDG for poverty reduction and environmental sustainability.

7 There was discussion in the Hassi Report 2008 about inclusion of afforestation in the post-2012 framework, but it appears that this issue is still excluded from projects.

References

ACP–EU (2009) *Joint ACP–EU Declaration on Climate Change and Development* Brussels, 28–29 May.

Adger, N., Huq, S., Brown, K., Conway, D. and Hulme, M. (2003) 'Adaptation to Climate Change in the Developing World', *Progress in Development Studies*, 3 (3): 179–95.

Ayers, J. and Dodman, D. (2010) 'Climate Change Adaptation and Development', *Progress in Development Studies*, 10 (2): 161–8.

Ayers, J., Huq, S. and Chandani, A. (2010) 'Assessing EU Assistance for Adaptation to Climate Change in Developing Countries: A Southern Perspective', in S. Oberthür and M. Pallemaerts (eds), *The New Climate Policies of the EU*, Brussels: VUB Press.

Bagoyoko, N. and Gibert, M. (2009) 'The Linkage between Security, Governance and Development: The European Union in Africa', *Journal of Development Studies*, 45 (5): 789–814.

Behrens, A. (2008) 'Financial Impacts of Climate Change: An Overview of Climate Change-Related Actions in the European Commission's Development Cooperation', CEPS Working Document 305/September 2008, Briefing note prepared for the European Parliament, Brussels: Centre for European Policy Studies.

Boko, M., Niang, I., Nyong, A., Vogel, C., Githeko, A., Medany, M., Osman-Elasha, B., Tabo, R. and Yanda, P. (2007) 'Africa', in M.L. Parry, O.F. Canziani, J.P. Palutikof, P.J. van der Linden and C.E. Hanson (eds), *Climate Change 2007: Impacts, Adaptation and Vulnerability, Contribution of Working Group II to the Fourth Assessment Report of the Intergovernmental Panel on Climate Change*, Cambridge: Cambridge University Press, pp. 433–67.

Bonaglia, F., Goldstein, A. and Petito, F. (2006) 'Values in European Union Development Cooperation Policy', in S. Lucarelli and I. Manners (eds), *Values and Principles in EU Foreign Policy*, London: Routledge.

Bossuyt, J. and Sherriff, A. (2010) 'What Next for the Joint Africa–EU Strategy? Perspectives on Revitalising an Innovative Framework: A Scoping Paper', Discussion Paper No. 94, ECDPM Maastricht.

Brande, E. (2008) 'Green Civilian Power Europe?', in J. Orbie (ed.), *Europe's Global Role: External Policies of the European Union*, Farnham: Ashgate, pp. 157–79.

Bretherton, C. and Vogler, J. (2006) *The European Union as a Global Actor*, 2nd edition, London: Routledge.

Burchell, J. and Lightfoot, S. (2004) 'Leading the Way? The European Union at the WSSD', *European Environment*, 14: 331–41.

Carbone, M. (2007) *The European Union and International Development: the Politics of Foreign Aid*, London: Routledge.

Carbone, M. (2008) 'Mission Impossible: The European Union and Policy Coherence for Development', *Journal of European Integration*, 30 (3): 323–42.

Carbone, M. (2010) 'The EU in Africa: Increasing Coherence, Decreasing Partnership', in F. Bindi (ed.), *The Foreign Policy of the European Union: Assessing Europe's Role in the World*, Washington, DC: Brookings Institution Press.

Charles, M., Ryanb, R., Oloruntobac, R., von der Heidtd, T. and Ryane, N. (2009) 'The EU–Africa Energy Partnership: Towards a Mutually Beneficial Renewable Transport Energy Alliance?', *Energy Policy*, 37: 546–55.

Collier, P., Conway, G. and Venables, T. (2008) 'Climate Change and Africa', *Oxford Review of Economic Policy*, 24 (2): 337–53.

Council of the European Union (2008), *The Africa–European Union Strategic Partnership*, Brussels: European Communities.

Curtin, J. (2010) *The Copenhagen Conference: How Should the EU Respond?*, Dublin: Institute of International and European Affairs, January.

Dai, X. and Diao, Z. (2010) 'Towards a New World Order for Climate Change: China and the European Union's Leadership Ambition', in R. Wurzel and J. Connelly (eds), *The EU as a Leader in International Climate Change Politics*, London: Routledge/UACES, pp. 252–68.

Dearden, S. (2008) 'Introduction: European Union Development Aid Policy: The Challenge of Implementation', *Journal of International Development*, 20: 187–92.

De Lopez, T., Ponlok, T., Iyadomi, I., Santos, S. and Mcintosh, B. (2009) 'Clean Development Mechanism and Least Developed Countries: Changing the Rules for Greater Participation', *The Journal of Environment and Development*, 18 (4): 436–52.

Egenhofer, C. (2006) *Policy Coherence for Development in the EU Council: Strategies for the Way Forward*, Brussels: Centre for European Policy Studies.

European Commission (2005a) *The 2005 Review of the EU Sustainable Development Strategy: Initial Stocktaking and Future Orientations*, COM (2005) 37, 9 February.

European Commission (2005b) *EU Strategy for Africa: Towards a Euro-African Pact to Accelerate Africa's Development*, COM (2005) 489, 12 October.

European Commission (2007) *Building a Global Climate Change Alliance between the European Union and Poor Developing Countries most Vulnerable to Climate Change*, COM (2007) 540, 18 September.

European Commission (2008) *One Year after Lisbon: The Africa–EU Partnership at Work*, COM (2008) 617, 17 October.

Farrell, M. (2005) 'A Triumph of Realism over Idealism? Cooperation between the European Union and Africa', *Journal of European Integration*, 27 (3): 263–83.

Faust, J. and Messner, D. (2005) 'Europe's New Security Strategy: Challenges for Development Policy', *European Journal of Development Research*, 17 (3): 423–36.

Fioramonti, L. and Poletti, A. (2008) 'Facing the Giant: Southern Perspectives on the European Union', *Third World Quarterly*, 29 (1): 167–80.

Flint, A. (2008) 'Marrying Poverty Alleviation and Sustainable Development? An Analysis of the EU–ACP Cotonou Agreement', *Journal of International Relations and Development*, 11: 55–74.

Geyer, R. and Lightfoot, S. (2010) 'The Strengths and Limits of New Forms of EU Governance: The Cases of Mainstreaming and Impact Assessment in EU Public Health and Sustainable Development Policy', *Journal of European Integration*, 32 (4): 339–56.

Hadfield, A. (2006) 'The Role of Energy in Sustainable Development: Greening the Environment and Securing Energy Supply', in M. Pallemaerts and A. Azmanova (eds), *The EU and Sustainable Development: Internal and External Dimensions*, Brussels: VUB Press.

Harmeling, S., Bals, C. and Burck, J. (2007) 'Adaptation to Climate Change in Africa and the European Union's Development Cooperation', Germanwatch Briefing Paper.

Harmeling, S. (2009) 'Adaptation to Climate Change in the Joint Africa–EU Strategy and the Copenhagen Climate Summit', Venro Working Paper.

Holden, P. (2006) 'Conflicting Principles in the Organisation of Aid Policy for Political Purposes: A Case Study of the European Union's Mediterranean Aid', *European Journal of Development Research*, 18 (3): 387–411.

Holland, M. (2008) 'The EU and the Global Development Agenda', *Journal of European Integration*, 30 (3): 343–62.

Holstenkamp, L. (2010) 'An Overview of European Programs to Support Energy Projects in Africa and Strategies to Involve the Private Sector', in P. Hoebink (ed.), *European Development Cooperation: In Between the Local and the Global*, Amsterdam: Amsterdam University Press, pp. 95–123.

Hoste, J.C. (2010) 'Where was United Africa in the Climate Change Negotiations?', available at: www.edc2020.eu/.../Jean_Christophe_Hoste_-_Where_was_united_Africa_in   _the_climate_change_negotiations_-_EDC_2020.pdf (accessed 30 June 2012).

Hout, W. (2010) 'Governance and Development: Changing EU Policies', *Third World Quarterly*, 31 (1): 1–12.

Huliaras, A. and Magliveras, K. (2008) 'In Search of a Policy: EU and US Reactions to the Growing Chinese Presence in Africa', *European Foreign Affairs Review*, 13: 399–420.

Kilian, B. and Elgström, O. (2010) 'Still a Green Leader? The European Union's Role in International Climate Negotiations', *Cooperation and Conflict*, 45 (3): 255–73.

Lamin, M. (2004) 'Climate Change and Development: The Role of EU Development Cooperation', *IDS Bulletin*, 35 (3): 62–5.

Langan, M. (2012) 'Normative Power Europe and the Moral Economy of Africa–EU Ties: A Conceptual Reorientation of "Normative Power"', *New Political Economy*, 17 (3): 243–70.

Lövbrand, E., Rindefjäll, T. and Nordqvist, J. (2009) 'Closing the Legitimacy Gap in Global Environmental Governance? Lessons from the Emerging CDM Market', *Global Environmental Politics*, 9 (2): 74–100.

Manners, I. (2002) 'Normative Power Europe: A Contradiction in Terms?', *Journal of Common Market Studies*, 40 (2): 235–58.

Manners, I. and Whitman, R. (2003) 'The "Difference Engine": Constructing and Representing the International Identity of the European Union', *Journal of European Public Policy*, 10 (3): 380–404.

Matthews, A. (2008) 'The European Union's Common Agricultural Policy and Developing Countries: The Struggle for Coherence', *Journal of European Integration*, 30 (3): 381–99.

Newell, P., Jenner, N. and Baker, L. (2009) 'Governing Clean Development: A Framework for Analysis', *Development Policy Review*, 27 (6): 717–39.

Oberthür, S. and Roche Kelly, C. (2008) 'EU Leadership in International Climate Policy: Achievements and Challenges', *The International Spectator*, 43 (3): 35–50.

Olsen, G.R. (2004) 'Challenges to Traditional Policy Options, Opportunities for New Choices: The Africa Policy of the EU', *The Round Table*, 93 (375): 425–36.

Pallemaerts, M. and Armstrong, J. (2009) 'Financial Support to Developing Countries for Climate Change Mitigation and Adaptation: Is the EU Meeting Its Commitments?', paper presented at the International Conference on the External Dimension of the EU's Sustainable Development Strategy, Brussels, 28 January.

Parks, B. and Roberts, J.T. (2006) 'Environmental and Ecological Justice', in M. Betsill, K. Hochstetler and D. Stevis (eds), *Palgrave Advances in International Environmental Politics*, Basingstoke: Palgrave.

Paterson, M. and Grubb, M. (1992) 'The International Politics of Climate Change', *International Affairs*, 68 (2): 293–310.

Peskett, L., Grist, N., Hedger, M., Lennartz-Walker, T. and Scholtz, I. (2009) 'Climate Change Challenges for EU Development Cooperation: Emerging Issues', EDC Brief No. 3, EADI, Bonn.

Raleigh, C. (2010) 'Political Marginalization, Climate Change, and Conflict in African Sahel States', *International Studies Review*, 69–86.

Redclift, M. and Sage, C. (1998) 'Global Environmental Change and Global Inequality', *International Sociology*, 13 (4): 499–516.

Scheipers, S. and Sicurelli, D. (2008) 'Empowering Africa: Normative Power in EU–Africa Relations', *Journal of European Public Policy*, 15 (4): 607–23.

Sicurelli, D. (2010) *The European Union's Africa Policies: Norms, Interests and Impact*, Farnham: Ashgate.

Stern, N. (2006) *The Economics of Climate Change: The Stern Report*, Cambridge: Cambridge University Press.

Storey, A. (2006) 'Normative Power Europe? Economic Partnership Agreements and Africa', *Journal of Contemporary African Studies*, 24 (3): 331–46.

Syzmanski, M. and Smith, M. (2005) 'Coherence and Conditionality in European Foreign Policy: Negotiating the EU–Mexico Global Agreement', *Journal of Common Market Studies*, 43 (1): 171–92.

Taylor, I. (2010) 'Governance and Relations between the European Union and Africa: The Case of NEPAD', *Third World Quarterly*, 31 (1): 51–67.

Van Der Grijp, N. and Etty, T. (2010) 'Mainstreaming Climate Change into EU Development Cooperation Policy', in J. Gupta and N. Van Der Grijp (eds), *Mainstreaming Climate Change in Development Cooperation: Theory, Practice and Implications for the European Union*, Cambridge: Cambridge University Press.

Wettestad, J. (2009) 'Interaction between EU Carbon Trading and the International Climate Regime: Synergies and Learning', *International Environmental Agreements*, 9: 393–408.

Yamin, F. (2005) 'The European Union and Future Climate Policy: Is Mainstreaming Adaptation a Distraction or a Part of the Solution?', *Climate Policy*, 5: 349–61.

13

The EU–Africa migration partnership: the limits of the EU's external dimension of migration in Africa

Tine Van Criekinge

The intensification of migratory movement between Africa and Europe since the early 2000s has encouraged renewed political engagement from the EU towards the continent. This engagement has mainly taken the form of migration dialogue between the European Union (EU) and migrant-sending countries in Africa, aiming to create channels for communication and cooperation between Europe and its southern neighbours. Dialogue with migration countries has become a crucial part of the external dimension of the EU's migration policy, or rather the integration of migration policy with traditional foreign policy domains such as development, trade and security, and the establishment of cooperation mechanisms between receiving and sending countries. Both the EU and African side recognise that through a coherent and coordinated policy of 'joint migration management', migration can be beneficial for both sides.[1] The EU's intensification of migration dialogue with migrant-sending countries is evidence to the changing dynamics shaping EU–Africa relations in the twenty-first century. The necessity for cooperation has given some African countries a renewed or increased strategic importance *vis-à-vis* Europe. Yet, to what extent has the EU been able to capitalise on this new area of engagement with Africa? Have these changing dynamics leaned in favour of an EU-wide approach towards the continent in duly engaging migrant-sending countries, and inducing cooperation on such a strategic area of importance for the EU?

This chapter explores the extent to which the EU's migration dialogue with African countries has impacted on its relations with the continent. Despite the rapid development of the external dimension of the EU's migration policy and its definite focus on the African continent, it is argued that the EU faces constraints in conducting migration dialogue and coherently implementing migration-related initiatives. These constraints have negatively affected the EU's ability to establish itself as an effective and relevant actor in the context of African migration, and has resulted in a policy that has been characterised as incoherent and weak. The first section establishes the major developments in EU–Africa migration dialogue. High-level dialogue particularly with Africa on joint migration management has become a priority item on the EU's political agenda. Although the EU has sensitised African countries to the importance of the migration phenomenon for the EU and for the countries' development, at the same time agendas are characterised by diverging interests and concerns. The second section reviews Africa's views on migration. Here the focus has largely been on making migration a positive tool for development. African governments generally agree with the need to coordinate migration agendas and have been willing to engage in dialogue with the EU provided this considers the concerns of both origin and destination countries. The last section analyses the EU's capacities and constraints in formulating an effective and coherent approach to cooperation with migrant-sending countries. Although the EU has endorsed a balanced and comprehensive approach, in practise it has employed a combination of repressive measures and incentive mechanisms meant to induce countries of origin to comply with readmission and migration control measures. Regardless of the EU's increased efforts in formulating an effective and comprehensive migration policy, the institutional and political constraints it faces tend to weaken its effectiveness.

Developments in the EU–Africa migration partnership

The EU's commitment to the formulation of a migration policy has grown rapidly since its initial foundation in the Maastricht Treaty. The gradual institutionalisation of the policy (Geddes, 2000) and the growth of high-level political discourse on migration have had significant implications for the EU's relations with migrant-sending countries. The African continent has witnessed an increasing willingness from the EU to engage in cooperation efforts geared at developing a joint strategy on migration. The Cotonou Agreement between the EU and the African, Caribbean and Pacific group of states (ACP), which in linking migration with development and inserting clauses on migration into the text, brought the issue to the forefront of the EU's relations with Africa (European Union, 2000). The insertion of a migration clause, Article 13, was amongst the most contentious issues during the Cotonou negotiations, highlighting the weight and importance of the issue for both the EU and the ACP. It defines the parameters of the EU–ACP dialogue on migration, and is essentially the result of a difficult compromise between the parties' different views and interests. While on the EU side, member states, under domestic pressure to reduce irregular migration stemming from Africa, strongly endorsed the integration of a readmission clause, instead the ACP was keen on securing the rights and protection of their migrants in the EU (Vanheukelom *et al.*, 2006: 6). Although the Article considers the EU and ACP's obligations towards migrants by committing to the right of fair treatment, importantly the Article also allows the EU to negotiate readmission agreements with individual countries and solicits cooperation in joint migration management.[2]

In the Maghreb, the EU has sought cooperation on migration in the context of the Euro-Mediterranean Association Agreements (AA), signed with all countries in the region aside from Libya. These agreements include chapters on 'cooperation in the prevention and control of illegal immigration', requiring dialogue on joint migration management, combating clandestine migration and readmission, as well as providing commitments regarding the treatment of migrants with regards to working conditions, remuneration, dismissal and social security (Van Criekinge, 2009: 184). In addition to the AAs, the European Neighbourhood Policy (ENP) offers another framework within which the EU has promoted migration dialogue with the North African region. The Arab Spring of 2011 forced the EU to revisit its approach towards region. In some ways, the EU has responded through changes introduced in the ENP as a result of a major policy review conducted in June 2011. EU action is envisioned in training the young and unemployed, visa facilitation for certain categories of migrants, and the opening of mobility partnership negotiations with Egypt, Morocco and Tunisia. The latter will provide the countries with temporary labour market access and increased assistance while requiring cooperation on readmission and irregular migration. However, issues such as true visa-free travel have been placed under a longer-term framework, with provisions on these remaining vague.

Shocking events in Ceuta and Melilla in 2005, in which several unarmed African migrants were killed while attempting to surmount blockades surrounding the two Spanish enclaves in Morocco, further incited a call for migration dialogue with Africa. In 2005, the Council adopted the *Global Approach to Migration: Priority Actions Focusing on Africa and the Mediterranean*, prompting action in three key areas: strengthening cooperation and action between the member states, increasing dialogue and cooperation with Africa, and promoting the creation of a framework for funding and implementation of a strategy on migration (Council of the European Union, 2005). In December 2006, the Council urged for the establishment of a comprehensive migration policy, which would include partnerships with key migration countries and regions. Euro-African Ministerial Conferences on Migration and

Development, which included government officials, civil society, academic and media responsible for migration issues in the region, were held in Rabat, Tripoli and Paris between 2006 and 2008, leading the parties to agree to a multi-annual programme of cooperation based on enhancing regular migration combating irregular migration and establishing synergies between migration and development. A Partnership on Migration, Mobility and Employment also became one of the strategic priority areas under the 2007 Joint EU–Africa Strategic Partnership, setting out an ambitious agenda for long- and short-term cooperation. A Second Action Plan was adopted in late 2010 at the EU–Africa Summit held in Tripoli, calling for a more enhanced dialogue and identifying a series of concrete actions to be undertaken between 2011 and 2013. Lastly, the *European Pact on Immigration and Asylum* (Council of the European Union, 2008) adopted by the Council in October 2008, again reaffirmed the EU's intent on creating comprehensive partnerships with countries of origin and transit.

The Global Approach to Migration also called for the EU diplomatic missions to be sent to key African countries to conduct dialogue based on Cotonou's Articles 13 and 8 (political dialogue). These missions have been led by EU delegations, embassy staff of Council Presidencies and interested member states, and Commission representatives, and ideally covers, 'a broad range of issues from institution and capacity building and effective integration of migrants to return and the effective implementation of readmission obligations, in order to establish a mutually beneficial cooperation in this field' (Council of the European Union, 2005: 5). Since 2006, missions have been sent to Cameroon, Cape Verde, Ethiopia, Ghana, Kenya, Libya, Mali, Mauritania, Nigeria, Senegal, South Africa and Tanzania.

Three main instruments have been used to incite cooperation with migrant-sending countries: financial and technical assistance, increased labour market access offers and the implementation of migration control mechanisms. Since 2005, the European Development Fund (EDF) has implemented some migration-related assistance. Until 2007, the Aeneas budgetary scheme provided specific and complementary financial and technical assistance to third countries in support of efforts to manage migration flows. This type of assistance continues under the new budget line, the Thematic Cooperation Programme with Third Countries in the Development Aspects of Migration and Asylum. The EU has also taken steps to increase opportunities for legal migration. In 2009, the Council approved the so-called 'Blue Card directive', aimed at recruiting highly skilled workers to the EU by offering a single work and residence visa; and the 'Rights directive', aimed at facilitating migrant integration by allowing access to a range of socio-economic benefits (European Union, 2009). With the entry into force of the Lisbon Treaty in 2009, migration policy has become even further communitarised. All EU decisions on asylum, immigration and integration will now be subject to qualified majority voting in the Council, and the European Parliament has been given joint decision-making power, potentially strengthening the EU's capacity to offer further labour market access towards migrants. The EU has also encouraged the implementation of circular migration schemes, to ensure that regular labour migration remains temporary. In 2008, the Council approved the creation of mobility partnerships with third countries, and has implemented pilot schemes in Africa and Eastern Europe.[3] The scheme offers temporary labour market access in exchange for cooperation on combating irregular migration and assistance on border security through Frontex, the EU's external border agency. Lastly, political engagement has been coupled with an increased use in security instruments to fight irregular migration. Since 2005, Frontex has carried out studies on surveillance systems and produced risk analysis reports on African migratory routes, and has implemented several border patrolling missions mostly in West Africa. Additionally, the *European Pact* also evidenced the EU's commitment to fighting irregular migration, by making the strengthening of border controls and ensuring readmission and return of irregular migrants priority areas for action.

Increased high-level dialogue between the EU and Africa and the variety of instruments used to consolidate this dialogue, indicate the EU's interest and intent to work in cooperation with migration countries in regulating and managing migration flows (Gnisci, 2008: 85). The intensification of the EU's efforts on the continent and the gradual establishment of a framework for cooperation and action have also increased pressure on African governments to consider how migration affects them. Although placing migration issues on the African agenda has been spurred largely by European pressure, at the same time the interests and issues of concern between the continents diverge. With migration bringing potential developmental benefits, it should come as no surprise that when cooperating with destination countries, African governments have focused on the positive aspects of the phenomenon. In general, African governments have been willing to engage with Europe provided the dialogue considers the concerns of both origin and destination countries. That said, interests in finding a common approach to migration management diverge somewhat between Africa and the EU. While developments in the external dimension of the EU's migration policy have tended to combine repressive measures with incentive instruments in order to incite cooperation from sending countries, African governments have instead focused largely on enhancing the developmental prospects that migration provides. The next section further explores how the African migration agenda has tended to focus more on a migration–development nexus, as opposed to migration control.

The African migration agenda: making migration work for development

With the growing interest on the EU side to place migration issues high on the agenda, the African side too now considers migration an important element on its political agenda. African governments now accept that migration cannot be seen in isolation of development policy, and, as such, they are aiming to ensure that migration does indeed contribute to development. The African migration agenda has focused on four major issues: remittances, mitigating brain drain and encouraging brain gain, assistance in fighting root causes and capacity-building in migration management.

Perhaps the most positive effect of migration on development is visible in remittances, or the private transfer of funds migrants send to their countries of origin. Although in Africa, Official Development Assistance (ODA) and Foreign Direct Investment (FDI) continue to represent the most significant flows of capital, remittances are steadily on the rise and are less volatile. In countries where remittances constitute an increasingly important source of revenue, governments have placed the facilitation and institutionalisation of remittance flows high on the national agenda. Secondly, the African migration agenda has focused on encouraging the positive link between migration and development through potential gains in human capital and promoting economic growth. The long-term loss of skilled migrants specialised in sectors of particular relevance to development has led to 'brain drain' becoming a major problem for Africa, where the rate of university graduates who have migrated to Europe is higher than in any part of the world (Katseli *et al.*, 2006: 19). Some African governments have begun exploring ways to encourage migrants to return and contribute to development efforts in their country of origin. Brain gain is also stimulated through increased engagement with the diaspora, which facilitates the forging of trade, investment and development links between origin and destination countries. Indeed, involving the diaspora in development processes is yet another item of high importance on the African migration agenda. Thirdly, African countries have also demanded more assistance in tackling 'root causes' which act as the main 'push factors' for migration, such as poverty, unemployment, political and economic instabilities, overpopulation and environmental degradation. To this extent, assistance in the creation of increased employment opportunities and schemes which provide incentives for potential migrants to stay through higher wages and better working conditions have played an important role in requests for assistance. Furthermore, because institutional capacity required to formulate and implement migration policies is relatively weak, governments have also been keen on assistance towards capacity-building and training of officials to deal with migration issues at the domestic level (Adepoju, 2008: 40). Thus, the need to coordinate migration initiatives and policies amongst relevant national stakeholders and institutionalising information exchanges and dialogue with destination countries is considered necessary for better management of migration flows.

A commitment towards developing the migration–development nexus is clear at the continental, regional and national levels. At the all-ACP level, the 2006 Brussels Declaration and Plan of Action outlined ACP commitment to developing a migration dialogue with the EU, as well as identifying the need for developing a holistic approach to migration (ACP, 2006). At the pan-African level, since 2006 the African Union (AU) has started to develop a common position on African migration policies and has also participated in migration dialogue and initiatives with the UN and the EU. The common position emphasises the challenges migration poses for the continent, providing guidelines for member states to implement common measures advocated by national, regional and international bodies. The framework focuses not only on the development aspect of migration, but also recognises the need to work with countries of destination to combat irregular migration flows (Gnisci, 2008: 97–8). At the regional level, several economic communities have begun formalising migration dialogue. East Africa held its first workshop on migration in May 2008, initiated namely under the auspices of the AU's migration framework and the EU–Africa migration dialogue. The region has launched a Regional Consultative Process on migration between the Inter-Governmental Authority on Development (IGAD) and strengthening cooperation on migration with North Africa. The Southern African Development Community (SADC) established the Migration Dialogue for Southern Africa in 2000, to facilitate regional cooperation and dialogue. The Economic Community of West African States (ECOWAS) adopted the Common Approach on Migration in January 2008, highlighting, firstly, the region's commitment to adopting a common legal framework and key principles on migration, and, secondly, setting out an action plan for implementing a framework on migration and development. Although the formulation of the Common Approach was largely influenced by EU pressure on the region to develop a coherent framework (Oucho, 2008: 96), the willingness of West African governments to engage in regional migration dialogue has been driven by the recognition that prospects for successful regional integration are strongly linked to both intra-regional and international migration dynamics. As such, ECOWAS has tended to focus on the linkages between migration, development and regional integration (Gnisci, 2008: 106).

Although there are some differences in the various emerging agendas on the continent, some common factors characterise the African position on migration. Firstly, the role of EU/European influence in the formulation of an African migration agenda cannot be understated, as increased salience of migration and much of what is happening at the African level has been a reaction to increased pressure from the international community, and especially Europe, for better joint management of migration (Zoomers *et al.*, 2008: 4). Secondly, the need to better coordinate and harmonise migration-related policies, country and regional frameworks and international efforts at joint migration management also features heavily on both agendas. The different African initiatives have demonstrated a clear commitment to dialogue and cooperation amongst relevant stakeholders. Indeed, dialogue has become ever-more frequent and intense, and, in general, African governments have been willing to engage provided the dialogue considers the concerns of both origin and destination countries. That said, while there has been a considerable acceleration of both EU and African efforts in developing a strategy of joint cooperation, the initial years following the *Global Approach* can be characterised mainly by 'agenda setting with Africa' (Bosch and Haddad, 2007: 17), or, rather, in conducting dialogue to find a common approach to how best to deal with migration issues. Thus, 'by furthering dialogue and co-operation with African partners to implement the global approach, a consensus has emerged – theoretically, if not for the moment practically – on the strategy linking "migration and development"' (Gnisci, 2008: 85).

The EU has indeed stepped up its efforts in migration dialogue with Africa. But although the EU's rhetoric on migration has clearly intensified, 'it now remains to be seen whether concrete implementation can match the political statements made and policy initiatives taken' (Bosch and Haddad, 2007: 17). The following section investigates whether the increased opportunities for dialogue and cooperation created by migration pressures generating from Africa towards the EU have indeed been effectively exploited by the EU.

The limits of the EU's external dimension of migration in Africa

Despite the manifest changing dynamics between the EU and Africa and the high-level political agenda generated by migration, the EU is constrained in its application of an effective migration policy towards Africa. Institutional and political constraints have weakened the effectiveness of the policy, and therefore the EU's ability to engage fully with countries of origin in seeking cooperation on migration. These are found namely with regard to the EU's capacity to engage with the countries, incoherence with other external policies and limited coordination and harmonisation amongst the various levels of policy- and decision-making. On the basis of in-depth interviews with relevant policy-makers and stakeholders in Brussels, as well as brief considerations of two country case studies, Senegal and Ghana,[4] it will become evident that these constraints severely limit the EU's effectiveness in migration dialogue and the implementation of policies in its relations with migrant-sending countries.

Capacity

The increased salience of migration issues on the EU–Africa agenda calls for relevant funding and policy expertise, requiring both sufficient financial and human resources. Yet the provision (or lack thereof) of adequate resources can be a significant impediment to progress. Thus, a very concrete constraint on the EU's ability to engage in a coordinated and coherent migration policy towards Africa is found in the limitations in resources which it is able and prepared to employ in executing the policy.

According to Bosch and Haddad, 'progress [in EU engagement with migrant-sending countries] can only be made if there are adequate resources. This is the case for the Commission, but also for member states, and includes particularly the EC delegations and member state embassies where officials frequently have to cover a whole range of issues in addition to migration' (2007: 16). Although delegation staff are expected to engage in migration dialogue with African governments, the challenge in effectively managing the policy lies partly in allocating sufficient financial and human resources and time to the policy, and in building up relevant policy expertise to handle an increasingly important profile (Interviews Brussels, Accra, Dakar, 2008, 2009).[5] EU migration-related programmes however are mostly managed through intermediary agencies such as the International Organisation for Migration (IOM) or the United Nations Development Programme (UNDP). Since African governments are therefore not directly working with the EU, the EU is sometimes perceived as either incapable in handling the profile, or other agencies have taken a more prominent leadership role in working with the governments. Indeed, the European Commission itself has acknowledged its limitations in terms of expertise and technical capacity in migration management. In 2008, it proposed enhancing expertise on migration management through the exchange of information and expertise, and the training of delegation and member states embassy staff (European Commission, 2008: 12); however, much of the EU's work continues to be outsourced (Interviews Brussels and Dakar, 2010).

This is clearly evidenced in the cases of Senegal and Ghana. In both countries, the IOM, and to a lesser extent the UNDP, have worked in close cooperation with the governments, and almost all EU migration-related projects have been coordinated and implemented by these organisations. In Senegal, the delegation of project implementation to other international organisations is perceived as failing to strengthen the government's own capacity in migration management and undermining ownership (Interviews Brussels and Dakar, 2008, 2009). Conversely, this delegation has especially strengthened the IOM's role in Ghana, as it has worked closely with relevant national and international stakeholders on the ground, rather than merely at the level of high-level dialogue, agenda-setting, or project formulation, as is the case for the EU. At the same time, delegating competences to the IOM has also contributed to the image of the EU as limited in its involvement in direct cooperation with the government, with the IOM instead seen as spearheading progress at the national level (Interviews Accra, 2008). This is particularly relevant for understanding the EU's effectiveness in engaging with African countries on migration, as the inability to undertake a strong leadership role at the national level has allowed other international organisations to surpass the EU despite the EU having driven the high-level agenda.

A similar capacity constraint is found in the limited financial resources that have been dedicated towards migration-related programmes to date. Although increased resources are now being invested in the policy, and specific financing instruments have been set up,[6] budgets have been criticised as being too modest, and therefore limit the effectiveness of migration-related initiatives (Roig and Huddleston, 2007: 378). The 10th EDF integrated financial instruments dedicated to migration in some West African Country Strategy Papers (CSPs). Regardless of the extensive migration discourse found in some of these CSPs, migration-related funding has been modest in comparison. Although Senegal's CSP, for example, claims the EU is committed to closer cooperation with the government on migration, only 4 million euro were allocated to migration-related assistance, or less than 1 per cent of the country's total allocation; for Ghana this was less than 0.5 per cent. In addition, the Commission itself notes that regardless of the various financial instruments that fund migration-related initiatives, problems remain in coordination amongst the different funding schemes, with many of the instruments currently in place considered limited or incomprehensive (Interviews Dakar, Accra, Brussels 2008, 2009).

Despite these capacity constraints evident in the EU's approach towards Africa, the policy is currently undergoing a major geographic expansion. Although the *Global Approach* initially prioritised relations with the Mediterranean and Africa, since 2007 it includes Eastern and South-Eastern regions bordering the EU, and extends cooperation as far as Asia, the Middle East, Latin America and the Caribbean. Certainly the most pressing migration movements towards Europe originate along the EU's borders, guaranteeing that at least in the short-term, the EU's concentration on these geographic areas will remain. Although geographic expansion is essentially in line with a more comprehensive approach to migration, at the same time 'such further broadening of the strategic horizon involves a risk of operational and financial over-stretching' (Pastore, 2007: 7), which in turn may undermine the EU's effectiveness in managing such a far-reaching policy. According to both European and African migration officials, the implementation of concrete policy initiatives in the field of migration can only be considered effective when the EU has the adequate resources and policy expertise necessary to translate rhetoric into actions (Interviews Dakar 2008, 2009). Nevertheless, the EU clearly faced capacity constraints in terms of the financial and human resources it can dedicate to implementing its migration policies in Africa.

Policy (in)coherence

As an area that overlaps with other EU external policies, especially development, trade, and security, and for joint migration management with countries of origin to be effective, the EU considers coherence amongst the relevant policies and policy-actors essential (European Commission, 2006: 4; 2008: 12). Yet this is one policy area, where the lack of coherence has been particularly evident and problematic. Trade policies and to a lesser extent development policies operate in separate policy spheres from migration, while the security field has instead adopted a leadership position in the formulation and implementation of the external dimension. The dominance of a security-led approach in the EU's migration policy has led several observers to point towards an 'overwhelming presence of the "security rationale" surrounding the debate concerning migration and development' (Chou, 2006: 2–3). The institutional set-up of the EU perpetuates this security-oriented approach and has led to a weakening of policy objectives in Africa.

There are essentially two approaches which the EU can take in linking migration and development: either to 'use development tools to reach migration goals such as tackling illegal immigration' or to 'utilise migration tools such as legal immigration to achieve development objectives' (Carrera and Chou, 2006: 141). Importantly:

> The former represents a more 'coercive approach' in the form of restricting or conditioning development aid if certain non-EU countries do not comply with member states' requests on migration management and the readmission of illegal immigrants. The latter can be characterised as a more 'open approach' which seeks to foster the potential of 'brain circulation', circular migration and the positive effects of remittances. (Carrera and Chou, 2006: 141; see also Chou, 2006, 2009b)

In 2005, the Council adopted the *Policy Coherence for Development* (PCD) Strategy, stressing the need for improvement in the coherence between 12 non-aid policy areas, including migration and development, in order to meet the Millennium Development Goals by 2015. Specifically, the EU sought to promote managed labour migration, improve remittance flows, turn the brain drain into brain gain, promote responsible recruitment practises, diaspora engagement and South–South migration management (European Commission, 2005: 15). The Commission's *Communication on Migration and Development*, released that same year, solicited similar actions, but added 'encouraging circular migration and facilitating return to the country of origin' as another priority area (European Commission, 2005). In other words, in making migration policy coherent with the EU's development policy, the EU opted for the 'open approach' by promoting migration as a positive factor for development in migration countries.

Although the EU has certainly demonstrated a willingness to engage with countries of origin, the growing trend in its approach towards the external dimension of migration has been the combination of repressive measures and incentive mechanisms soliciting closer cooperation. Concessions given in the field of legal migration have been coupled with enhanced cooperation in combating irregular migration and requiring countries to sign readmission agreements. The pilot mobility partnership schemes, for example, go beyond simply offering temporary labour market access and increased assistance, by also requiring countries to cooperate on readmission and in the fight against irregular migration. The EU's diplomatic missions to several African countries have also employed this approach, with dialogue intended to cover 'a broad range of issues from institution and capacity building and effective integration of legal migrants to return and the effective implementation of readmission obligations' (Council of the European Union, 2005) as well as increased assistance and political engagement.

The integration of migration profiles and readmission clauses in agreements with third countries and legislation on irregular migration 'enhances the EU's capacity to control and reduce unwanted immigration to its territory' (Lavenex, 2002: 162). These control elements are further coupled with instruments to induce third countries to cooperate with the EU in the first place, in that 'trade and aid are increasingly made conditional on the reduction of push factors and the readmission of persons staying illegally in the contracting party' (Lavenex, 2002: 162). Thus, the EU's approach tends to be restrictive and weighted towards a politics of control (Boswell, 2003; Geddes, 2009; Lavenex, 2002, 2007; Lavenex and Kunz, 2008; Pastore, 2007; Sterkx, 2008). Indeed, the *European Pact on Immigration and Asylum* (Council of the European Union, 2008) clearly highlights the use of a carrot and stick approach, with the Council insisting that EU-level and bilateral migration agreements should include clauses on legal migration and development but also on irregular immigration and readmission. In effect, the Pact commits the EU to undertaking not just closer cooperation with sending countries, but also to organising legal migration policies in accordance with each member state's needs and reception capabilities, to ensure that irregular migrants are sent back to their countries of origin or transit, and to increase the effectiveness of border controls (Council of the European Union, 2008). Thus, while the EU seemingly promotes a development-friendly approach, recent policy developments and the combined used of coercive and incentive mechanisms point towards a much more restrictive approach.

Thus, policy-making in migration continues to be dominated largely by the security field with limited attempts at better coordination with other relevant policies. Even in light of the changing discourse on migration in recent years through such developments as the *Global Approach*, mobility partnerships and increased political dialogue with migrant-sending countries in Africa, the EU has continued to focus on migration control rather than overcoming obstacles for greater coherence. According to Lavenex and Kunz, 'barriers towards greater policy coordination are sustained by the institutional set-up of policy-making in the EU' (2008: 453). The external dimension of migration is governed mainly by the

High-Level Working Group (HLWG) on Migration and Asylum within the General Affairs and External Relations Council (GAERC), composed of national justice and security officials. This has perpetuated command of the migration field by one group and poses as a barrier for linking migration to development and other complementary policies (Lavenex and Kunz, 2008: 453–4; Chou, 2006: 17).[7] Thus migration and development policies 'are marked since an early phase by a certain strategic fuzziness, by intrinsic political ambiguities, overlapping competences, policy incoherencies and bureaucratic competition (including, at European level, turf battles within the European Commission)' (Pastore, 2007: 3). For the most part, coordination on migration policy within the Commission has often been insufficient. In fact, the Commission's justice and home affairs (JHA) branch has taken a leadership role in the formulation of external dimension of migration, often without adequate consultation with DG Development or DG RELEX on external policy coherence (Egenhofer, 2006: 28; Sterkx, 2008: 127).

Migration dialogue with African countries is thus embedded in a security-oriented approach in that the purpose is still migration control. This is sharply in contrast to a development-oriented perspective, where the purpose is not to achieve migration goals through migration control, but to achieve development goals through migration management. This is evidenced by the EU's intense engagement with Senegal as opposed to Ghana, largely because the former is more closely aligned with the EU's strategic interests in migration management. Senegal's geographic position as a major point of departure for both regular and irregular and mostly low-skilled migrants has placed it amongst the more relevant sending-countries with which the EU is interested in collaborating on migration (Adepoju, 2009). Because the EU's main objective has been to seek cooperation on migration control (through a framework of migration management), it should come as no surprise that Senegal has become a priority partner country. Ghana, much unlike Senegal, is neither geographically nor demographically (Ghana exports mostly high- to medium- skilled migrants) placed as a strategic country (Adepoju, 2009). Migration tends to affect Ghana more adversely (in terms of brain drain) rather than the EU (in terms of irregular or low-skilled migrants), resulting in much less intense involvement on the part of the EU in Ghana as opposed to Senegal (Interviews, Accra and Brussels, 2008). Indeed, according to the Senegalese government and IOM officials, the EU began its interactions by pushing a migration control agenda rather than joint migration management. The fundamental interest in conducting dialogue was mainly to find a common position on stemming irregular migration and implementing control measures, and to find agreement on readmission and border controls. Thus, initial dialogue and policy actions in Senegal can be characterised by a systematic divergence in interests and views (Interviews, Brussels and Dakar, 2008).

Certainly, with the coming into force of the Lisbon Treaty in 2009, the dominant position of JHA, and perhaps the security-oriented approach has been challenged, as JHA has officially ceased to exist as an EU pillar and was fully integrated into Community decision-making mechanisms. Yet better linking of migration policies with other relevant areas will depend on the extent to which the European External Action Service (EEAS) will be willing and capable in formulating a more coherent approach to migration dialogue with third countries by collaborating with its counterpart in the Commission, DG EuropAid Development and Cooperation (DEVCO). On the other hand, a stronger institutional link between migration and development is not very likely given that many provisions in the Treaty actually strengthen the security-oriented approach. Firstly, Lisbon reinforces the Commission's legal competence to negotiate readmission agreements, thus augmenting its possibility for increasing dialogue with third countries for the purposes of signing readmission agreements. Secondly, the dominance of justice and home affairs officials in formulating the orientation of the external dimension of migration policy will continue, in that institutional practises and decision-making in this realm remain unchanged.

The dominance of a security-oriented approach in migration constrains the EU's capacity to engage fully with African countries, largely because the EU's interests and policy objectives diverge from most of theirs. African countries are instead seeking a more 'open' approach to migration, in which the link between development and migration is more manifest. The case of Senegal is especially relevant here, because despite the EU's efforts to have the government sign a readmission agreement and then later to negotiate a mobility partnership, the Senegalese government has continued to resist. This is evident from the very beginnings of the EU's attempts at establishing dialogue, and has continued into present-day negotiations. In 2006, the government delayed the employment of the first FRONTEX mission, until guarantees were made as to the treatment of intercepted migrants and additional funding. This issue was eventually settled when in August 2006 the Senegalese government was allowed to participate in the missions. In 2007, readmission talks also stalled. The government resisted the inclusion of non-nationals in any such agreement unless sufficient guarantees could be made regarding funding for capacity-building to manage the influx of repatriated migrants.[8] In March 2009, negotiations on a mobility partnership also reached a stalemate, 'as a result of Senegalese dissatisfaction with what they were to receive in return for the EU's terms' (Chou, 2009a: 10). Senegalese officials claim that the government has little interest in negotiating the partnership unless it clearly provides for increased development and capacity-building assistance and improved labour market access for Senegalese migrants (Interviews, Dakar, 2009, 2010). The case of Senegal has been particularly instructive for the Commission's approach towards third countries. It has since recognised that the decision to embark on negotiations for mobility partnership without the country's formal expression of interests, needs and expectations at an early stage has not worked (European Commission, 2009: 3), and has instead served to push countries like Senegal towards adopting resistant tactics and the continual formulation of high-level demands that the Commission is often unable to concede to. This is exacerbated by a lack of coordination between the Commission and the member states, which has led to further constraints on the EU's ability to engage with migrant countries in Africa.

Coordination between the Commission and the member states

While member states can agree on the benefits of developing a common migration policy or a common position on migration, separate national policy agendas restrict an effective and practical approach at the EU level (Sterkx, 2008: 126–8). Although the EU's migration policy is indeed becoming increasingly communitarised, EU member states have kept a firm grip on the external aspects of the policy, especially in relation to cooperation with third countries. This has created tensions between the supranational and inter-governmental levels of policy-making and implementation. These tensions, in turn, limit the EU's ability to promote its policy objectives abroad, as a lack of coordination between the different levels has weakened the potential effectiveness of engaging with migration countries.

Thus, policy coherence has suffered not only because of the institutional set-up that has perpetuated a security agenda rather than a developmental one, but just as importantly because of the member states' role in shaping and orienting the policy according to their preferences, and for providing 'the main impetus for the incorporation of migration policies into EU external relations' (Higazi, 2005: 5). Not only do member states steer policy orientations, their approach has leaned towards securitisation. According to Sterkx (2008), member states have pushed for a security approach to migration rather than for policy coherence in external cooperation, evidencing the extent to which they tend to be concerned first and foremost with migration management for the purposes of migration control. Commission interviewees note that member states have adopted a much more restrictive approach towards third countries as opposed to what the Commission might prefer. This has led to a constant struggle for compromise on how the policy should evolve, and, more specifically, to what extent it should adopt a 'coercive' as opposed to an 'open' approach to migration (Interviews Brussels 2008, 2009). Thus, while the EU's objective has been to pursue 'a rather innovative (at least on paper) common migration policy compared to the tradition of its member states', the policy 'has encountered considerable difficulties in shaking off the prevailing approach at the level of its member states, which often manifests itself in an opposite manner, that is, as biased towards restrictiveness and weighted towards controls, sectoral, reactive and essentially unilateral' (CeSPI/SID 2006: 5). Indeed, Gnisci reiterates this point well, when she writes that, 'the EU's global approach seeks to be integrated, balanced, negotiated and consensual. It shows that migration is now one of the strategic domains of negotiation between the Community and third countries. From the operational perspective, however, the 'control' issue is overriding and liberating national agendas from it is often difficult' (2008: 85).

In addition, delegating competence in migration matters to the EU level is delicate because of the careful balance to be struck between cooperation amongst different member states and between member states and the Commission on the one hand, and maintaining national sovereignty over migration matters on the other (Brady, 2008: 18). Indeed, while cooperation and coherence in migration policy is encouraged, some member states have been apprehensive towards delegating further competence to the EU. Even those member states that have been the main architects of the EU's emerging migration policy (namely France, Spain and, to a lesser extent, the Netherlands and the UK) have nevertheless continued to operate separate bilateral migration schemes with third countries, often leading to duplication, surpassing, or undermining of EU policy (Sterkx, 2008: 126; Interviews Dakar and Accra, 2008). For example, although the Commission has been given the mandate to negotiate readmission agreements with third countries on behalf of the EU, member states also conduct bilateral talks with migration countries. Accordingly, 'third countries do not understand this situation of parallel negotiations, and take advantage of it, which often results in the deferral of Community negotiations' (Sterkx, 2006: 126).

This overlap of efforts in third countries goes further than readmission agreements however, and extends into the EU's efforts at becoming more active in migration and development. A limited set of EU countries, namely France, Spain and, to lesser extent, Italy, are managing African migration flows through 'second-generation migration agreements'.[9] These agreements apply a carrot and stick approach to managing migration flows, by giving labour access, financial or technical assistance or investment opportunities to migration countries, in exchange for their cooperation in fighting irregular migration and readmission. Bilateral agreements thus address both the individual member states' labour needs as well as some of the root causes of migration in third countries, while providing incentives for closer cooperation in controlling irregular migration (Panizzon, 2008). Although this approach is similar to that of the EU approach, coordination between the member states and the Commission (or in this case the delegation) is often inadequate. This often perpetuates an image of the EU as unable to adequately deal with migration matters, as opposed to individual member states, whose bilateral efforts are seen as more relevant (Interviews Dakar, 2009).

Some relevant examples can serve to better illustrate this point. In Senegal, constraints faced by the EU in conducting migration dialogue with the government have been coupled with a rather unique level of intense bilateral engagement from interested EU member states. Faced with this state of affairs, the Senegalese government has opted to engage bilaterally while undermining and bypassing EU engagement until agreement can be reached that would substantially concede towards Senegalese demands for increased funding, labour market access and capacity-building measures. Bilateral engagement on migration in Senegal is not a new phenomenon, but has intensified since 2006, with the signing of bilateral migration agreements between Senegal and France and Spain. These agreements employ an incentive-based strategy for migration management by offering increased labour mobility, development assistance and/or legal migration opportunities in exchange for cooperation on migration control (Panizzon, 2008: 55). Bilateral agreements provide an alternative and more comprehensive means for the government to engage with Europe in the face of limited progress at the EU level. Intense bilateral engagement has not helped to strengthen the EU's position in the country however. Both the Senegalese government and the EU delegation admit that coordination between member states and the Commission has been lacking. Each member state is seen as pursuing its own bilateral interests and political agendas, which are often much more appropriate in relation to Senegalese preferences (Interviews, Brussels and Dakar, 2008, 2009). The EU delegation, while constrained by a lack in capacity, expertise and human resources necessary to carry-out EU-wide dialogue with Senegal, sees its efforts surpassed by individual member states that are able to offer concrete labour market access in return for increased Senegalese cooperation on migration control. Thus, while Senegal has adopted a resistant approach towards EU efforts, it has instead engaged intensively in bilateral efforts to develop a more comprehensive approach to migration management. Similarly, in Ghana, the government also recognised that the EU is constrained by a rather narrow mandate for conducting dialogue on migration matters. The issues that interest the government, such as increased labour migration opportunities or temporary and circular migration schemes, essentially go beyond the EU's competences (Interviews Accra, 2008). At the practical level, therefore, the EU's role in Ghana has been rather limited and constrained to either providing funding, or merely supporting member state initiatives (CEC–GoG, 2006). The fact that under Lisbon, legal migration is now a Community competence is unlikely to change this situation, as member states will continue to determine individual quotas for migrants allowed to enter their territories.

Bilateral agreements have also been particularly relevant in the North African region. While the migration pacts or agreements negotiated between individual EU countries with certain North African countries are now terminated as a consequence of the Arab Spring, a lack of a comprehensive EU response to migration pressures originating from the region has made a unilateral or bilateral response from individual member states again the preferred option. Member states immediately commenced new negotiations, with Italy agreeing to a new agreement on border security with Libya in April 2012. Although migration remains a key area in which the EU is expected to engage with North African countries in the aftermath of the uprisings, a limited and vague EU response has characterised the EU's initial post-crises engagement with the region. EU migration policies and debates have remained mainly reactionary, with true policy-planning taking a backseat in relation to events such as the economic crisis and the Arab Spring.

A lack of coordination weakens the effectiveness and strength of the EU's total efforts in third countries. Coherence and coordination in the EU's migration policy is undermined because of 'the exclusion of key institutional actors who prefer the comprehensive approach from the decision-making process … [and] the isolation of decision-making power within an institutional setting which favours the coercive strategy' (Chou, 2006: 3). Inadequate consultation on the part of the security branch with other relevant policy actors has led to the dominance of a restrictive and limited approach being adopted at the EU level. Furthermore, tensions between the supranational and intergovernmental levels that characterise the EU's migration policy have led to further perpetuating a security-oriented approach over a more comprehensive approach in migration cooperation with countries of origin. The fact that member states have continued operating migration policies at the bilateral level, while the EU level remains inadequately equipped to implement the policy, has undermined the total EU effort in engaging with third countries.

Conclusion

This chapter has traced the evolution of the EU's migration policy towards Africa. It was argued that, although the EU has become increasingly active in promoting and instigating dialogue and policy initiatives at the African level, EU and African interests in migration do not always converge. Moreover, the chapter has outlined the extent to which the EU is constrained in fully engaging in migration dialogue with external actors due to three main reasons.

Firstly, the EU is limited in its capacity to engage fully with governments of migrant-sending countries in Africa. The increased salience of migration issues on the EU–Africa agenda calls for relevant funding and policy expertise, requiring both sufficient financial and human resources. Yet, the provision (or lack thereof) of adequate resources can be a significant impediment to progress.

Secondly, the EU is constrained with regards to policy coherence. Although the EU has started to move towards increased coherence between migration and development, as demonstrated by the evolution of dialogue and policy initiatives, at the same time migration and development are linked only insofar as they create incentive mechanisms meant to induce countries to comply with joint cooperation in tackling irregular migration. Thus, the migration dialogue continues to be embedded in a security-oriented approach in that the purpose is still migration control, rather than reaching development goals through joint cooperation and policy coherence. The dominance of a security-oriented approach in migration constrains the EU's capacity to engage fully with countries of origin, largely because the EU's interests and policy objectives diverge from those of the countries of origin, which are instead seeking a more 'open' approach to migration.

Thirdly, the multilevel structure of the EU poses a constraint on effective coordination amongst the different actors involved in the EU's emerging migration policy towards Africa. Although decision-making and policy implementation are to be coordinated, namely between the delegations and headquarters in Brussels and between the Commission and EU member states, coordination and coherence is often lacking, and, more importantly, these gaps have consequences for the effectiveness of the EU's policy objectives projected abroad. The Arab Spring and the continuing crisis in the Eurozone have further impacted coordination between the EU and member states' levels. These events have given rise to more negative perceptions over the EU's ability to adequately handle European migration policies. As a consequence, political leaders across Europe have kept a firm grasp on migration-related policies with more European integration in migration unlikely.

In recent years, a definite resurgence of Africa has occurred on both the EU and international agenda. Indeed, this has been confirmed, at least in theory, by the coming into force of the EU–Africa Strategic Partnership. In implementing the partnership, the EU recognised the regained strategic and economic importance of the continent, and confirmed its intention to remain a crucial actor in the developments that are now re-shaping Africa's relations with external actors. One such development has been the increased migratory movement between the African continent and Europe, consequently spurring increased engagement at the EU level with migrant-sending countries in Africa. Despite the increased engagement however, the EU has proven limited in its ability to narrow the gap between capabilities and expectations, between security interests and development prospects and in coordination at different levels of governance comprising the external dimension of migration policy towards Africa. Successful engagement with African countries, and therefore the EU's ability to reach its preferences, will certainly require moving far beyond the rhetoric.

Acknowledgements

The author thanks Karen E. Smith, Gorm Rye Olsen, and Mary Farrell for their feedback on earlier versions of this piece and the University of London Central Research Fund, the LSE Department of International Relations and the Robert Schuman Centre for Advanced Studies (EUI) for their funding.

Notes

1 Joint migration management has been used by the EU and various international organisations to describe the dialogue and policies being negotiated and implemented between countries of origin, transit and destination in order to maximise the benefits and minimise the disadvantages of migration for all actors concerned. Management refers to policies that aim to reduce irregular migration, promote the rights and protection of migrants, reduce economic pressures that influence outward migration and regulate labour migration.

Countries work together to find a common or burden-sharing approach towards migration.

2 Tampere European Council in 1999 established that readmission clauses would henceforth be included in all agreements with third countries. The insertion of this clause in Cotonou was met with fraught resistance from the ACP, on the grounds that the obligation to readmit third country nationals was incompatible with international law (*Statewatch Bulletin*, 2000). The issue was left unsettled until the final round of negotiations, where the EU and ACP eventually agreed that readmission clauses would be negotiated at the bilateral level between individual countries and the EU. In 2010, during the negotiations for a revised Cotonou Agreement, the ACP once again resisted the reinsertion of a readmission clause in the Agreement. Talks continue only at a bilateral level.

3 Pilot partnerships were launched with Cape Verde and Moldova in 2008 and with Georgia in 2009. Since 2009, negotiations started with Senegal, and in 2010 with Ghana. In 2011, partnerships were proposed with Egypt, Morocco and Tunisia.

4 Both countries provide a significant source of migrants towards Europe, contributing to both regular and irregular migration flows. The EU sent its first diplomatic mission to Senegal to initiate dialogue under Cotonou's Article 13 in 2006. A similar mission was sent to Ghana in 2007.

5 This chapter draws on a number of interviews conducted in Brussels with officials from the European Commission, and in Accra and Dakar with officials from the governments of Ghana and Senegal between April 2008 and September 2010.

6 The 2004–08 budget for the Aeneas programme was 250 Million euros, while its follow-up programme (2007–13) has increased the budget to 384 Million euros.

7 Established in December 1998, the HLWG's main task is the implementation of cross-pillar programmes in the main countries of origin and transit. The HLWG draws up action plans, for selected countries identified as heavy migrant sending countries, in which it stipulates short- and long-term measures for cooperation with the countries in the areas of foreign policy, development policy, economic and humanitarian aid, conflict prevention and combating illegal migration and organised crime. The HLWG is said to have a leading role in the EU's migration policies with an external dimension (Council of the European Union, 2002; Lindstrøm, 2005).

8 Roig and Huddleston (2007) argue that third-country willingness to sign readmission agreements is dependent on the integration of incentives such as visa facilitation or even EU membership prospects. Since these incentives cannot be offered to many migrant-sending countries, agreements are likely to stall.

9 As opposed to 'first-generation' schemes which can be classified mainly as guest worker programmes or working-holiday maker schemes.

References

ACP (2006) *Brussels Declaration on Asylum, Migration and Mobility*, ACP/28/025/06 Final, Brussels, 13 April.

Adepoju, A. (2009) 'Migration Management in West Africa within the Context of ECOWAS Protocol on Free Movement of Persons and the Common Approach on Migration: Challenges and Prospects', in M. Tremolieres (ed.), *Regional Challenges of West African Migration: African and European Perspectives*, Paris: OECD.

Adepoju, A. (2008) 'Perspectives on International Migration and National Development in Sub-Saharan Africa', in A. Adepoju *et al.* (eds), *International Migration and National Development in sub-Saharan Africa*, Leiden: Koninklijke Brill NV.

Bosch, P. and Haddad, E. (2007) 'Migration and Asylum: An Integral Part of the EU's External Policies', *Forum Natolinskie*, 3 (11): 1–19.

Boswell, C. (2003) 'The "External Dimension" of EU Immigration and Asylum Policy', *International Affairs*, 79 (3): 619–38.

Brady, H. (2008) 'EU Migration Policy: An A–Z', CER Briefing, Centre for European Reform, London.

Carrera, S. and Chou, M.-H. (2006) 'Fiche on EU Migration Policy', in C. Egenhofer (ed.), *Policy Coherence for Development in the EU Council: Strategies for the Way Forward*, Brussels: Centre for European Policy Studies.

Centro Studi di Politica Internazionale (CeSPI) and the Society for International Development (SID) (2006) 'European Migration Policy on Africa – Trends, Effects and Prospects', Policy Paper for the Colloquium on Migration and Development in Africa – Scenarios and Proposals.

Chou, M-H. (2006) 'EU and the Migration–Development Nexus: what prospects for EU-wide policies?', COMPAS Working Paper No. 37, University of Oxford.

Chou, M-H. (2009a) 'European Union Migration Strategy towards West Africa: the Origin and Outlook of "Mobility Partnerships" with Cape Verde and Senegal', paper presented at the EUSA Conference, Los Angeles, April 2009.

Chou, M-H (2009b) 'The European Security Agenda and the "External Dimension" of EU Asylum and Migration Cooperation', *Perspectives on European Politics and Society*, 10 (4): 541–59.

Commission of the European Communities (CEC)–Government of Ghana (GoG) (2006), *Ghana Migration Profile*, Accra, October.

Council of the European Union (2005) *A Global Approach to Migration: Priority Actions Focusing on Africa and the Mediterranean*, 15744/05, Brussels, 13 December.

Council of the European Union (2008) *European Pact on Immigration and Asylum*, 13440/08, Brussels, September.

Egenhofer, C. (ed.) (2006) *Policy Coherence for Development in the EU Council: Strategies for the Way Forward*, Brussels: Centre for European Policy Studies.

European Union (2000) 'Partnership Agreement between the Members of the African, Caribbean and Pacific Group of States of the One Part, and the European Community and its Member States, of the Other Part', signed in Cotonou on 23 June 2000, *Official Journal of the European Communities*, L 317/3, 15 December.

European Commission (2005) *Migration and Development: Some Concrete Orientations*, COM (2005) 390, 1 September.

European Commission (2006) *A Global Approach to Migration One Year On: Towards a Comprehensive European Migration Policy*, COM (2006) 735, 30 November.

European Commission (2007) *Strategy Paper for the Thematic Programme of Cooperation with Third Countries in the Areas of Migration and Asylum 2007–2010*, Brussels: EuropeAid.

European Commission (2008) *Strengthening the Global Approach to Migration: Increasing Coordination, Coherence and Synergies*, COM (2008) 611, 8 October 2008.

European Commission (2009) *Mobility Partnerships as a Tool of the Global Approach to Migrationm*, SEC (200() 1240, 18 September.

Geddes, A. (2000) *Immigration and European Integration*, Manchester: Manchester University Press.

Geddes, A. (2009) 'Migration as a Foreign Policy? The External Dimension of EU Action on Migration and Asylum', Swedish Institute for European Policy Studies Report No. 2, Stockholm.

Gnisci, D. (2008) *West African Mobility and Migration Policies of OECD Countries*, West African Studies series, Geneva: OECD Secretariat.

Higazi, A (2005) 'Integrating Migration and Development Policies: Challenges for ACP-EU Cooperation', Discussion Paper 62, ECPDM, Maastricht.

Katseli, L., Lucas, R. and Xenogiani, T. (2006) Effects of Migration on Sending Countries: What Do We Know?, OECD Development Centre Working Paper No. 250, Paris: OECD.

Lavenex, S. (2002) 'EU Trade Policy and Immigration Control', in S. Lavenex and E. Uçarer (eds), *Migration and the Externalities of European Integration*, Oxford: Lexington Books.

Lavenex, S. (2007) 'Shifting Up and Out: The Foreign Policy of European Immigration Control', in V. Guiraudon and G. Lahav (eds), *Immigration Policy in Europe: The Politics of Control*, Oxon: Routledge.

Lavenex, S. and Kunz, R. (2008) 'The Migration-Development Nexus in EU External Relations', *Journal of European Integration*, 30 (3): 439–57.

Oucho, J. (2008) 'African Brain Drain and Gain, Diaspora and Remittances: More Rhetoric than Action', in A. Adepoju *et al.* (eds), *International Migration and National Development in Sub-Saharan Africa*, Leiden: Koninklijke Brill NV.

Panizzon, M. (2008) 'Labour Mobility as a Win-Win Model for Trade and Development in the Case of Senegal', NCCR Trade Regulation Working Paper No. 7, June.

Pastore, F. (2007) 'Europe, Migration and Development: Critical Remarks on an Emerging Policy Field', CeSPI Working Paper, August, CeSPI, Rome.

Roig, A. and Huddleston, T. (2007) 'EC Readmission Agreements: A Re-Evaluation of the Political Impasse', *European Journal of Migration and Law*, 9: 363–83.

Statewatch Bulletin (2000) 'Lomé Convention used to impose repatriation on the world's poorest countries', 10 (2).

Sterkx, S. (2008) 'The External Dimension of EU Asylum and Migration Policy: Expanding Fortress Europe?', in J. Orbie (ed.), *Europe's Global Role: External Policies of the European Union*, Aldershot: Ashgate.

Van Criekinge, T. (2009) 'The Integration of Migration Issues in the EPAs', in J. Orbie and G. Faber (eds), *Beyond Market Access for Economic Development: EU–Africa Relations in Transition*, London: Routledge.

Vanheukelom, J., Mackie, J. and Bossuyt, J. (2006) 'Political Dimensions: Introductory Note', *ECDPM seminar on The Cotonou Partnership Agreement: What role in a changing world?*, Maastricht: ECDPM.

Zoomers, A., Adepoju, A. and van Naerssen, T. (2008) 'International migration and national development: An introduction to policies in sub-Saharan Africa', in A. Adepoju et al. (eds), *International migration and national development in sub-Saharan Africa*, Leiden: Koninklijke Brill NV.

14

Work in progress: the social dimension of EU–Africa relations

Jan Orbie

Since the early 2000s, the European Union (EU) has explicitly committed itself to promoting the social dimension of globalisation.[1] The emergence of this new external policy objective reflects broader trends such as the post-Washington Consensus in the development sphere and the resurrection of the International Labour Organisation (ILO) as the core institution for global social governance.[2] More specifically, the ILO's social dimension of the globalisation agenda provided an alternative for the failed labour–trade linkage at the World Trade Organisation (WTO). Instead of a binding linkage between trade relations and social standards, which had been resisted fiercely by developing countries fearing hidden protectionism through lofty social objectives, EU and international policy-makers opted for a broader and softer approach to international social issues (Orbie and Tortell, 2008).[3]

The first European Commission communication on *Promoting Core Labour Standards and Improving Social Governance in the Context of Globalisation* was published in July 2001 and was followed by many other Commission and Council documents on the EU's global role in relation to labour standards, decent work, employment and social cohesion (e.g. European Commission, 2004, 2006, 2007a, and subsequent Council conclusions). At the same time, the EU has strengthened its cooperation with the ILO through annual high-level meetings since 2001 and other activities (e.g. EC-ILO, 2007, 2008, 2010). The European Commission played an active role in the World Commission on the Social Dimension of Globalization, established by the ILO in 2002.

Social objectives also became integrated into the EU's main development policy strategies. The 'social dimension of globalization, promotion of employment and decent work' formed one of the 12 areas of the *Policy Coherence for Development* document (European Commission, 2005: 13–14). Equally, the European Consensus on Development – presented as an alternative to the Washington Consensus in EU relations with the global South – included a separate section on 'social cohesion and employment' (European Union, 2005: 15; see also Michel, 2006). In 2007, European civil society organisations started to mobilise more actively on this topic by creating a European Working Group on Social Protection and Decent Work in Development Cooperation. In the same year, the Commission issued a communication on EU development cooperation and employment (European Commission, 2007a) and the Council invited the Commission to prepare a communication on social protection in EU development cooperation (Council, 2007). However, it took five years before the Commission started a public consultation process which should ultimately result in a communication on this issue. Also, the social dimension did not appear among the five 'priority areas' identified for the EU's policy coherence for development agenda (European Commission, 2009a). Thus, in the subsequent report the social dimension was only touched upon in the context of trade and migration (European Commission, 2011c).

Thus, while the social dimension has become increasingly important, it clearly does not (yet) figure on the top of the EU's development agenda. This chapter examines whether and to what extent a social dimension has been integrated into EU–Africa relations. We argue that the social dimension has been virtually non-existent and that decent work and labour standards objectives have been framed through, and overshadowed by, other considerations. Interestingly, these conclusions seem to apply even more to EU relations with sub-Saharan Africa (SSA) compared with Latin America or Asia. The chapter is structured as follows. The next section provides some conceptual clarification on how we define and approach the 'social dimension' in development policies as well as the historical-institutionalist perspective that will be inform further analysis. Then, we elaborate on the social dimension of EU–Africa relations in empirical terms, looking at political agreements, budgetary commitments and trade arrangements. Finally, we formulate the conclusions and point to the position of African governments in this story.

Defining and approaching the social dimension of globalisation

Before analysing the social dimension of EU–Africa relations, some conceptual issues need to be clarified. What do we mean by the 'social dimension'? The World Commission on the Social Dimension of Globalization (2004) presents an all-embracing description which includes not only decent work objectives but also broader goals such as freedom, prosperity, security, health, education and shelter for a liveable environment, as well as human dignity, sustainable development and democratic governance. The concept of 'decent work' as defined on the ILO website is equally vague, encompassing a wide range of 'aspirations of people in their working lives' such as a fair income, security in the workplace, social protection, freedom of expression and equal opportunities. The conceptual fuzziness around the concept makes it difficult to assess the EU's activities and progress in this area. The European Commission (2009b: 112) acknowledged that 'a clear understanding of the notion of "social dimension of globalization, employment and decent work" is lacking'. However, a more profound study of EU and ILO documents reveals that the 'social dimension' mainly refers to the four pillars of the decent work agenda (labour standards, employment, social protection and social dialogue), and that most emphasis has been placed on the promotion of the core labour standards (CLS) (Orbie and Tortell, 2008: 2–4). These CLS were identified in the 1998 ILO Declaration on Fundamental Rights and Principles at work and include the freedom of association, elimination of child labour, elimination of forced labour and non-discrimination. Each of these four CLS corresponds with two ILO Conventions.

Our definition of social dimension as 'the decent work agenda with emphasis on the CLS' implies that we will not classify EU policies in relation to water, health, education and infrastructure under the 'social dimension'. This is despite the fact that initiatives in these areas may well foster development, and have often been included by the EU and others in the conceptual dustbin of the 'social dimension' (see below).[4] Another delineation of this study is that we only consider policies that aim to *directly* address the social dimension. EU policies in other areas, such as trade, infrastructure or migration management, may also impact (positively or negatively) on the social dimension. In fact, the EU often labels initiatives in these areas as 'social' because of the expected long-term effects on social development. For example, the EU's migration management policies towards Africa have often been framed through the objective of employment, for example the EU's support of migration centers in West Africa was presented as an example in the Commission communication on employment and development (e.g. European Commission, 2007a: 17). Another example concerns the expected positive consequences of the Economic Partnership Agreements (EPAs) on social systems in African countries, although critics argue that these would be devastating (for an analysis, see d'Achon and Gérard, 2010). While indirectly a whole range of EU policies may affect social development, we only study those initiatives that directly concern the social dimension as defined above. That said, we will also look at how 'social' objectives are framed by the EU in its relations with SSA.

Behind these choices lies a criticism of the more traditional, developmental view which assumes that economic growth and poverty reduction should come first and will eventually improve decent work and labour standards. Despite some convergence of views, this 'chicken egg' discussion continues to divide approaches by the EU and ILO to the social dimension of globalisation (Interview, November 2009). However, also within the EU institutions, divergences exist between the development ministries and the Commission's DG Development (now DevCo) on the one hand, and social ministries and the DG Employment and Social Affairs on the other hand. While the former advocate a growth and poverty oriented approach, the latter prioritise the promotion of decent work objectives.

Since different views on the social-development nexus seem to prevail depending on the institutional setting within the EU, this chapter will start from a historical-institutionalist perspective. We consider the EU as a compartmentalised external policy actor, whose external relations are to a large extent shaped by the interplay of the different institutional sub-units (or policy sub-systems) which have their own preferences on various dimensions of foreign policy. The EU's foreign policy system is notorious for its compartmentalisation, facilitating stove-piping and hindering a coherent approach between the various sub-systems. While authors have illustrated this point before (see e.g. Elgström and Pilegaard, 2008, on the trade–development nexus, and Olsen, 2009, on the development–security nexus), the social–development nexus has barely been researched. From this institutionalist perspective, it may be expected that the most ambitious EU global social policy comes from the Directorate General for (DG) Employment and Social Affairs. However, the outcome can also be expected to be limited, since this DG is mainly concerned with the intra-European dimension. The capacities in terms of personnel and budget at the international unit of DG Employment and Social Affairs are limited. To the extent that this DG engages in global social policies, it concerns the general policy formulation and the EU's policies towards Latin America and Asia. Officials at this DG confirmed that they are working 'more generally' on the EU's advancement of the social dimension of globalisation and refer to the people at DG Development to look at how this is implemented in Africa because this is 'not their domain' (Interview, December 2009). In contrast, we expect that the DG Development will favour a more traditionalist approach (see above), whereas DG Trade will be more focused on economic liberalisation rather than directly addressing decent work and labour standards issues. Both policy spheres would respectively assume that economic growth and liberalisation will (indirectly) foster the social dimension, which should not be tackled up-front. Given their nature as *horizontal* external policy objectives, social objectives would need to be mainstreamed into EU trade and development-related initiatives. In fact, the Lisbon Treaty introduces a 'horizontal social clause' which can be seen as an attempt to mainstream social policies in the all areas of EU policy (Bruun *et al.*, 2012: 4). However, given the limited institution-alisation of strong pro-social forces in EU external relations, we expect that the social dimension of EU–Africa relations will be limited.

Taking these conceptual and theoretical considerations into account, the next section will delve into the empirics of EU–Africa relations and scrutinise the various EU documents and initiatives in relation to SSA. The social dimension of this relationship will be examined by looking respectively into the political agreements, budgetary commitments and trade arrangements.

Tracking the social dimension of EU–Africa relations

Political agreements

The social dimension has not only been mentioned in the EU's own policy documents such as the European Consensus on Development (see above), it also features in political agreements between the EU and Africa. However, if we look at the EU–ACP Cotonou Agreement and the Joint Africa–EU Strategy (JAES), it becomes clear that references to the social dimension are barely implemented and largely overshadowed by other issues. The Cotonou Agreement mentions the CLS several times. The preamble states that the EU and the African, Caribbean and Pacific (ACP) group are 'anxious to respect basic labour rights, taking account of the principles laid down in the relevant conventions of the International Labour Organization'. In Article 9 the 'fundamental social rights' are explicitly considered as part of the 'essential elements' of the agreement, standing on an equal footing with the principles of human rights, democracy and the rule of law. Thus, the incorporation of the ILO's CLS into the 'essential elements' is an important confirmation of their human rights status. This implies not only that these labour standards can be the subject of a structured political dialogue between the EU and an ACP country, but also that substantial violations of these principles could lead to consultations and, eventually, 'appropriate measures' in line with Article 96 procedures. In addition, Article 25 of the Cotonou Agreement on 'social sector development' mentions health and education as well as social rights, social dialogue, systems of social protection and security (European Union, 2000).[5] Furthermore, Article

50 reaffirms the parties' commitment to promote the ILO core labour standards and to enhance cooperation in social affairs.

However, the social provisions of Cotonou have not been implemented. Reflecting the concerns of the international trade union movement and some members of the European Parliament, Article 50 has not been used to discuss the implementation of CLS with the ACP countries. Research by Kerremans and Gistelinck (2009: 316) shows that the promotion of CLS through this article has been overwhelmed by other foreign policy and trade priorities such as weapons of mass destruction, the International Criminal Court, the fight against terrorism, EPAs, climate change and the political situation of individual ACP countries. Similarly, the possibility of including CLS in the political conditionality system has not been tested in practice. Article 96 consultations have mainly dealt with violations of democratic principles, good governance and the rule of law. This confirms that the EU prioritizes the first generation of civil and political human rights over the second generation of social and economic human rights (see, e.g., Deacon, 1999: 25–8; Clapham and Bourke Martignoni, 2006: 291), even if a clear-cut distinction is increasingly difficult to make (Manners, 2009: 787–8). In particular, it has been argued that EU only applies Article 96 sanctions in cases of flawed elections or coup d'état (Del Biondo, 2011: 668).

An internal European Commission evaluation in 2007 also concluded that 'we could have done more' through the Cotonou Agreement.[6] The Commission has developed social sector support programmes towards Latin America, Asia and the European neighbourhood, but not towards the ACP (European Commission, 2007a: 21). However, the JAES launched in at the end of 2007 provided a new opportunity for high-level dialogue on social issues. The First Action Plan for the implementation of the Africa–EU Strategy includes eight areas for strategic partnership between 2008–10. The social dimension does not constitute a separate partnership but is incorporated in partnership 7 on 'Migration, mobility and employment'. Priority action three is to implement and follow up the Ouagadougou Declaration and Action Plan on Employment and Poverty Alleviation in Africa, agreed within the African Union in 2004 (African Union, 2004a, 2004b). A close reading of this partnership shows that the main focus is on labour *market* and labour *mobility* rather than the social dimension as defined above. Throughout the Action Plan, 'labour' is mentioned seven times, but it is *always* framed in the context of 'labour market' and 'labour mobility', never in the context of labour rights, social dialogue or social protection. 'Decent work' principles are mentioned, but not the ILO's

CLS; even the ILO is not explicitly mentioned among the actors to be involved.[7]

In line with our historical-institutionalist perspective, this particular framing of the employment dimension through a migration and mobility lens may not be surprising because the European side of the partnership on 'Migration, Mobility and Employment' is not only managed by DG Development, but also closely followed by DG Justice and Home Affairs and by the ministries of home affairs in the EU member states. Even apart from this particular framing of the employment dimension of the JAES through a migration and mobility lens, it should also be noticed that the implementation of the employment provisions was long delayed. Again, the European Commission (2009a: 108) had to acknowledge that 'limited progress has been registered under this heading at this early stage'. It took more than two years after the signing of the JAES before a first meeting of the employment partnership was organised. As admitted by the European co-chair of this meeting, until then the main focus of the Migration, Mobility and Employment partnership had been on migration and mobility (Matres-Manso, 2010). In 2010, the employment dimension of the JAES was finally launched and an informal working group on social protection was created, even if there are no clear budgets and benchmarks on what to achieve. The working group came to the conclusion that social protection should be prioritised within the JAES and that support of the development of national social protection systems should be a major objective. The Second Action Plan for 2011–13 adopted at the Africa-EU Summit in Tripoli (November 2010) also pays more attention to decent work and employment.

Thus, social issues have not featured high on the political agenda of EU–Africa relations. On a lower level, some initiatives have been taken. For example, in 2009 a regional seminar on the integration of decent work in EU development aid was organised between the EU and the ACP countries in cooperation with the ILO. In 2010, the EU also started organising Social Protection Training Courses for its own personnel in the headquarters and delegations, which will possibly lead to a higher involvement and prioritisation of these issues on the European side (ERD, 2010: 176). However, contrary to stated intentions (EC–ILO, 2007: 3, 2008: 7, 2010: 3), the EU has not yet applied the ILO 'toolkit for mainstreaming of employment and decent work in UN operations' to EU programmes (ILO, 2007). Moreover, at the technical level of the delegations there is still a tendency to think of poverty reduction in a more narrow sense while neglecting the social dimension as such (Interviews, July 2010).

Budgetary commitments

While the JAES sets an overarching framework for EU–Africa relations, it does not have a separate budget and continues to rely on existing financial frameworks. Discussions on a Pan-African financial support programme and the alignment of existing instruments to the JAES have barely made progress (Bossuyt and Sherriff, 2010: 7). Since 2001, the multi-annual programming of European Community aid has been based on the country strategy paper (CSP) mechanism. Each strategy paper contains a national indicative programme (NIP) which indicates the focal areas where EU resources will be spent. CSPs and NIPs are developed between the EU (all relevant Commission DGs and member states) and partner governments.

An analysis of 48 CSPs with SSA countries reveals that the social dimension of globalisation and decent work is usually referred to in the country analysis. However, except for South Africa, 'social cohesion and employment' is never mentioned as one of the focal areas of cooperation. 'Human and social development' appears as a focal area in only seven CSPs with mostly Southern African countries (Angola, Botswana, Côte d'Ivoire, Lesotho, Namibia, Swaziland, Zambia), and these objectives are typically framed in the context of health, education and sometimes water. The social dimension (as defined above) is neglected in the focal areas. When considering the subsequent text of CSPs, NIPs, as well as the indicators, it becomes clear that employment and social services are mentioned, albeit overshadowed by other concerns. The objective of employment is framed through the perspective of education (professional training with a view to integration in the labour market, e.g. in cases of Benin and Gabon) while social services form part of commitments in terms of infrastructure (transport initiatives in order to reach basic social services, e.g. in cases of Comoros, Eritrea, Ethiopia, Togo). Also, the language in the documents illustrates the traditional developmental view that social objectives will be reached through economic development and should not be pursued as a goal as such.[8] Importantly, this not only reflects EU preferences but also those of African partner governments (Interview, November 2010). While the commitments of CSPs/NIPs with ACP countries are disbursed through the European Development Fund (EDF), the EU's thematic budget lines are also an important source of aid. An examination of the 'Investing in People' (or 'Human and Social Development') budget line reveals the same pattern as identified above: the bulk of the budget (almost 60 per cent) is spent on health and less than 8 per cent goes to the social dimension.[9] The EU admits that the social dimension has 'received relatively limited attention' in EU programmes and attributes this to implementation problems as well as partner governments' priority for growth over social issues.[10]

In addition, social rights could be promoted through the European Instrument for Democracy and Human Rights (EIDHR) budget line. Indeed, the ILO CLS are widely considered equivalent to human rights, and the core labour conventions on freedom of association and the right to organise are linked to democratic participation. An overview of EIDHR projects between 2000 and 2006 (EuropeAid, n.d./a) reveals that several projects in relation to CLS have been implemented, but that this has barely been the case in SSA countries. Again, it becomes clear that the EU's external social policies have been more ambitious towards non-African regions. For example, the EIDHR funded projects on child labour in Brazil, Morocco, Palestine, India and Egypt, and initiatives to promote CLS and corporate social responsibility have taken place in China, Indonesia, Pakistan and Tajikistan. Among the projects for the support of trade union movements, four are directed at SSA countries (two projects in Nigeria, one in Ethiopia and one in Burundi;[11] besides projects in Ukraine, Kazakhstan, Tunisia, China, three projects in Colombia and two in Palestine). A similar pattern emerges when considering the EIDHR projects for January 2007–April 2009. Nine projects concern the promotion of CLS, in addition to six projects which focus explicitly on child labour and children's rights. However, only two of these, namely projects on the strengthening of trade unions in Rwanda and the DRC, would be implemented in SSA (EuropeAid, nd/b). In conclusion, the number of projects directly targeted at labour rights and decent work has been limited, even when compared to other regions, such as Latin America and Asia.[12]

However, there are other entry points through which social expenditures can be provided. EU general budget support can be used by African governments for social sector spending. More specifically, the 'MDG contracts' which have been signed between the Commission and eight African countries (Burkina Faso, Ghana, Mali, Mozambique, Rwanda, Uganda, Zambia and Tanzania) offer a longer term, more predictable budget for the realisation of the Millennium Development Goals (MDGs). However, these MDG contracts have not been published in full, and according to the Commission website the variable component rewards performance against MDG-related results 'notably in health, education and water',[13] thus again not directly addressing the social dimension. The same conclusion applies to the sector budget support provided by the EU. Only about 1.8 per cent of the 10th EDF was spent on social cohesion and employment, of which only 0.4 per cent went through sector budget support.[14]

Trade relations between the EU and SSA countries have been significantly reformed. While the old Lomé regime evaporated the spirit of the so-called New International Economic Order in the 1970s, the post-Lomé system reflects the neo-liberal international environment by abandoning the interventionist parts of Lomé and introducing reciprocal trade liberalisation (Orbie, 2007). Also the geographical reach of the trade arrangements has changed: while the Lomé trade arrangements applied to the ACP group as a whole, EPAs are being negotiated with several ACP sub-regions. However, the EU has experienced serious difficulties to achieve its objective to negotiate reciprocal, comprehensive and regional trade arrangements. Only the Caribbean group signed a full EPA with the EU. Trade relations with SSA have become more complicated than ever: several countries signed 'interim' EPAs while others reverted back to preferential market access under the Generalised System of Preferences (GSP) and its 'Everything but Arms' (EBA) variant for the least-developed countries (LDCs). In turn, South Africa continues to export under its free trade agreement with the EU. What about the social dimension in this hotchpotch of EU–Africa trade arrangements?

Interestingly, the social dimension of the CARIFORUM EPA is quite innovative and ambitious compared with other bilateral agreements concluded by the EU. For example, while Article 50 of the Cotonou Agreement is largely developmentalist and non-committal, the agreement with the Caribbean group commits the parties not to lower their level of social protection, both in legislation and in implementation, for reasons of competitive advantage in trade and investment (Article 193). Although the dispute settlement procedure under the CARIFORUM agreement does not allow for trade concessions or financial compensations in cases of violation of the social provisions, negative panel outcomes may have an important 'blaming and shaming' effect in the long run and contribute to the EU's normative power through trade (cf. Manners, 2009; Orbie, 2011).[15] However, it is doubtful that social provisions in the EPAs with African countries will be equally far-reaching. Firstly, Caribbean governments are more responsive to EU demands for a social clause in trade agreements. In fact, the Dominican Republic, which is the most important member of the CARIFORUM, had already made similar commitments through the CAFTA-DR trade agreement with the US (Kerremans and Gistelinck, 2009: 317–19). In contrast, African countries are less enthusiastic about any linkage between trade and social provisions (Interviews, November 2009). This reluctance may reflect different levels of economic development, but it is also based on the fear that social provisions will be misused by the EU for protectionist purposes. Secondly, since DG Trade is leading the EPA negotiations, more emphasis will be put on liberalisation and market enhancing rules rather than human rights or redistributive issues which do not belong to DG Trade's core mandate (Faber and Orbie, 2009a: 368–70). A notable institutional evolution in this context is that the 'sustainable development' unit is no longer a separate unit within DG Trade (it has been enlarged to include sanitary and phytosanitary standards). Thirdly, most African ACP countries have only few incentives to continue negotiations towards full EPAs because they already receive free access to the European market through the interim EPAs or through the EBA system.

Indeed, about half of the African ACP countries have opted for EBA as an alternative to EPAs. Under this special GSP arrangement, LDCs are allowed to export duty-free and quota-free to the EU. Therefore, there is no room for the EU to grant additional trade preferences to EBA beneficiaries which comply with CLS. On the other hand, trade preferences from GSP/EBA beneficiaries which have seriously and systematically violated CLS can be withdrawn. Under this system, only two countries have lost their preferential trade access to the European market: Burma (since 1997) and Belarus (since 2007) because of practices of forced labour and violation of freedom of association respectively. These GSP trade sanctions have typically coincided with strong condemnations against core labour standards by the competent expert bodies of the ILO (Orbie and Tortell, 2009: 675–6). However, the EU has never initiated investigations against violations of labour standards by African EBA/GSP beneficiaries.

There is also a positive side to the GSP social conditionality system. The GSP-plus – officially the 'special incentive arrangement for sustainable development and good governance' – allows the EU to grant additional trade preferences to developing countries that have ratified and effectively implemented the ILO core labour conventions (besides a range of other multilateral agreements). Despite some flows, this is a relatively successful linkage between EU trade policies and social conditionality, which has effectively spurred on the ratification of ILO conventions in a number of countries, such as El Salvador, Bolivia and Ecuador (Orbie and Tortell, 2009). As part of the strengthening of the strategic partnership between the Commission and the ILO in the area of development, input from the ILO in the Commission's external human rights policy and GSP-plus trade decisions has increased (EC-ILO, 2010: 1). However, this system has only been applied to countries in Latin America, Asia and the European neighbourhood, but not to African countries because these have either opted for EPAs or for EBA.

Only three African and non-LDC ACP countries could potentially apply for the GSP-plus incentives: Nigeria, Gabon and Congo. Both Nigeria and Gabon were unsuccessful in applying for these additional trade preferences in 2008 because they had not ratified the Genocide Convention and the ILO Convention on Child Labour respectively. It remains to be seen whether the 'carrot' of GSP-plus incentives will stimulate these countries to ratify and implement the relevant conventions and apply again at a later stage.[16]

On a more general level, and still in line with a historical-institutionalist approach, it could be argued that the limited social dimension of EU–Africa trade relations also stems from path-dependencies. Previous trade arrangements that were established at a time when the social dimension had not yet reached the EU external policy agenda continue to constrain the room for maneuver of current trade initiatives. For example, the EBA regulation, which went into force in 2002 but goes back to a Council decision in 1997 (Faber and Orbie, 2009b: 771), provides completely free market access for African LDCs without any (social) strings attached. Thus, for these countries a social incentive system like GSP-plus cannot apply (see above). Equally, the Cotonou Agreement was signed in 2000, building on the 1996 Green Paper and the 1998 Council mandate which had already outlined the main objective of the EPAs, namely WTO compatibility. As a result, a 'light EPA' is sufficient and there is no urgent need to negotiate more ambitious trade arrangements to include a social chapter. Thus, even if social issues have become more important in EU–Africa relations, constraints stemming from earlier EU decisions make it difficult to introduce these social commitments in the short run. In addition, the absence of a consensus in the WTO on labour rights and the reluctance of developing countries in this regard make it extremely difficult to depart from the existing situation where the social dimension is by default subordinated to trade-related issues.

Conclusion

Since the beginning of this millennium, the European Union has taken a more systematic and a more assertive approach to advancing the social dimension of globalisation, although its policy in this area still constitutes 'a patchwork of ideas and initiatives rather than a full-fledged strategy' (Orbie and Tortell, 2008: 21), and although 'decent work is not yet an organizing principle behind project programming' (Andrieu *et al.*, 2008: 2). This chapter showed that the latter is all the more true for the social dimension of EU policies towards sub-Saharan Africa; so far the social dimension has been virtually non-existent (e.g. non-implementation of Cotonou Article 50; limited social projects under Investing in People and EIDHR; no social conditionality through trade). Our analysis of the political agreements, budgetary commitments and trade arrangements of EU–Africa relations makes clear, to the extent that initiatives under the banner of 'social dimension of globalisation' have been advanced, that this was almost exclusively framed through and overshadowed by other priorities, such as health, education, labour market, economic growth, and migration management (e.g. in the CSPs and the JAES). Although further research into this would be needed, it also seemed that these conclusions apply less to EU relations with Latin America and Asia, where the social dimension tends to be advanced more prominently.

In line with the expectations outlined in the introduction and based on the historical-institutionalist perspective, the chapter also confirms the importance of path-dependencies, especially for explaining the social dimension of EU trade policies towards Africa, and of the compartmentalised nature of the EU's foreign policy system. The institutional setting of EU–Africa cooperation reinforces the bias away from the social dimension as an objective in itself. For example, the partnership on Migration, Mobility and Employment in the context of the JAES has been dominated by the Justice and Home Affairs sub-system, while the impact of development and social affairs sub-systems remained limited. This could at least partly explain why the employment dimension has been addressed only marginally, and only through a migration lens. However, more generally, the Commission's DG Development does play the first fiddle in EU–Africa relations, for example through the negotiation of CSPs with African partner countries, whereas the role of DG Employment and Social Affairs has been marginalised in relation to SSA. This sheds light on the EU's rather traditionalist, developmental approach to the social dimension. Indeed, interviews confirm that DG Development officials tend to hold the view that economic growth and poverty reduction should be first and foremost promoted, which will eventually trickle down to social progress. In addition, they seem to favour a 'sectorial' approach focusing on specific projects in areas such as education or health, without engaging in a broader view on the decent work objectives as promoted by the ILO. In the same vein, EU officials involved in development cooperation with Africa tend to consider the ILO as a technical institution which could be involved in the implementation of specific projects, without recognising its political relevance as the main forum for global social governance. As a consequence, according to an ILO official, the potential for cooperation and synergies between both institutions has not yet been fully exploited

(Interview, July 2010). On the other hand, a comparison of interviews at DG Development over several years (March 2007, December 2009 and November 2010) suggests that these officials have come to take a more nuanced approach to the 'chicken–egg' problem between social and economic development. It has been increasingly recognised that the social dimension should be addressed more straightforwardly and not only indirectly through a growth-oriented approach.

This observation qualifies our initial expectations. How can we explain these evolving views within the Commission's development sub-system? It should be noted that this European shift mirrors the position of other international institutions, but perhaps with some delay. Initially launched by the ILO, the social dimension and decent work agendas had received support from several other international institutions, such as the UN Economic and Social Council, the World Bank, the OECD-DAC, the G8 and the G20, by the end of the 2000s. Since the UN MDG+5 World Summit in 2005, decent work has become one of the targets within the MDGs. Similarly, the upcoming initiative on social protection in EU development cooperation refers to the renewed impetus at the global level on this issue, for example in the UN, the ILO and the G20 (European Commission, 2011b: 2). Given its role as a norm-taker in development issues (see e.g. Farrell, 2008), it does not come as a surprise that the EU's DG Development is following these international trends. In addition, there is the realisation that economic growth in Asia and Latin America has not significantly improved social equality and formal employment. Furthermore, interviewees at the ILO and the Commission's DG Development indicated that the global economic crisis, with its social consequences in Africa, has increased donors' interests in the social–development nexus.[17]

Another finding that emerged from the analysis is the importance of political preferences in the African partner countries themselves. We mostly looked at the EU side based on the implicit assumption that EU–Africa relations are asymmetric and thus that its outcomes reflect European preferences. However, this seems to be a case where EU preferences, or at least those of the EU development policy sub-system, largely correspond with the preferences of African governments. African policy-makers have also been reluctant towards an ambitious social dimension, not just in trade but also in aid policies. Just like their European counterparts, they tend to favour traditional aid paradigms which prioritise growth and poverty reduction over social issues. Again, there is an institutionalist dimension to it: negotiators involved in EU–African development policies tend to come from the ministry of finance, whereas social ministries (and also trade unions) are usually in a weaker position within their political system and less involved in relations with the EU. One interviewee stressed that ACP officials and ministries involved in development cooperation continue to 'harp on the same string' which is an economic and poverty-oriented approach to the social dimension (Interviews, November 2009). In addition, African policy-makers are more inclined to consider the EU's emphasis on labour standards as a neo-colonial agenda which interferes with their sovereignty. All these factors are less prominent in for example Latin America, where social ministries and trade unions often hold a stronger position. The leftist revival in Latin America has also been accompanied with the emergence of social protection schemes, with the Bolsa Familia in Brazil as the most famous example.

This is not to say that social issues are considered unimportant in African countries. On the contrary, as amply illustrated in Deacon's (2009) account of social cooperation in Western and Southern Africa, the numerous meetings, declarations, intentions and Action Plans that have been formulated reflect an unquestionable commitment to social development.[18] However, the author argues that implementation has been limited due to a lack of political will as well as capacity problems. Interestingly, it seems that these social commitments suffer from the same biases as the EU approach: an analysis of the NEPAD Action Plans makes clear that the focus is mostly on health and education as the major social topics, and on economic growth as the main strategy to advance social progress (Deacon, 2009).

Further research should further explore this side of the EU–Africa relationship. In any case, it is clear that despite a growing interest in the social dimension, it has not yet been firmly established and prioritised in EU development cooperation with Africa. However, if the EU really aims to become an ambitious global social actor, there are even more thorny challenges that have not been addressed in this chapter. For instance, social protection should be promoted as part of a wider vision on 'transformative' social policy, which is not only concerned with resource transfers and risk management but also addresses discriminatory legal and social practices that produce or exacerbate vulnerability (Adesina, 2010). In addition, it would ideally be embedded within a universal approach to 'global welfare state (re)building', which involves redistribution within and across countries (Deacon and Cohen, 2011).

There is clearly still a long way to go towards such a global, transformative social policy. The Agenda for Change (European Commission, 2011a) illustrates Europe's ambiguity on the social–development nexus; while social protection is explicitly mentioned, this is again exclusively in the context of health, education and migration.

In addition, the document seems to herald a new emphasis on growth and investment as central development topics. Although this is not a regression to the view of the 1980s when cutting social spending was deemed necessary for developing countries' growth strategies, the references to social protection are still limited and were only included after hard-edged debates within the Commission. Ultimately, this ambiguity mirrors the balancing-act that the EU is also exercising internally in relation to the Europe 2020 strategy and the reaction to the economic crisis between the preservation and extension of a so-called European social model on the one hand and the prerogatives of the European market project on the other. Both internally and externally, it has been challenging to incorporate ambitious social objectives into the EU's economic strategies.

Notes

1 When referring to the EU, we mean the former European Community. Thus, we focus on the European Commission as a donor and not at the EU member states' aid policies.

2 Also noticeable is the UN World Summit for Social Development which took place (Copenhagen, 1995).

3 This research partly builds on a Jean Monnet Information and Research Activity on the European Union's global social role, see www.eu-sdg.ugent. be. This chapter is partly based on seven interviews with officials at DG Trade, DG Development and DG Social Affairs and Employment, as well the ILO. Interviewees have requested anonymity.

4 For the same reason, we do not explicitly focus on gender and civil society, which have increasingly appeared on the EU development agenda. However, existing studies on these topics (for example, Lister and Carbone, 2006; Debusscher, 2011) come to similar conclusions as ours.

5 The Cotonou revisions in 2005 and 2010 added small changes to Article 25 in the field of health and education, but not on the social dimension as such.

6 DG Trade official discussion on 'The social dimension of EU trade policies', Ghent, 22 November 2007. Transcripts are available from http://www.eu-sdg.ugent.be (accessed 1 November 2008).

7 The core ILO conventions are briefly mentioned under partnership 2 on 'democratic governance and human rights' but they are not referred to in the section on employment.

8 Specifically focusing on social protection, the ERD study also found that 'direct support to social protection in Sub-Saharan Africa is relatively scarce', although it points the EU's €100 million funding of the Productive Safety Net Programme in Ethiopia as a major exception (ERD, 2010: 153).

9 See, for example, Table 6.1 in the Mid-Term Review, available online at http://ec.europa.eu/development/icenter/repository/investing_people_mid-term_review.pdf (accessed 11 July 2012).
10 Ibid., pp.18–19.
11 According to the European Commission (2007b: 21) the project aimed at strengthening the political role and the effectiveness of trade unions in Burundi is one of the EIDHR success stories.
12 For example, the EUROsociAL Regional Programme which aims to increase social cohesion in Latin America, *inter alia* through employment policies and in cooperation with the ILO.
13 See http://ec.europa.eu/development/how/aid/mdg-contract_en.cfm.
14 EDF10 – indicative sector breakdown, national indicative programmes (A-envelopes), DEV-C/1, 27/08/2008 (unpublished document).
15 Trade Sustainability Impact Assessments (Trade SIAs) also pay 'increasing attention' to employment and social impacts of trade agreements (Europa, 2012).
16 For more information about the social dimension of EU trade policies, see the special issue of *European Foreign Affairs Review*, 2009, 14.
17 For example, an *ad hoc* 500 million euro V-FLEX mechanism was created in 2009 aimed at helping the most vulnerable ACP countries safeguard social spending in a context of economic crisis.
18 At the level of the African Union, the 2004 Ouagadougou Declaration and Action Plan on employment and poverty alleviation in Africa (2004), the Livingstone Call for Action on social protection (2006) and the Social Policy Framework for Africa (2008) are worth mentioning. At the country level, various 'green shoots of institutionalized social protection' programmes have emerged in SSA (Niño-Zarazúa *et al.*, 2010). It should also be noted that most African countries have ratified all the ILO core conventions.

References

Adesina, J. (2010) 'Rethinking the Social Protection Paradigm: Social Policy in Africa's Development', commissioned background paper for the European Report on Development (ERD), European University Institute, Florence.

African Union (2004a) *Declaration on Employment and Poverty Alleviation in Africa*, Assembly of the African Union Third Extraordinary Session on Employment and Poverty Alleviation, EXT/ASSEMBLY/AU/3 (III), Ouagadougou, Burkina Faso, 8–9 September.

African Union (2004b) *Plan of Action for Promotion Of Employment and Poverty Alleviation*, Assembly of the African Union Third Extraordinary Session on Employment and Poverty Alleviation, EXT/ASSEMBLY/AU/4 (III) Rev. 3, Ouagadougou, Burkina Faso, 8–9 September.

Andrieu, J.B., Bell, S., Gibbons, S. and Newitt, K. (2008) *Bilateral Relations and Co-operation Activities in the Area of Employment and Decent Work between EU Member States or Relevant International Organizations, on the One Hand, and Selected Emerging Economies, Neighbourhood Countries and Strategic Partners of the EU on the Other*, London: Ergon Associates.

Bossuyt, J. and Sherriff, A. (2010) 'What Next for the Joint Africa–EU Strategy? Perspectives on Revitalising an Innovative Framework: A Scoping Paper', ECDPM Discussion Paper, 94.

Bruun, N., Lörcher, K. and Schömann, I. (2012) 'Introduction', in N. Bruun, K. Lörcher and I. Schömann (eds), *The Lisbon Treaty and Social Europe*, Oxford: Hart Publishing, pp. 1–15.

Clapham, A. and Bourke Martignoni, J. (2006) 'Are We There Yet? In Search of a Coherent EU Strategy on Labour Rights and External Trade', in V. Leary and D. Warner (eds), *Social Issues, Globalisation and International Institutions*, Leiden: Martinus Nijhoff Publishers, pp. 233–310.

Council of Ministers (2007), Conclusions of 21 June 2007 on Promoting Employment through EU Development Cooperation.

d'Achon, E. and Gérard, N. (2010) 'Les Accords de Partenariat Economique et le travail decent: Quels enjeux pour l'Afrique de l'ouest et l'Afrique centrale?', BIT Document de travail de l'Emploi, 60.

Deacon, B. (1999), 'Socially Responsible Globalization: A Challenge for the EU', available at: www.ose.be/files/deaconOK.pdf (accessed 24 December 2010).

Deacon, B. (2009) 'Regional Social Policies in Africa: Declarations Abound', in B. Deacon, M.C. Macovei, L. Van Langenhove and N. Yeates (eds), *World-Regional Social Policy and Global Governance*, London and New York: Routledge, pp. 162–87.

Deacon, B. and Cohen, S. (2011) 'From the Global Politics of Poverty Alleviation to the Global Politics of Welfare State (Re)Building', *Global Social Policy*, 11 (2): 233–49.

Debusscher, P. (2011) 'Mainstreaming Gender in European Commission Development Policy: Conservative Europeanness?', *Women's Studies International Forum*, 34 (1): 39–49.

Del Biondo, K. (2011) 'Democracy Promotion Meets Development Cooperation: The EU as a Promoter of Democratic Governance in Sub-Saharan Africa', *European Foreign Affairs Review*, 16 (5): 659–72.

Elgström, O. and Pilegaard, J. (2008) 'Imposed Coherence: Negotiating Economic Partnership Agreements', *Journal of European Integration*, 30 (3): 363–80.

Europa (2012) *MDG contract*, available at: http://ec.europa.eu/europeaid/what/millenium-development-goals/contract_mdg_en.htm (accessed 2 November 2012).

EuropeAid (nd/a), 'EIDHR 2000–2006: Ambitious in scope ... Global in reach, More than € 731 Million in EIDHR Funding Supporting more than 2400 Projects In more than 140 Countries.

EuropeAid (nd/b), 'EIDHR Compendium January 2007–April 2009: Promoting Democracy and Human Rights Worldwide', available at http://ec.europa.

eu/europeaid/what/human-rights/documents/eidhr_compendium_en.pdf (accessed 24 December 2010).

ERD (2010) *Social Protection for Inclusive Development: A New Perspective in EU Co-operation with Africa*, The 2010 European Report on Development.

European Commission (2001) *Promoting Core Labour Standards and Improving Social Governance in the Context of Globalisation*, COM (2001) 416, 18 July.

European Commission (2004) *The Social Dimension of Globalisation: the EU's Policy Contribution on Extending the Benefits to all*, COM (2004) 383, 18 May.

European Commission (2005) *Policy Coherence for Development: Accelerating Progress towards Attaining the Millennium Development Goals*, COM (2005) 134, 12 April.

European Commission (2006) *Promoting Decent Work for All: The EU Contribution to the Implementation of the Decent Work Agenda in the World*, COM (2006) 249, 24 May.

European Commission (2007a) *Promoting Employment through EU Development Cooperation*, SEC (2007) 495, 13 April.

European Commission (2007b) *Furthering Human Rights and Democracy across the Globe*, Luxembourg: Office for Official Publications of the European Communities.

European Commission (2009a) *Policy Coherence for Development – Establishing the Policy Framework for a Whole-of-the-Union Approach*, COM (2009) 458, 15 September.

European Commission (2009b) *Commission Staff Working Paper Document Accompanying the EU 2009 Report on Policy Coherence for Development*, SEC (2009) 1137, 17 September.

European Commission (2011a) *Increasing the Impact of EU Development Policy: An Agenda for Change*, COM (2011) 637, 13 October.

European Commission (2011b) *Public Consultation: Social Protection in EU Development Cooperation*.

European Commission (2011c) *EU 2011 Report on Policy Coherence for Development*, SEC (2011) 1627, 15 December.

European Union (2000) 'Partnership Agreement between the Members of the African, Caribbean and Pacific Group of States of the One Part, and the European Community and its Member States, of the Other Part', signed in Cotonou on 23 June 2000, *Official Journal of the European Communities*, L 317/3, 15 December.

European Union (2005) *European Union Joint Statement of 20 December 2005 on 'The European Consensus'*.

European Commission and International Labour Office (EC-ILO) (2007) *Joint Conclusions of the 6th High-Level Meeting*, 21 November.

European Commission and International Labour Office (EC-ILO) (2008) *Joint Conclusions of the 7th High-Level Meeting of 2 December*.

European Commission and International Labour Office (EC-ILO) (2010) *Joint conclusions of the 8th High-Level Meeting of 2 February*.

Faber, G. and Orbie, J., (2009a) 'Of Potholes and Roadblocks: The Difficult Path to Development Relevant EPAs', in G. Faber and J. Orbie (eds), *Beyond Market Access for Economic Development: EU–Africa Relations in Transition*, London and New York: Routledge, pp. 361–72.

Faber, G. and Orbie, J. (2009b) '"Everything but Arms": Much More than Appears at First Sight', *Journal of Common Market Studies*, 47 (4): 767–87.

Farrell, M. (2008) 'Internationalising EU Development Policy', *Perspectives on European Politics and Society*, 9 (2): 225–40.

International Labour Organization (ILO) (2007) *Toolkit for Mainstreaming Employment and Decent Work*, Geneva: International Labour Office.

Kerremans, B. and Gistelinck, M.M. (2009) 'Labour Rights in EPAs: Can the EU-CARIFORUM EPA be a Guide?', in G. Faber and J. Orbie (eds), *Beyond Market Access for Economic Development: EU–Africa Relations in Transition*, London and New York: Routledge, pp. 304–21.

Lister, M. and Carbone, M. (2006) *New Pathways in International Development: Gender and Civil Society in EU Policy*, Aldershot: Ashgate.

Manners, I. (2009) 'The Social Dimension of EU Trade Policies: Reflections from a Normative Power Perspective', *European Foreign Affairs Review*, 14 (5): 785–802.

Matres-Manso, J. (2010) 'Speech at the Workshop on Employment, Social Protection and Decent Work in Africa – Sharing Experience on the Informal Economy', Dakar (Senegal), 30 June.

Michel, L. (2006) 'Speech at the ILO Governing Body', GB.295/WP/SDG/1, March.

Niño-Zarazúa, M., Barrientos, A., Hulme, D. and Hickey, S. (2010) 'Social Protection in Sub-Saharan Africa: Getting the Politics Right', paper presented at the Conference 'Experiences and Lessons from Social Protection Programmes across the Developing Word: What Role for the EU?', ERD 2010, Paris, 17–18 June.

Olsen, G.R. (2009) 'The Missing Link: EPAs, Security and Development Interventions in Africa', in G. Faber and J. Orbie (eds), *Beyond Market Access for Economic Development: EU–Africa Relations in Transition*, London and New York: Routledge, pp. 342–58.

Orbie, J. (2007) 'The European Union and the Commodity Debate: From Trade to Aid', *Review of African Political Economy*, 34 (112): 297–311.

Orbie, J. (2011) 'Promoting Labour Standards through Trade: Normative Power or Regulatory State Europe?', in R. Whitman (ed.), *Normative Power Europe: Empirical and Theoretical Perspectives*, Basingstoke: Palgrave, pp. 161–86.

Orbie, J. and Tortell, L. (2008) 'From the Social Clause to the Social Dimension of Globalization', in J. Orbie and L. Tortell (eds), *The European Union and the Social Dimension of Globalization: How the EU Influences the World*, London and New York: Routledge, pp. 1–26.

Orbie, J. and Tortell, L. (2009) 'The New GSP+ Beneficiaries: Ticking the Box or Truly Consistent with ILO Findings?', *European Foreign Affairs Review*, 14 (5): 663–81.

World Commission on the Social Dimension of Globalization (2004) *A Fair Globalization: Creating Opportunities for All*, Geneva: ILO.

Part IV

Conclusion

15

European policies, African impact and international order: (re)evaluating the EU–Africa relationship

Michael Smith

As noted in several chapters in this volume, the European Union (EU)'s status as an international actor is contestable and (frequently) contested. There is a wide range of approaches to the external policies of the EU, ranging from those based on neo-Realism and its related perspectives (for example Hyde-Price, 2006, 2007), through to those based essentially on Liberal or neo-Liberal institutionalism (such as M.E. Smith, 2003, 2004) to those centred on ideas, values and the EU's identity as an international presence and force (including Aggestam, 2008; Sjursen, 2007; Whitman, 2011). No less contested, as demonstrated in this volume as a whole, is the nature of the EU's engagement with and impact on its key international partners, and in this case Africa.[1] Is the relationship essentially a projection of, and in some ways an artefact of, the EU's development of a foreign policy apparatus and its search for an international identity and role(s)? Is it a partnership among equals, or an asymmetric relationship carrying with it at least some of the characteristics of neo-colonialism? Is it a contribution to global order and governance and an encapsulation of desirable global values, or an instrumental relationship in which both 'sides' play games to maximise material advantage and avoid excessive commitment? Is it a distinctive if not unique kind of partnership, or one that has competitors and to which there are live alternatives?

All of these questions are dealt with at various points and in various ways in this volume, and the intention here is certainly not to duplicate them. Rather, it is to undertake an overall evaluation of the issues and evidence addressed in the book, and to place these within a somewhat broader understanding of the EU's external policies. In pursuing these aims, this Conclusion relates closely to the two key areas identified by Maurizio Carbone in the Introduction to the volume: (1) the internal development of the EU, its institutions and its policies and (2) the external contextual forces and factors that surround the EU's Africa policies and shape them as they are projected into the continent.

Issues and evidence from EU–Africa relations

The issues and evidence emerging from this volume can be explored in four areas: firstly, institutional issues within the EU; secondly, issues of EU external policy implementation; thirdly, issues surrounding the impact of the EU's policies towards Africa; and, finally, issues relating to the links between EU–Africa relations and the broader world order.

Institutional issues within the EU

It is almost a given in the study of EU external policy-making to assert that 'institutions matter' and that the nature of internal institutional arrangements has a profound influence on the ways in which the EU can conduct its international relations. In the context of EU–Africa relations, this set of drivers and constraints takes a distinctive form, but is none-theless powerful. One expression of the institutional factor is the pres-ence of internal institutional bargains that have successively shaped and re-shaped the EU's development policies and its policies towards Africa (which are not, of course, identical). As pointed out by Carbone in the Introduction, the framework for the EU's development policies had been renegotiated during the period since 2000, but in many ways that renegotiation reflects the impact of bargains made elsewhere and earlier (for example, in the negotiation of the Lomé and then the Cotonou Agreements, or more distantly and indirectly the deals done with central and east European countries in the course of their entry into the Union). Such evidence also supports in a broader context the arguments made by Orbie in chapter 14 about path dependency: he is concerned with the ways in which successive institutional bargains have constrained the development of a 'social dimension' relating to employment in the EU's development and Africa policies, but many of the arguments he makes can be applied in different ways to the ways in which the EU–Africa relationship has become part of the internal institutional 'game' within the EU. In the mid-2000s, the agreement of the European Consensus on Development and the establishment of a more formal division of labour between the various actors in the EU's development policies seemed to bring together a lot of the threads through which EU development policy has evolved, but it did not resolve some of the key institutional problems relating to EU–Africa relations. The Lisbon Treaty has more recently generated a further set of institutional challenges, not least through the creation of the European External Action Service (EEAS), which has taken responsibility for policy formation in the development field (and in EU–Africa relations). This has been a disruptive influence, at least in the short term, since it has required the transfer of responsibilities and of people from the Commission to the EEAS, the creation of a new division of responsibilities in the field of development, and the confrontation of apparently disparate organisational cultures in the formation and the delivery of development assistance. There is thus an institutional problem of 'ownership' of development policy and Africa policy in the EU's central institutional framework, and one that has not been resolved.

The problem of internal 'ownership' translates also into a problem familiar both to scholars and to practitioners in the field of EU external relations: that of coherence, consistency and coordination (see for example Nuttall, 2005; Gebhard, 2011). Many of the chapters in this volume refer more or less directly to the fragmentation of EU policy-making and to the consequent issues of coordination and consistency between the actions of different arms of the EU machine. Carbone addresses what could be summarised as the fragmentation of the 'development community', whilst Chris Stevens in chapter 9 points out the 'different worlds' inhabited by DG Development (renamed DG DEVCo in January 2011) and DG Trade when it comes to the negotiation of the Economic Partnership Agreements (EPAs) required by the Cotonou Agreement. But of course, this problem is not limited to the 'development community' – it pervades the EU's external policy-making and implementation, and is a central issue in the EU's policies towards Africa in particular. Thus, we see in chapter 13 that there is a lack of coordination on migration, with the 'justice and home affairs' parts of the machine often dominating the 'development' parts. Equally, we see that there is a high level of policy incoherence in the domains of energy (chapter 11 by Hadfield) and climate change policy (chapter 12 by Lightfoot). In both of these areas, there can be seen not only the clash of bureaucratic interests, but also the juxtapositioning of those institutions that are well established and well embedded, and those such as energy that are at an earlier stage of institutional consolidation. It is unclear whether the implementation of the Lisbon Treaty has on balance made this problem more or less severe, but clearly there are likely to be costs and inefficiencies occurring in the process of re-shaping the institutional framework.

The discussion so far has assumed that the Brussels institutions are insulated from other influences. This, of course, is far from the truth, and the reality is that what goes on in Brussels reflects only one level of a multilevel and multidimensional policy process. At the centre of this process are the member states, and it is fundamental to the EU's Africa policies that they are almost entirely shared between the member states and the Brussels institutions. The mixed nature of policy-making, especially in the area of development but also in areas bearing on security, environment, migration, energy and human rights, is at the core of the EU's dilemmas in relation to Africa. In chapter 7, Carbone points out that the changing membership of the EU has had a far-reaching effect on the willingness of member states to pay for Africa policy; some of the new member states are only just coming to terms with having a development policy at all, and others (along with some of the older-established member states) are reluctant to provide or sign up to additional resource provision either because of domestic demands or because of objections to aspects of the EU's involvement in Africa. At the same time, a number of member states with sophisticated and well-established development policies are reluctant to abandon them in favour of an EU policy framework. The problem is not simply one of resources, though. As indicated by Van Criekinge in chapter 13, in areas such as migration the member states can and do pursue essentially parallel negotiations with African countries, despite the attempts to prioritise the Euro-African partnership; this is also to be remarked in areas such as energy, where there is no firm set of obligations to give priority to EU-level policies, and where strong strategic and commercial interests come into play. Member states are thus at the core of the multilevel policy-making dilemmas faced by the EU, but they are not its only component. The chain extends upwards to the institutions of global governance, which on issues such as migration, climate change and human rights can play a major shaping role. It also extends downwards to the roles played by EU delegations in African countries, which as we will see later can have major impacts not only on the framing of policy but also on its implementation.

Internal institutional forces can thus have major shaping effects on the formulation and the direction of the EU's Africa policies. There is also a major and less tangible effect on the way in which the EU defines 'Africa'. To put it simply, there is evidence in the earlier chapters that the EU actually finds itself dealing with several Africas, defined partly by the institutional frameworks within which negotiations and transactions are undertaken and partly also by the ways in which the organisational cultures and self-perceptions produced within the EU play a role in shaping its policy orientation. At the 'European level', there thus appear to be several definitions of Africa. There is the 'big Africa' of the strategic partnership with the African Union (AU), but alongside it there are others: the 'regionalised Africa' of the EPAs (which as we shall see later does not necessarily coincide with African conceptions of regions and regionalism); the 'least developed Africa' represented in the Everything But Arms (EBA) regulation and its application; the 'neighbourhood Africa' reflected in the EuroMed agreements and in the Union for the Mediterranean (UfM); 'energy producing Africa' representing the emerging and established oil producers; and no doubt others depending on geopolitical, cultural and other constructions of the continent. But this is not all, a number of the member states have their 'Africas' as well, reflecting colonial and other historical ties and exercising a powerful influence on their willingness or reluctance to enter into EU-level commitments as opposed to 'special relationships'.

External policy issues

When it comes to the external projection of EU policies towards Africa, as has already been noted it is logical to assume that they are affected by the kinds of institutional complexity and fragmentation that have already been identified. One area in which these effects can be seen is that of 'target identification'. By this I mean not only the selection of appropriate partners or target countries, but also the selection of the issues on which policy will focus and the ways in which such policies are developed. It is easy to assert that such policy development will not necessarily be linear or always appropriate; and less easy to understand exactly what this means in terms of the EU's Africa policies. To some extent, targets are set by external institutional frameworks, as for example in the case of the Millennium Development Goals (MDGs) to which the EU is an enthusiastic adherent. There is evidence in several chapters of this volume that the MDGs form an important normative context, and to a certain extent an operational context, for EU policies. Equally, however, there is evidence of other influences, emanating both from within and outside the Brussels institutions. The thematic priorities and partnerships established under the Joint Africa–EU Strategy (JAES) draw upon but are not the same as the MDGs. At the same time, different targets are implied by the Cotonou Agreement and especially by the EPAs, with strong overtones of political conditionality as well as economic reciprocity. Yet other targets are identified both in terms of countries and in terms of activities by the EBA regulation, or by the various sectoral arrangements set up under Lomé and largely continued under Cotonou – subject of course to the broader dictates of World Trade Organisation (WTO) legality. In the area of security, as pointed out by Whitman and Haastrup in chapter 4, there is further differentiation by types of mission and mandate, and by the extent of local 'ownership' of operations. The result of these and other arrangements is a highly differentiated set of relationships between the EU and its external targets, whether these are defined in terms of countries or of issues. The influence of global governance institutions is evident, as are the interests of member states, mediated through the Brussels institutions or running parallel to them.

Given this complex, differentiated set of relationships, and the issues relating to resources and responsibilities outlined in the previous section, it is to be expected that the implementation of the EU's Africa policies will raise distinctive problems. The EU is often said to be very good at setting out general frameworks and norms, but less good at concrete implementation of policies in the field, and this is a particular problem in relation to Africa. Söderbaum in chapter 2 identifies 'field operations' as a key area in which the EU's general policy frameworks are tested and often found wanting (and of course this was also one of the driving perceptions behind the re-shaping of EU development policies dealt with by Carbone in chapter 7). Once the problem of acquiring resources has been addressed – and, as we have seen, it is not an easy one to resolve – the delivery of those resources in the right forms at the right places and at the right times presents an entirely different challenge. In large degree, this challenge falls to the EU delegations and project teams in African countries, who are at the end of a lengthy implementation chain which contains many possibilities for 'leakage' of authority, resources and control. This means that there is a substantial 'principal-agent' problem in the implementation of the EU's Africa policies where they address concrete problems of development and where they attempt what Stephan Keukeleire (2003) has described as 'structural diplomacy'. This form of diplomacy entails getting under the skin of target societies in the attempt to re-shape internal economic, social and political structures, and it is one of the implicit aims of many of the EU's development strategies. But it carries with it major problems of coordination between EU agencies, and of evaluation of the results of policy – not to mention its impact on the target societies (see below).

Discussion of the implementation of EU policies naturally raises another question: what (or who) are these policies for? As noted earlier, there are issue of internal 'ownership' of the EU's Africa policies, but here it is important to focus on what might broadly be described as the motivations behind them. It is frequently argued, not least by some EU officials, that the EU is a 'force for good' in the global arena, exercising influence through the use of its normative power rather than through coercive or other traditional means. This argument – that the EU is not only different but also better – is particularly relevant to the framing of the EU's development policies, where it is argued that the EU's links with Africa amongst others are unique in terms of their long-term nature, their institutionalisation and their permeation by a normative approach. The earlier chapters in this volume, though, show a rather mixed picture. It appears that the EU's normative commitment to African development is not the whole of the story. To be sure, that commitment is there, and the EU's adherence to the MDGs and other normative frameworks is unquestionably part of its external policy projection in Africa. But it is also clear that as Olsen points out in chapter 3, there is a complex interplay of interests going on when the EU engages with Africa (in whichever of the forms Africa takes – see above). Söderbaum is more explicit in chapter 2, pointing out that there is a basic tension between two sets of norms in the practice of EU Africa policy: on the one hand there is the liberal, progressive and holistic version of development expressed in many of the EU's normative statements, but on the other there is a more hard-edged neo-liberal commitment to deregulation, the retreat of the state and the pursuit of balanced budgets. The EU's relationship to the state in Africa is also ambiguous, as pointed out by Taylor in chapter 5, and by Crawford in chapter 8. Some of the implications of this ambiguity are dealt with in the following section on the impact of the EU's policies, but it is important here to note that the ambiguity generates a number of contradictions in the EU's motivations for partnership with Africa. Most notably, several chapters draw attention to the 'securitisation' of the EU's Africa policy, reflecting the new strategic interest attached to emerging African oil producers, the construction of African migration to the EU as a threat to be countered with (at times) coercive measures and the development of the so-called 'development/security nexus'. As Whitman and Haastrup point out in chapter 4, the EU's strategic security aims in Africa are set by a variety of forces and processes, including a specific interpretation of the 'security–development nexus' and a range of preferences reflecting the interests of member states and other organisations. This is a challenge to the normative basis of the EU's Africa policies, and brings into play a number of tensions surrounding the EU's conception of its own identity and 'strategic culture'. The challenge is faced by other international actors, but it might be argued that they have not exposed themselves to it in quite the same way as the EU, through the proclamation of its distinctiveness in this area and its commitment to a norms-based model of development. As noted earlier, this issue also has important links to the internal institutional tensions attending the EU's Africa policies, with some parts of the EU institutions committed to a securitised model and others to a more normative perspective.

Discussion of the tensions between normative and material interests in the EU's Africa policies leads naturally to consideration of what might be termed the 'identity problem'. It was argued earlier that external policy can be seen as playing an important role in the generation and propagation of an external identity for the EU, and Africa policy is certainly no exception to this. The need to project a certain image of the EU into (or onto) Africa is central to the ways in which the EU's policies have developed, but it is not unproblematic. Farrell points out in chapter 6 that the EU has promoted itself as encouraging regionalism in Africa, and that this reflects a powerful EU self-conception. But at the same time, the EU has pursued policies, such as those centred on the EPA negotiations, that act counter to what might be seen the legitimate regional aspirations of African countries (see for example chapter 9). As Söderbaum points out in chapter 2, there is a strong commitment to increasing the visibility of the EU, but it is not clear what version of the EU is being promoted, and also not clear what the results of this self-promotion might be. Additionally, there is a sense that the EU actually has several identities in dealing with Africa, and that this reflects the differentiation and complexity referred to earlier in the chapter. There is the EU of comprehensive development and of partnership, but this exists alongside the EU of hard-nosed trade negotiations, of resistance to African migration, of exacting environmental and other standards and of hard-edged rather than soft power. There is no doubt that the EU has set itself to become a power in Africa, but it remains unclear what the identity of this 'power' is.

External impact

It is apparent from the argument so far that the EU's policies and roles in respect of Africa are open to debate and that they demonstrate a number of important tensions arising from the framing and the execution of policy. Thus far, however, the argument has avoided talking very much about the impact of those policies in Africa itself and on African countries. In this section, I will discuss four aspects of the impact question: firstly, the question of 'ownership' and the relationship of the EU's policies to African interests; secondly, the ways in which the EU's development policies and other activities relate to each other in the African context; thirdly, the nature of Africa as a 'test bed' for EU policies; and, finally, the presence in the African context of competitors for the EU with different priorities and ways of operating.

A number of the chapters in this volume touch on the important question of 'ownership'. We have seen that this is an issue within the

European institutions, but here we are concerned with the extent to which the EU's Africa policies are 'owned' either by the EU or Africa, and the implications this carries with it for African interests and approaches to the relationship. On the face of it, the Joint Africa–EU Strategy, as pointed out by several contributors to the volume, is one of equality in which the strategy is jointly owned by the EU and its African partners in the shape of the AU. The rhetoric is that of equality and shared ownership, throughout the eight thematic partnerships on which the JAES centres. But this is at odds with some of the other ways in which the EU acts in relation to Africa. Whilst the JAES is jointly owned, the Cotonou Agreement, and more particularly the negotiation of the EPAs, does not demonstrate such an even-handed relationship. As Stevens shows in chapter 9, there is a sense in which the EU's approach to the EPA negotiations can be seen as coercive, setting peremptory demands and threatening dire consequences if they are not met. Other evidence of a similar type might be extracted from chapter 10, where the fisheries partnerships favoured by the EU do not appear to be even-handed or mutual in their operation, or from chapter 13, where the issue of migration seems to have evoked distinctly coercive forms of behaviour on the part of the EU towards certain African countries. Chapter 4 argues that the EU's approach to African security issues reflects a consistent commitment to a number of central preferences, but it is not difficult to conceive of situations in which these preferences conflict and in which the imperatives of specific security challenges subvert the apparent overall consistency. This of course links with another issue, that of multilateralism. Whilst the EU has declared itself in many contexts in favour of 'effective multilateralism', many of its policies towards Africa seem to adopt a bilateral or what might be termed a 'selective multilateral' approach, reflecting the differentiation already noted in the motivations of EU policy-makers. Not only this, but some of the EU's policies, particularly the EPAs, seem to fly in the face of existing African efforts at regional integration, and thus to be positively disruptive in their effects.

This issue of ownership in turn relates to the differences of interest that exist between the EU and its African partners. Taylor in chapter 5 of this volume makes the important point that the nature of the African partners themselves – specifically in terms of the characteristics of African statehood – coupled with the varying amounts of agency available to them within EU–Africa relations can radically affect the extent to which they express and can pursue their own interests, and resist the demands of the EU. It seems clear that African countries can manoeuvre within the EU–Africa partnership and within the broader global system to pursue their interests, and that they can either avoid, resist or reinterpret the pressure exerted by the EU so as to gain advantage. Many studies of the EU's development policies and the EU–Africa relationship have a tendency to assume that 'Africa' is a kind of passive recipient of the benefits bestowed or to the demands made by the EU, but this is an oversimplification; as we shall see later, the changing nature of the global context and the ways in which it penetrates Africa have given new opportunities for manoeuvre, and when this is added to the sheer variety of African states themselves, it means that the EU is confronted with an active, diverse and dynamic partner. The chapters in this volume dealing with the EPAs, migration, human rights and the social dimension, among others, show in detail how some of the manoeuvring and re-configuration can work.

   One of the reasons why the African countries can find space to avoid, resist or re-configure EU policies lies in the nature of those policies themselves, and the ways in which they interact within the African context. To some extent, these interactions are rooted in the nature of the EU's institutional framework and policy-making processes; as we have seen, these produce a competition and elements of fragmentation in EU policy-making that is then reproduced in the field. In different areas of policy, the EU's strategies espouse not only different aims but also different modes of implementation, and this presents both a challenge and an opportunity to African countries. Thus, for example, in chapter 3 of this volume Olsen explores the 'gap' between development policy and security policy, whilst in chapter 9 Stevens notes the different approaches and modes of operation of DG DEVCo and DG Trade. Whilst such differences are often discussed in terms of the sub-optimality of EU policy-making, it is important to be aware of the ways in which they impact on the EU's African partners. These partners are confronted not only with complexity and often confusion in EU policy, but also with opportunities that they can exploit to balance between different areas and to benefit from the uncoordinated character of EU activities. Most recently, of course, as pointed out by a number of contributors, the linkage (or lack of it) between development and security in EU external relations – despite the Union's declaratory commitment to a specific version of the development–security nexus as noted in chapter 4 – has provided new challenges and opportunities for African countries, although by the nature of the linkage, this can be a high-risk arena within which to manoeuvre. The agency available to African states, as pointed out by Taylor in chapter 5, can as easily be used to resist reform as it can to undertake changes in line with EU conditionality or other forms of demand. Such African responses, as we shall see later, are also assisted by the availability of alternatives to partnership with the EU, either at the global or at the African level.

For the EU, the issues dealt with above – of ownership, relationships to African interests and African agency – are given additional importance by the ways in which Africa has been seen as a 'test bed' for EU policies. This perception lies behind the declaratory commitments noted by Whitman and Haastrup in chapter 4 of this volume, and they draw attention to the ways in which Africa has generated not only the security challenges to which the EU has responded, but also the 'space' within which to respond. The EU has had, if not a free run in sub-Saharan Africa particularly, then a significant amount of freedom within which to experiment, producing initiatives not only in the area of security but also in a number of areas connected with economic and social activity. These initiatives often carry with them the rhetoric of partnership and of shared values, but it can also be argued that they reflect the kinds of European norms that are unlikely to gain much traction in the African context. Thus, as Orbie shows in chapter 14, norms relating to employment and workers' rights have not made much of an impression in the African arena, reflecting a lack of attention and resources both in the EU and in Africa itself. More significant have been the impacts of various forms of conditionality, for example as expressed in the Cotonou Agreement and more sharply in the EPAs; sub-Saharan Africa has been the largest single arena in which such conditionality has been pursued. In the broader area of development policy, Africa has been the region in which development models, instruments and actions have been tested and refined by the EU. But the question remains: is Africa to be seen as a passive 'test-bed' or as a dynamic region in which EU policies can be diverted, subverted or resisted?

The idea that Africa provides a kind of empty space onto which EU norms, practices and policies can be inscribed is clearly open to challenge because of the agency available to African countries and the 'gaps' in EU policies themselves. But there is an important additional force to be considered in this context: the emergence of substantial competitors to the EU when it comes to Africa policy. Several of the chapters in this volume point out that both the established and the emerging 'great powers' are developing a keen interest in Africa as a source of raw materials, as a security challenge and as a platform on which they can further their ambitions in an increasingly multi-polar world. As a result, the status of the EU as an African power – if not *the* African power – has come under pressure in ways not foreseen during the 1980s and 1990s. In chapter 3, Olsen points out that the USA and China are key emerging competitors to the EU, and that they present themselves in ways that are at times radically different from those in which the EU operates. Thus the USA is interested in Africa primarily as a source of vital raw materials and as a potential security challenge, especially given the increasing presence of China. China, on the other hand, is interested in access to raw materials for economic reasons, and presents itself as a 'responsible great power' in the African arena. In this way, China appears to be more of a challenge to the EU on its traditional territory, since it can offer development assistance in what appears to be a more pragmatic way and also link this to a broader model of world order that appeals to at least some African governments. But as pointed out in chapter 5, China is not the only actual or potential rival: Brazil, Russia, Mexico, Turkey and Iran are also active to a greater or lesser extent in cultivating African partners. The net result of this changing landscape is that the EU's positioning of itself as not only the most desirable partner for Africa but also a 'different' type of power with distinctive norms and approaches to partnership has come under pressure, and that its status as *the* external power in Africa will increasingly be questioned within and outside the region.

External order

One of the central elements in the EU's external policies has always been the generation or the maintenance of order (Smith, 2007). This is implicit in the EU's role as a kind of 'trading state' in the world arena, interested in maintaining a stable and accommodating environment within which to pursue its prosperity and in avoiding the kinds of confrontational politics that might attend other kinds of world order projects. The EU has succeeded over a long period in establishing a strong model of order in Europe itself, and it has set about preserving that through initiatives in relation to other regions and to the world arena in general. It is thus important to look at the EU's relationship with Africa as part of this search for order, and to relate it to the broader processes of world order and global governance. In this part of the chapter, I will focus on four aspects of the question: firstly, the extent to which the EU has tried to 'order' Africa, especially in terms of governance; secondly, the ways in which a new geopolitics of order has affected the EU's relationship with Africa; thirdly, the effects of the EU's inter-regional policies on the establishment of regional order; and, finally, the linkages between the EU–Africa relationship and broader issues of world order.

As noted at many points in this volume, the EU has approached Africa with a set of preconceptions about what constitutes political order at the national level, expressed in terms such as 'good governance' and related to the rule of law and the promotion of liberal democracy. As Taylor shows in chapter 5, this conception of domestic political order is often far removed from the reality of governance in Africa, even when there is rhetorical acceptance of the EU's norms. There is a profound ambiguity, as noted above, about the role of the state and the nature of rule in Africa as seen from Brussels, which is compounded by the engagement of the EU with a wide variety of domestic regimes and at a variety of levels from field operations to grand strategy. The EU has also been committed to promoting regional order in Africa, but again here there are tensions and ambiguities; it was noted earlier that the promotion of EPAs with their 'constructed' sub-regions in Africa was essentially disruptive of established efforts at regional integration, and this non-coincidence of images of regional order makes negotiation and the 'sharing' of concepts of regionalism significantly more difficult. In a number of ways, the EU is pursuing in Africa a form of the 'external governance' identified by scholars especially in relation to the European 'neighbourhood' (Schimmelfennig and Sedelmeier, 2004; Lavenex, 2004; Lavenex and Wichmann, 2009) but without the proximity or the incentives provided by the possibility of accession to the Union. Not surprisingly, there are problems, and these are demonstrated in the EU's efforts to promote its vision of human rights in Africa (detailed by Crawford in this volume), by the tensions between the development of state capacity and the priorities of the EU's development model also noted in chapter 6, and by the underlying tension between a unified and a differentiated view of Africa itself (see above).

The picture is complicated further by the re-birth of geopolitics (and its close relative, geo-economics) in contemporary Africa. We have already noted that the EU is no longer in a position of dominance as a partner for African countries (although, as will be seen below, its partnership with the African Union especially does give it a distinctive position). The growing interest in Africa on the part of the USA, China and other 'powers' has produced a situation in which a version of the 'great game' is being played out, generating the potential for conflict and for linkages between external power and interests and the dynamics of African regional order. In contrast to most other regions of the world, the EU has taken an active military role in the newly emerging African order, although this has been strictly confined both by the interests of member states and by the normative and institutional leverage provided by the Treaties – for example over the 'Petersberg tasks' and related commitments to peace-keeping, reconstruction and post-conflict reconciliation. The JAES, as pointed out by Soderbaum in chapter 2, has a peace and security dimension, expressed not only in rhetoric but also in the development of the African Peace Facility (see also chapters 3 and 6). But as Whitman and Haastrup point out in chapter 4, a variety of rationales have been used to justify EU action in specific circumstances, and the ground rules for EU action are constantly being renegotiated. In addition to specific interventions, the EU has a more general interest in those areas that have become increasingly 'securitised', two of the key examples being energy and migration; here the dictates of geopolitics and geo-economics are emerging as the EU experiments with new ideas of regional order and intervention, and this dynamism represents a considerable challenge to the EU's adaptive capacity (see chapters 11 and 13). If this is geopolitics or geo-economics, clearly it is being played with a variety of maps and shifting boundaries between issues as well as between actors. Inevitably, the jury is still out on the extent to which the EU, as a distinctive type of actor, can enter into the contemporary version of African power politics and power balancing.

Linked to the notions of order in Africa and the new geopolitics of Africa is the EU's interest in inter-regionalism. The pursuit of partnerships across regions with the global arena is built into the DNA of the EU, and relates strongly to some of the issues of identity as well as interest dealt with earlier in this chapter. True to form, the EU has attempted to construct a comprehensive inter-regional partnership with Africa in the shape especially of the JAES and its partnership with the African Union. This is one of the truly distinctive aspects of the EU's relationship with Africa: the attempt to build and institutionalise a long-term partnership at the inter-continental level, and to build into it both the normative and the material dimensions of the EU's engagement with international order. As Farrell points out in detail in chapter 6 of this volume, this policy is linked fundamentally to the desire by the EU to diffuse its normative position on key aspects of African affairs, and to provide an 'architecture' within which the relationship between the EU and Africa can be negotiated and adapted to a changing global arena. But two points must be made about this effort. Firstly, the JAES, as shown in many chapters in this volume, is actually only part of a complex, multilayered and differentiated set of EU–African partnerships, some of which are in tension with each other. The product of the EU's African engagement is a form of 'complex inter-regionalism' (Hardacre and Smith, 2009; Hardacre, 2010), in which different layers of activity, different institutional contexts and different forms of bilateral, multilateral and 'bi-multilateral' relationships coexist. As a result, the EU and its African partners face significant issues in relation to coordination, management and linkage, arising from the 'externalities' of actions taken at the different levels and in different areas of activity. This problem can be spotted in several of the chapters in this volume, especially those dealing with energy, climate change, human rights and related areas; it also provides part of the explanation for the limitations Farrell describes in the outcomes of the EU's and its partners' search for inter-regional cooperation.

Inter-regional order is of course closely linked to world order. In the EU's perspective, the two are not just linked, they play into each other and developments at one level can be functional or dysfunctional at the other. It seems clear from several of the chapters in this book that the EU is interested in the ways in which the EU–Africa partnership can form part of a broader approach to global order. Partly this is normative, implying that the EU–Africa partnership is a means of generating traction for the EU's values and priorities at the global level. Partly also it is linked with key issues of EU identity and its leaders' desire to see the EU not only as a 'different power' but also as a 'better power' in the global arena. But the EU's interest is also pragmatic and instrumental, especially when it comes to the pursuit of EU interests in global contexts. Thus, we find in chapter 12 that the EU has made significant efforts to recruit African support for joint initiatives on climate change. It is also suggested that the EU might use the African partners as support for its positions in the World Trade Organisation, or in human rights bodies, and it is clear from chapter 14 that at least potentially the EU sees Africa as a test-bed for actions in the social dimension that might then be reinforced at the global level. But there are two key limitations surrounding the pursuit of these aspirations. The first is that as pointed out in chapter 14 by Orbie, the global regimes to which the EU's efforts are linked might be weak and the results of their standard-setting thus unpredictable. In the African context, such weakness is likely to lead to wide variations in compliance and the embedding of global norms, whether via the EU or otherwise. The second limitation is linked to this: that on the evidence provided in this volume, African governments (often abetted by non-state actors in the commercial and industrial sectors) are inclined to sign up to EU norms at the level of rhetoric, but to ignore them – often because they have to – at the level of compliance and practical action. Whilst the JAES represents a joint commitment on the part of the EU and its African partners to a large number of global norms, these are more often honoured in the breach than in the observance. And this can lead to a major gap between the EU's normative aspirations and the reality of its partnership with African countries. When this is taken along with the governance, geopolitical and inter-regional issues outlined in this part of the chapter, it is clear that there is no linear relationship between the EU's commitment to a particular version of 'order' in its partnership with Africa and the contribution made by the partnership to broader world order or global governance.

Conclusions

It is clear from the issues and evidence dealt with here that the EU's relationship with Africa presents in some ways a microcosm of the EU's external policy dilemmas in the early part of the twenty-first century. The institutional base for EU policy is shifting, is often contested and sets up a series of tensions between internal institutional development and external policy demands. The projection of EU policies is affected not only by the internal institutional uncertainties but also by the nature of the context within which the policies are projected and implemented. Implementation does not always lead to the anticipated impact, and raises further questions about the extent to which African countries are willing targets of the EU's institutional, material and normative activities. These three sets of problems – institutions, implementation and impact – in turn have ambiguous effects for the EU's capacity to contribute to wider world order or to situate Africa within its vision of the 'good world'.

Africa, however, is not simply a test-bed for EU policies and a test for them. Given the re-emergence of geo-political and geo-economic competition within the continent, and its likely intensification in the coming decade, this is an area in which the entire set of EU claims to international actorness and influence is likely to be at issue. This book provides a rich seam of evidence on the extent to which the EU is likely to be able to survive this existential challenge. It is clear that the EU's capacity to do so is partly a function of the internal logic of European integration, but no less clearly a function of external challenges and the ways in which the EU can develop a conception of its own role in dealing with them, both at the conceptual and normative level and at the level of operational effectiveness.

Note

1 In this chapter, in common with most of the authors in this volume, I will essentially focus on sub-Saharan Africa, although as several chapters recognise, the North African countries present their own distinct challenges and opportunities for EU policies.

References

Aggestam, L. (2008) 'Ethical Power Europe?', special issue of *International Affairs*, 84 (1).

Gebhard, C. (2011) 'Coherence', in C. Hill and M. Smith (eds), *International Relations and the European Union*, 2nd edition, Oxford: Oxford University Press, pp. 101–27.

Hardacre, A. (2010) *The Rise and Fall of Interregionalism in EU External Relations*, Dordrecht: Republic of Letters Publishing.

Hardacre, A. and Smith, M. (2009) 'The EU and the Diplomacy of Complex Interregionalism', *The Hague Journal of Diplomacy*, 4 (2): 167–88.

Hyde-Price, A. (2006) 'Normative Power Europe: A Realist Critique', *Journal of European Public Policy*, 13 (2): 29–44.

Hyde-Price, A. (2007) *European Security in the 21st Century: The Challenge of Multipolarity*, London: Routledge.

Keukeleire, S. (2003) 'The European Union as a Diplomatic Actor: Internal, Traditional and Structural Diplomacy', *Diplomacy and Statecraft*, 14 (3): 31–56.

Lavenex, S. (2004) 'EU External Governance in a Wider Europe', *Journal of European Public Policy*, 11 (4): 680–700.

Lavenex, S. and Wichmann, N. (2009) 'The External Governance of EU Internal Security', *Journal of European Integration*, 31 (1): 83–102.

Nuttall, S. (2005) 'Coherence and Consistency', in C. Hill and M. Smith (eds), *International Relations and the European Union*, Oxford: Oxford University Press, pp. 91–112.

Schimmelfennig, F. and Sedelmeier, U. (2004) 'Governance by Conditionality: EU Rule Transfer to the Candidate Countries of Central and Eastern Europe', *Journal of European Public Policy*, 11 (4): 669–87.

Sjursen, H. (ed.) (2007) *Civilian or Military Power? European Foreign Policy in Perspective*, London: Routledge.

Smith, M. (2007) 'The European Union and International Order: European and Global Perspectives', *European Foreign Affairs Review*, 12 (4): 437–56.

Smith, M.E. (2003) *Europe's Foreign and Security Policy: The Institutionalisation of Cooperation*, Cambridge: Cambridge University Press.

Smith, M.E. (2004) 'Toward a Theory of EU Foreign Policy-Making: Multi-Level Governance, Domestic Politics and National Adaptation in Europe's Common Foreign and Security Policy', *Journal of European Public Policy*, 11 (4): 740–58.

Whitman, R. (ed.) (2011) *Normative Power Europe: Empirical and Theoretical Perspectives*, Basingstoke: Palgrave/Macmillan.

Index

Lightning Source UK Ltd.
Milton Keynes UK
UKHW02f1936191217
314784UK00006B/760/P